The Circuit Designer's Companion

The Circuit Designer's Companion

Fourth Edition

Peter Wilson

Newnes is an imprint of Elsevier
The Boulevard, Langford Lane, Kidlington, Oxford OX5 1GB, United Kingdom
50 Hampshire Street, 5th Floor, Cambridge, MA 02139, United States

Notices
Knowledge and best practice in this field are constantly changing. As new research and experience broaden our understanding, changes in research methods, professional practices, or medical treatment may become necessary.

Practitioners and researchers must always rely on their own experience and knowledge in evaluating and using any information, methods, compounds, or experiments described herein. In using such information or methods they should be mindful of their own safety and the safety of others, including parties for whom they have a professional responsibility.

To the fullest extent of the law, neither the Publisher nor the authors, contributors, or editors, assume any liability for any injury and/or damage to persons or property as a matter of products liability, negligence or otherwise, or from any use or operation of any methods, products, instructions, or ideas contained in the material herein.

Library of Congress Cataloging-in-Publication Data
A catalog record for this book is available from the Library of Congress

British Library Cataloguing-in-Publication Data
A catalogue record for this book is available from the British Library

ISBN: 978-0-08-101764-7

For information on all Newnes publications visit our website at
https://www.elsevier.com/books-and-journals

Working together
to grow libraries in
developing countries

www.elsevier.com • www.bookaid.org

Publisher: Mara Conner
Acquisition Editor: Tim Pitts
Editorial Project Manager: Charlotte Kent
Production Project Manager: Lisa M. Jones
Designer: Mark Rogers

Typeset by TNQ Books and Journals

Contents

Introduction

INTRODUCTION TO THE FOURTH EDITION

Since the third edition of this book, there has been an explosion in "maker" projects with UAVs, robots, autonomous systems, mobile electronics, 3D printing, which has also required a similar level of creativity and capability in the electronics field. Industry has also kept moving at an incredibly rapid pace, and yet university or college students still are not exposed to many of the practical skills and techniques that engineers tend to accumulate over many years.

Many engineers (at all stages of their career) have contacted me since the last revision to suggest further enhancements, additions, and modifications, and therefore after 5 years the time is right to undertake another revision of this book. I hope I have captured most of the many suggestions I have received and added some significant level of updating to the book and, in particular, extended the practical and tutorial aspects of the book, this being one of the key suggestions. I sincerely hope you enjoy reading the book as much as I have in writing it, finding it interesting and useful.

Professor Peter Wilson
University of Bath, 2016

INTRODUCTION TO THE THIRD EDITION

When I was first approached to produce a third edition of *The Circuit Designer's Companion*, I was at first reluctant to "mess with it." It is rare to have a companion book that is not just a textbook, or a handbook, but is seen in many respects to contain all the essential information that a "real" circuit designer needs not only to produce a working circuit but also to enable that designer to understand all the related topics that make the design robust, tolerant to noise and temperature, and able to operate in the system that it was designed for. This book is a rare example of just that, and there is no other comparable text that will provide such a broad range of design skills to be passed on to the next generation of circuit designers.

It is interesting to note that in the 21st anniversary of the original edition of this book, there is no diminution of demand for analog and mixed signal design skills; however, most universities and colleges still teach a syllabus in Electronics that is dominated by digital design techniques. The comment made by Tim in the introduction to the first edition that analog electronics were "hard" and there was a reluctance to embark on analog electronics could have been written this year, rather than two decades ago! During the revision of this book, it was also interesting to note that much of the content was still completely valid in today's electronic systems, albeit some of the individual technology elements have of course moved on, with many of the fundamental concepts being essentially the same.

The third edition of the book has really been an exercise of revision rather than revolution, and I have tried to keep the philosophy the same as the original author intended. As with the second edition, the aim has been to update the technological aspects in the book, expand some sections and offer a slightly different personal perspective to hopefully further enhance the book. I am very grateful to Tim Williams for allowing me to make these revisions and for his discussions about the books' previous editions. I also acknowledge the advice, teaching, and knowledge of many friends and colleagues over the past three decades, which have provided much insight into the art of analog electronics including my father, Tom Wilson, Frank Fisher while at Ferranti, professor Alan Mantooth at the University of Arkansas, and Dr. Neil Ross and Dr. Reuben Wilcock, at the University of Southampton. I must also thank my wife, Caroline, who has tolerated my fascination with Electronics for many years. I hope that further generations of electronic designers will find this edition useful and that the book will continue to provide the assistance and help to circuit designers that the previous editions have done over the last two decades.

Dr. Peter Wilson
July 2011

INTRODUCTION TO THE SECOND EDITION (TIM WILLIAMS, 2004)

The first edition was written in 1990 and eventually, after a good long run, went out of print. But the demand for it has remained. There followed a period of false starts and much pestering, and finally the author was persuaded to pass through the book once more to produce this second edition. The aim remains the same, but technology has progressed in the intervening 14 years, and so a number of anachronisms have been corrected and some sections have been expanded. I am grateful to those who have made suggestions for this updating, especially John Knapp and Martin O'Hara, and I hope it continues to give the same level of help that the first edition evidently achieved.

INTRODUCTION TO THE FIRST EDITION (TIM WILLIAMS, 1990)

Electronic circuit design can be divided into two areas: the first consists in designing a circuit that will fulfill its specified function, sometimes, under laboratory conditions; the second consists in designing the same circuit so that every production model of it will fulfill its specified function, and no other undesired and unspecified function, always, in the field, reliably over its lifetime. When related to circuit design skills, these two areas coincide remarkably well with what engineers are taught at college—basic circuit theory, Ohm's law, Thévenin, Kirchhoff, Norton, Maxwell, and so on—and what they learn on the job—that there is no such thing as the ideal component, that printed circuits are more than just a collection of tracks, and that electrons have an unfortunate habit of never doing exactly what they are told.

This book has been written with the intention of bringing together and tying up some of the loose ends of analog and digital circuit design, those parts that are never mentioned in the textbooks and rarely admitted elsewhere. In other words, it relates to the second of the above areas.

Its genesis came with the growing frustration experienced as a senior design engineer, attempting to recruit people for junior engineer positions in companies whose foundations rested on analog design excellence. Increasingly, it became clear that the people I and my colleagues were interviewing had only the sketchiest of training in electronic circuit design, despite offering apparently sound degree-level academic qualifications. Many of them were more than capable of hooking together a microprocessor and a few large-scale functional block peripherals but were floored by simple questions such as the nature of the p—n junction or how to go about resistor tolerancing. It seems that this experience is by no means uncommon in other parts of the industry.

The colleges and universities can hardly be blamed for putting the emphasis in their courses on the skills needed to cope with digital electronics, which is after all becoming more and more pervasive. If they are failing industry, then surely it is industry's job to tell them and to help put matters right. Unfortunately it is not so easy. A 1989 report from Imperial College, London, found that few students were attracted to analog design, citing inadequate teaching and textbooks as well as the subject being found "more difficult." Also, teaching institutions are under continuous pressure to broaden their curriculum, to produce more "well-rounded" engineers, and this has to be at the expense of greater in-depth coverage of the fundamental disciplines.

Nevertheless, the real world is obstinately analogue and will remain so. There is a disturbing tendency to treat analog and digital design as two entirely separate disciplines, which does not result in good training for either. Digital circuits are in reality only overdriven analog ones, and anybody who has a good understanding of analog principles is well placed to analyze the more obscure behavior of logic devices. Even apparently simple digital circuits need some grasp of their analog interactions to be designed properly, as Chapter 6 of this book shows. But also, any product that interacts with the outside world via typical transducers must contain at least some analog circuits for signal conditioning and the supply of power. Indeed, some products are still best realized as all-analog circuits. Jim Williams, a well-known American linear circuit designer (who bears no relation to the author of the first two editions of this book), put it succinctly when he said "wonderful things are going on in the forgotten land between ONE and ZERO. This is Real Electronics."

Because analog design appears to be getting less popular, those people who do have such skills will become more sought-after in the years ahead. This book is meant to be a tool for any aspiring designer who wishes to develop these skills. It assumes at least a background in electronics design; you will not find in here more than a minimum of basic circuit theory. Neither will you find recipes for standard circuits, as there are many other excellent books that cover those areas. Instead, there is a serious treatment of those topics that are "more difficult" than building-block electronics: grounding, temperature effects, electromagnetic compatibility, component sourcing and characteristics, the imperfections of devices, and how to design so that someone else can make the product.

I hope the book will be as useful to the experienced designer who wishes to broaden his or her background as it will to the neophyte fresh from college who faces a first job in industry with trepidation and excitement. The traditional way of gaining experience is to learn on the job through peer contact, and this book is meant to enhance rather than supplant that route. It is offered to those who want their circuits to stand a greater chance of working first time every time, and a lesser chance of being completely redesigned after 6 months. It does not claim to be conclusive or complete. Electronic design, analog or digital, remains a personal art, and all designers have their own favorite tricks and their own dislikes. Rather, it aims to stimulate and encourage the quest for excellence in circuit design.

I must here acknowledge a debt to the many colleagues over the years who have helped me toward an understanding of circuit design, and who have contributed toward this book, some without knowing it: in particular Tim Price, Bruce Piggott, and Trevor Forrest. Also to Joyce, who has patiently endured the many brainstorms that the writing of it produced in her partner.

Grounding and Wiring

The Circuit Designer's Companion. http://dx.doi.org/10.1016/B978-0-08-101764-7.00001-3

1.1 GROUNDING

A fundamental property of any electronic or electrical circuit is that the voltages present within it are referenced to a common point, conventionally called the ground. This term is derived from electrical engineering practice, when the reference point is often taken to a copper spike literally driven into the ground. This point may also be a connection point for the power to the circuit, and it is then called the 0-V (nought-volt) rail, and ground and 0 V are frequently (and confusingly) synonymous. Then, when we talk about a 5-V supply or a −12-V supply or a 2.5-V reference, each of these is referred to the 0 V rail.

At the same time, ground is *not* the same as 0 V. A ground wire connects equipment to earth for safety reasons and does not carry a current in normal operation. However, in this chapter the word "grounding" will be used in its usual sense, to include both safety earths and signal and power return paths.

Perhaps the greatest single cause of problems in electronic circuits is that 0 V and ground are taken for granted. The fact is that in a working circuit there can only ever be one point which is truly at 0 V; the concept of a "0-V rail" is in fact a contradiction in terms. This is because any practical conductor has a finite nonzero resistance and inductance, and Ohm's law tells us that a current flowing through anything other

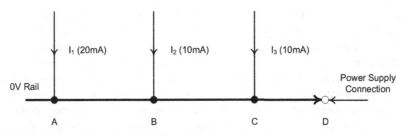

FIGURE 1.1 Voltages Long the 0-V Rail.

than a zero impedance will develop a voltage across it. A working circuit will have current flowing through those conductors that are designated as the 0-V rail and therefore, if any one point of the rail is actually at 0 V (say, the power supply connection) the rest of the rail will *not* be at 0 V. This can be illustrated with the example in Fig. 1.1.

Assume the 0-V conductor has a resistance of 10 mΩ/in. and that points A, B, C, and D are each 1 in. apart. The voltages at points A, B, and C referred to D are defined in Eqs. (1.1)–(1.3), respectively.

$$V_C = (I_1 + I_2 + I_3) \times 10 \text{ m}\Omega = 400 \text{ μV} \tag{1.1}$$

$$V_B = V_{C+}(I_1 + I_2) \times 10 \text{ m}\Omega = 700 \text{ μV} \tag{1.2}$$

$$V_A = V_{B+}(I_1) \times 10 \text{ m}\Omega = 900 \text{ μV} \tag{1.3}$$

Now, after such a trenchant introduction, you might be tempted to say well, there are millions of electronic circuits in existence, they must all have 0-V rails, they seem to work well enough, so what is the problem? Most of the time there is no problem. The impedance of the 0-V conductor is in the region of milliohms, the current levels are milliamps, and the resulting few hundred microvolts drop does not offend the circuit at all. 0 V plus 500 μV is close enough to 0 V for nobody to worry.

The difficulty with this answer is that it is then easy to forget about the 0-V rail, and assume that it is 0 V under all conditions, and subsequently be surprised when a circuit oscillates or otherwise does not work. Those conditions where trouble is likely to arise are as follows:

- where current flows are measured in amps rather than milli- or microamps
- where the 0-V conductor impedance is measured in ohms rather than milliohms
- where the resultant voltage drop, whatever its value, is of a magnitude or in such a configuration as to affect the circuit operation

When to Consider Grounding?

One of the attributes of a good circuit designer is to know when these conditions need to be carefully considered and when they may be safely ignored. A frequent complication is that you as circuit designer may not be responsible for the circuit's layout, which is handed over to a layout draftsman (who may in turn delegate many routing decisions to a software package). Grounding is always sensitive to layout, whether of discrete wiring or of printed circuits, and the designer must have some knowledge of and control over this if the design is not to be compromised.

The trick is always to be sure that you know where ground return currents are flowing, and what their consequences will be; or, if this is too complicated, to make sure that wherever they flow, the consequences will be minimal. Although the above comments are aimed at 0 V and ground connections, because they are the ones most taken for granted, the nature of the problem is universal and applies to any conductor through which current flows. The power supply rail (or rails) is another special case where conductor impedance can create difficulties.

1.1.1 GROUNDING WITHIN ONE UNIT

In this context, "unit" can refer to a single circuit board or a group of boards and other wiring connected together within an enclosure such that you can identify a "local" ground point, for instance the point of entry of the mains earth. An example might be as shown in Fig. 1.2. Let us say that printed circuit board (PCB) 1 contains input signal conditioning circuitry, PCB2 contains a microprocessor for signal processing, and PCB3 contains high-current output drivers, such as for relays and for lamps. You may not place all these functions on separate boards, but the principles are easier to outline and understand if they are considered separately. The power supply unit provides a low-voltage supply for the first two boards and a higher power supply for the output board. This is a fairly common system layout and Fig. 1.2 will serve as a starting point to illustrate good and bad practice.

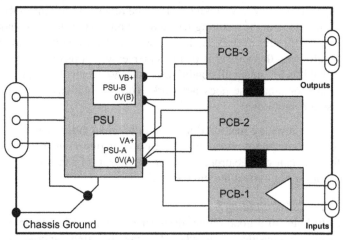

FIGURE 1.2 Typical Intraunit Wiring Scheme.

1.1.2 CHASSIS GROUND

First of all, note that connections are only made to the metal chassis or enclosure at one point. All wires that need to come to the chassis are brought to this point, which should be a metal stud dedicated to the purpose. Such connections are the mains safety earth (about which more later), the 0-V power rail, and any possible screening and filtering connections that may be required in the power supply itself, such as an electrostatic screen in the transformer. (The topic of power supply design is itself dealt with in much greater detail in Chapter 7.)

The purpose of a single-point chassis ground is to prevent circulating currents in the chassis.[1] If multiple ground points are used, even if there is another return path for the current to take, a proportion of it will flow in the chassis (Fig. 1.3); the proportion is determined by the ratio of impedances which depends on frequency. Such currents are very hard to predict and may be affected by changes in construction, so that they can give quite unexpected and annoying effects: it is not unknown for hours to be devoted to tracking down an oscillation or interference problem, only to find that it disappears when an inoffensive-looking screw is tightened against the chassis plate. Joints in the chassis are affected by corrosion, so that the unit performance may degrade with time, and they are affected by surface oxidation of the chassis material. If you use multipoint chassis grounding then it is necessary to be much more careful about the electrical construction of the chassis.

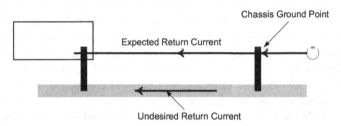

Chassis Ground Point

Expected Return Current

Undesired Return Current

FIGURE 1.3 Return Current Paths With Multiple Ground Points.

1.1.3 THE CONDUCTIVITY OF ALUMINUM

Aluminum is used throughout the electronics industry as a light, strong, and highly conductive chassis material—only silver, copper, and gold have a higher conductivity. You would expect an aluminum chassis to exhibit a decently low bulk resistance, and so it does, and is very suitable as a conductive ground as a result. Unfortunately, another property of aluminum (which is useful in other contexts) is that it oxidizes very readily on its surface, to the extent that all real-life samples of aluminum are

---

[1]But, when RF shielding and/or a low-inductance ground is required, multiple ground points may be essential. This is covered in Chapter 8.

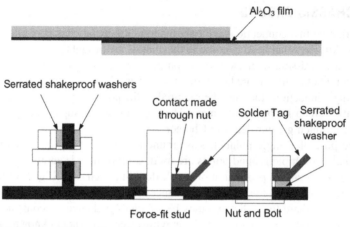

FIGURE 1.4 Electrical Connections to Aluminum.

covered by a thin surface film of aluminum oxide (Al_2O_3). Aluminum oxide is an insulator. In fact, it is such a good insulator that anodized aluminum, on which a thick coating of oxide is deliberately grown by chemical treatment, is used for insulating washers on heat sinks.

The practical consequence of this quality of aluminum oxide is that the contact resistance of two sheets of aluminum joined together is unpredictably high. Actual electrical contact will only be made where the oxide film is breached. Therefore, whenever you want to maintain continuity through a chassis made of separate pieces of aluminum, you must ensure that the plates are tightly bonded together, preferably with welding or by fixings that incorporate shakeproof serrated washers to dig actively into the surface. The same applies to ground connection points. The best connection (since aluminum cannot easily be soldered) is a force-fit or welded stud (Fig. 1.4), but if this is not available then a shakeproof serrated washer should be used underneath the nut, which is in contact with the aluminium.

Other Materials

Another common chassis material is cadmium- or tin-plated steel, which does not suffer from the oxidation problem. Mild steel has about three times the bulk resistance of aluminum so does not make such a good conductor, but it has better magnetic shielding properties and it is cheaper. Die-cast zinc is popular for its light weight and strength, and ease of creating complex shapes through the casting process; zinc's conductivity is 28% of copper. Other metals, particularly silver-plated copper, can be used where the ultimate in conductivity is needed and cost is secondary, as in radio frequency (RF) circuits. The advantage of silver oxide (which forms on the silver-plated surface) is that it is conductive and can be soldered through easily. Table 1.1 shows the conductivities and temperature coefficients of several metals.

Table 1.1 Conductivity of Metals

Metal	Relative Conductivity (Cu = 1, at 20°C)	Temperature Coefficient of Resistance (/°C at 20°C)
Aluminum (pure)	0.59	0.0039
Aluminum alloy:		
Soft-annealed	0.45–0.50	
Heat-treated	0.30–0.45	
Brass	0.28	0.002–0.007
Cadmium	0.19	0.0038
Copper:		
Hard drawn	0.895	0.00382
Annealed	1.0	0.00393
Gold	0.65	0.0034
Iron:		
Pure	0.177	0.005
Cast	0.02–0.12	
Lead	0.7	0.0039
Nichrome	0.0145	0.0004
Nickel	0.12–0.16	0.006
Silver	1.06	0.0038
Steel	0.03–0.15	0.004–0.005
Tin	0.13	0.0042
Tungsten	0.289	0.0045
Zinc	0.282	0.0037

1.1.4 GROUND LOOPS

Another reason for single-point chassis connection is that circulating chassis currents, when combined with other ground wiring, produce the so-called "ground loop," which is a fruitful source of low-frequency magnetically induced interference. A magnetic field can only induce a current to flow within a closed-loop circuit. Magnetic fields are common around power transformers—not only the conventional 50-Hz mains type (60 Hz in the United States) but also high-frequency (HF) switching transformers and inductors in switched-mode power supplies—and also other electromagnetic devices: contactors, solenoids, and fans. Extraneous magnetic fields may also be present. The mechanism of ground-loop induction is shown in Fig. 1.5.

Lenz's law tells us that the emf induced in the loop is given in Eq. (1.4)

$$V = -10^{-8} \times A \times n \times \frac{dB}{dt} \tag{1.4}$$

where A is the area of the loop in cm^2, B is the flux density normal to it in microtesla assuming a uniform field, and n is the number of turns ($n = 1$ for a single-turn loop).

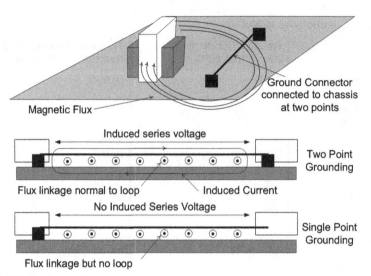

Ground Connector
connected to chassis
at two points

Magnetic Flux

Induced series voltage

Two Point
Grounding

Flux linkage normal to loop   Induced Current

No Induced Series Voltage

Single Point
Grounding

Flux linkage but no loop

FIGURE 1.5 The Ground Loop.

As an example, take a 10 μT 50 Hz field as might be found near a reasonable-sized mains transformer, contactor or motor, acting at right angles through the plane of a 10 cm² loop that would be created by running a conductor 1 cm above a chassis for 10 cm and grounding it at both ends. The induced emf is derived as described in Eqs. (1.5)–(1.7).

$$V = -10^{-8} \times 10 \times \frac{d}{dt}(10 \times \sin 2\pi \times 50 \times t) \tag{1.5}$$

$$V = -10^{-8} \times 10 \times 1000\pi \times \cos \omega t \tag{1.6}$$

$$V = 314 \text{ μV peak} \tag{1.7}$$

Magnetic field induction is usually a low-frequency phenomenon (unless you happen to be very close to a high-power radio transmitter), and you can see from this example that in most circumstances the induced voltages are low. But in low-level applications, particularly audio and precision instrumentation, they are far from insignificant. If the input circuit includes a ground loop, the interference voltage is injected directly in series with the wanted signal and cannot then be separated from it. The cures are as follows:

- Open the loop by grounding only at one point.
- Reduce the area of the loop (A in the equation above) by routing the offending wire(s) right next to the ground plane or chassis, or shortening it.
- Reduce the flux normal to the loop by repositioning or reorienting the loop or the interfering source.
- Reduce the interfering source, for instance by using a toroidal transformer.

1.1.5 POWER SUPPLY RETURNS

You will note from Fig. 1.2 that the output power supply 0-V connection (0 V(B)) has been shown separately from 0 V(A) and linked only at the power supply itself. What happens if, say for reasons of economy in wiring, you do not follow this practice but instead link the 0-V rails together at PCB3 and PCB2, as shown in Fig. 1.6?

The supply return currents I_{0V} from both PSB/PCB3 and PSA/PCB2 now share the same length of wire (or track, in a single-PCB system). This wire has a certain nonzero impedance, say for dc purposes it is R_S. In the original circuit this was only carrying $I_{0V(2)}$ and so the voltage developed across it can be described using Eq. (1.8):

$$V_S = R_S \times I_{0V(2)} \tag{1.8}$$

However, in the economy circuit

$$V_S = R_S \times \left(I_{0V(2)} + I_{0V(3)} \right) \tag{1.9}$$

This voltage is in series with the supply voltages to both boards and hence effectively subtracts from them.

Putting some typical numbers into the equations:

$I_{0V(3)} = 1.2$ A with a V_{B+} of 24 V because it is a high-power output board.
$I_{0V(2)} = 50$ mA with a V_{A+} of 3.3 V because it is a microprocessor board with
　　　　some CMOS logic on it.

Now assume that, for various reasons, the power supply is some distance remote from the boards, and you have without thinking connected it with 2 m of 7/0.2 mm equipment wire, which will have a room temperature resistance of about 0.2 Ω. The voltage V_S will be

$$V_S = 0.2 \times (1.2 + 0.05) = 0.25 \text{ V} \tag{1.10}$$

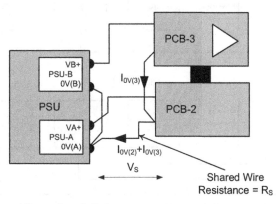

FIGURE 1.6 Common Power Supply Return.

which will drop the supply voltage at PCB2 to 3.05 V, less than the lower limit of operation for 3.3-V logic, *before* allowing for supply voltage tolerances and other voltage drops. One wrong wiring connection can make your circuit operation borderline! Of course, the 0.25 V is also subtracted from the 24-V supply, but a reduction of about 1% on this supply is unlikely to affect operation.

Varying Loads

If the 1.2 A load on PCB3 is varying—say several high-current relays may be switched at different times, ranging from all off to all on—then the V_S drop at PCB2 would also vary. This is very often worse than a static voltage drop because it introduces noise on the 0-V line. The effects of this include unreliable processor operation, variable set threshold voltage levels, and odd feedback effects such as chattering relays or, in audio circuits, low-frequency "motor-boating" oscillation.

For comparison, look at the same figures but applied to Fig. 1.2, with separate 0-V return wires. Now there are two voltage drops to consider: $V_{S(A)}$ for the 3.3-V supply and $V_{S(B)}$ for the 24-V supply. $V_{S(B)}$ is 1.2 A times 0.2 Ω, substantially the same (0.24 V) as before, but it is only subtracted from the 24-V supply. $V_{S(A)}$ is now 50 mA times 0.2 Ω or 10 mV, which is the only 0-V drop on the 3.3-V supply to PCB2 and is negligible.

The rule is: always separate power supply returns so that load currents for each supply flow in separate conductors (Fig. 1.7).

Note that this rule is easiest to apply if different power supplies have different 0-V connections (as in Fig. 1.2) but should also be applied if a common 0 V is used, as shown above. The extra investment in wiring is just about always worth it for peace of mind!

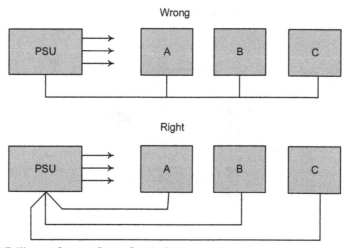

FIGURE 1.7 Ways to Connect Power Supply Returns.

Power Rail Feed

The rule also applies to the power rail feed as well as to its return, and in fact to any connection where current is being shared between several circuits. Say the high-power load on PCB3 was also being fed from the +5-V supply V_{A+}, then the preferred method of connection is two separate feeds (Fig. 1.8).

The reasons are the same as for the 0-V return: with a single feed wire, a common voltage drop appears in series with the supply voltage, injected this time in the supply rail rather than the 0-V rail. Fault symptoms are similar. Of course, the example above is somewhat artificial in that you would normally use a rather more suitable size of wire for the current expected. High currents flowing through long wires demand a low resistance and hence a thick conductor is required. If you are expecting a significant voltage drop then you will take the trouble to calculate it for a given wire diameter, length, and current. See Table 1.3 for a guide to the current-carrying abilities of common wires. The point of the previous examples is that voltage drops have a habit of cropping up when you are *not* expecting them.

Conductor Impedance

Note that the previous examples, and those on the next few pages, tacitly assume for simplicity that the wire impedance is resistive only. In fact, real wire has inductance as well as resistance and this comes into effect as soon as the wire is carrying ac, increasing in significance as the frequency is raised. A 1-m length wire of 16/0.2 equipment has a resistance of 38 mΩ and a self-inductance of 1.5 μH. At 4 A dc the voltage drop across it will be 152 mV. An ac current with a rate-of-change of 4 A/μs will generate 6 V across it. Note the difference! The later discussion of wire types includes a closer look at inductance.

1.1.6 INPUT SIGNAL GROUND

Fig. 1.2 shows the input signal connections being taken directly to PCB1 and not grounded outside the PCB. To expand on this, the preferred scheme for two-wire

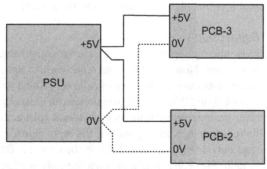

FIGURE 1.8 Separate Power Supply Rail Feeds.

single-ended input connections is to take the ground return directly to the reference point of the input amplifier, as shown in Fig. 1.9A.

The reference point on a single-ended input is not always easy to find: look for the point from which the input voltage must be developed for the amplifier gain to act on it alone. In this way, no extra signals are introduced in series with the wanted signal by means of a common impedance. In each of the examples in Fig. 1.9 of bad input wiring, getting progressively worse from (B) to (D), the impedance X−X acts as a source of unwanted input signal due to the other currents flowing in it as well as the input current.

Connection to 0 V Elsewhere on the Printed Circuit Board

Insufficient control over PCB layout is the most usual cause of arrangement (B), especially if autorouting layout software is used. Most CAD layout software assumes that the 0-V rail is a single node and feels itself free to make connections to it at any point along the track. To overcome this, either specify the input return point as a separate node and connect it later, or edit the final layout as required. Manual layout is capable of exactly the same mistake, although in this case it is due to lack of communication between designer and layout draftsman.

Connection to 0 V Within the Unit

Arrangement (C) is quite often encountered if one pole of the input connector naturally makes contact with the metal case, such as happens with the standard Bayonet Neill−Concelman (BNC) coaxial connector, or if for reasons of connector economy a common ground conductor is shared between multiple input, output, or control signals that are distributed among different boards. With sensitive input signals, the latter is false economy; and if you have to use a BNC-type connector, you can get versions with insulating washers, or mount it on an insulating subpanel in a hole in the metal enclosure. Incidentally, taking a coax lead internally from an uninsulated BNC socket to the PCB, with the coax outer connected both to the BNC shell and the PCB 0 V, will introduce a ground loop (see Section 1.1.4) unless it is the only path for ground currents to take. But at radio frequencies, this effect is countered by the ability of coax cable to concentrate the signal and return currents within the cable, so that the ground loop is only a problem at low frequencies.

External Ground Connection

Despite being the most horrific input grounding scheme imaginable, arrangement (D) is unfortunately not rare. Now, not only are noise signals internal to the unit coupled into the signal path but also all manner of external ground noise is included. Local earth differences of up to 50 V at mains frequency can exist at particularly bad locations such as power stations, and differences of several volts are more common. The only conceivable reason to use this layout is if the input signal is already firmly tied to a remote ground outside the unit, and if this is the case it is far better to use a differential amplifier as in Fig. 1.9E, which is often the only workable solution for low-level signals and is in any case only a logical development of the correct

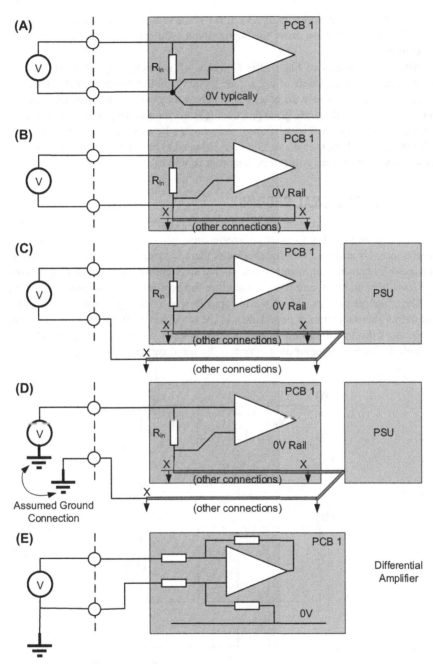

FIGURE 1.9 Input Signal Grounding.

approach for single-ended signals (A). If for some reason you are unable to take a ground return connection from the input signal, you will be stuck with ground-injected noise.

All of the schemes of Fig. 1.9B—D will work perfectly happily if the desired input signal is several orders of magnitude greater than the ground-injected interference, and this is frequently the case, which is how they came to be common practice in the first place. If there are good practical reasons for adopting them (for instance, connector, or wiring cost restrictions), and you can be sure that interference levels will not be a problem, then do so. But you will need to have control over all possible connection paths before you can be sure that problems will not arise in the field.

1.1.7 OUTPUT SIGNAL GROUND

Similar precautions need to be taken with output signals, for the reverse reason. Inputs respond unfavorably to external interference, whereas outputs are the cause of interference. Usually in an electronic circuit there is some form of power amplification involved between input and output, so that an output will operate at a higher current level than an input, and there is therefore the possibility of unwanted feedback.

The classical problem of output-to-input ground coupling is where both input and output share a common impedance, in the same way as the power rail common impedances discussed earlier. In this case the output current is made to circulate through the same conductor as it connects the input signal return (Fig. 1.10A).

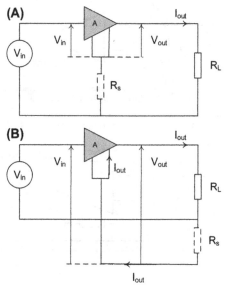

FIGURE 1.10 Output-to-Input Coupling.

A tailor-made feedback mechanism has been inserted into this circuit, by means of R_S. The input voltage at the amplifier terminals is supposed to be V_{in}, but actually it is as defined in Eq. (1.11):

$$V_{in'} = V_{in} - (I_{out} \times R_S) \tag{1.11}$$

Redrawing the circuit to reference everything to the amplifier ground terminals (Fig. 1.10B) shows this more clearly. When we work out the gain of this circuit, it turns out to be as shown in Eq. (1.12):

$$\frac{V_{out}}{V_{in}} = \frac{A}{\left(1 + \left[A \times \dfrac{R_S}{(R_L + R_S)}\right]\right)} \tag{1.12}$$

which describes a circuit that will oscillate if:

$$A \times \frac{R_S}{(R_L + R_S)} < -1 \tag{1.13}$$

In other words, for an inverting amplifier, the ratio of load impedance to common impedance must be less than the gain, to avoid instability. Even if the circuit remains stable, the extra coupling due to R_S upsets the expected response. Remember also that all the above terms vary with frequency, usually in a complex fashion, so that at high frequencies the response can be unpredictable. Note that although this has been presented in terms of an analog system (such as an audio amplifier), any system in which there is input−output gain will be similarly affected. This can apply equally to a digital system with an analog input and digital outputs which are controlled by it.

Avoiding the Common Impedance

The preferable solution is to avoid the common impedance altogether by careful layout of input and output grounds. We have already looked at input grounds, and the grounding scheme for outputs is essentially similar: take the output ground return directly to the point from which output current is sourced, with no other connection (or at least, no other susceptible connection) in between. Normally, the output current comes from the power supply so the best solution is to take the return directly back to the supply. Thus the layout of PCB3 in Fig. 1.2 should have a separate ground track for the high-current output as in Fig. 1.11A, or the high-current output terminal could be returned directly to the power supply, bypassing PCB3 (B).

If PCB3 contains only circuits that will not be susceptible to the voltage developed across R_S, then the first solution is acceptable. The important point is to decide in advance where your return currents will flow and ensure that they do not affect the operation of the rest of the circuits. This entails knowing the ac and dc impedance of any common connections, the magnitude and bandwidth of the output currents and the susceptibility of the potentially affected circuits.

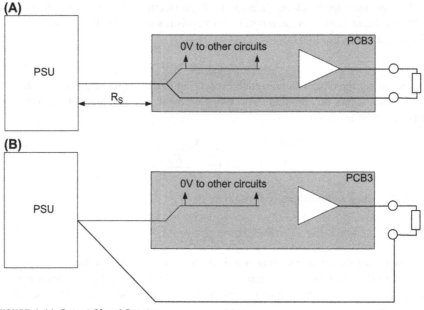

FIGURE 1.11 Output Signal Returns.

1.1.8 INTERBOARD INTERFACE SIGNALS

There is one class of signals we have not yet covered, and that is those signals which pass within the unit from one board to another. Typically these are digital control signals or analog levels that have already been processed, so are not low level enough to be susceptible to ground noise and are not high current enough to generate significant quantities of it. To be thorough in your consideration of ground return paths, these signals should not be left out: the question is what to do about them?

Often the answer is nothing. If no ground return is included specifically for interboard signals then signal return current must flow around the power supply connections and therefore the interface will suffer all the ground-injected noise V_n that is present along these lines (Fig. 1.12). But, if your grounding scheme is well thought out, this may well not be enough to affect the operation of the interface. For instance, 100 mV of noise injected in series with a CMOS logic interface, which has a noise margin of 1 V will have no direct effect. Or, ac noise injection onto a dc analog signal that is well filtered at the interface input will be tolerable.

Partitioning the Signal Return

There will be occasions when taking the long-distance ground return route is not good enough for your interface. Typically these are as follows:

- where high-speed digital signals are communicated, and the ground return path has too much inductance, resulting in ringing on the signal transitions;

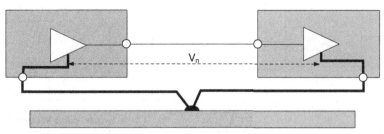

FIGURE 1.12 Interboard Ground Noise.

- when interfacing precision analog signals which cannot stand the injected noise or low-voltage dc differentials.

If you solve these headaches by taking a local interboard ground connection for the signal of interest, you run the risk of providing an alternative path for power supply return currents, which nullifies the purpose of the local ground connection. A fraction of the power return current will flow in the local link (Fig. 1.13), the proportion depending on the relative impedances, and you will be back where you started.

If you really need the local signal return, but are in trouble with ground return currents, there are two options to pursue:

Option 1: Separate the ground return (Fig. 1.14) for the input side of the interface from the rest of the ground on that PCB. This has the effect of moving the ground noise injection point inboard, after the input buffer, which may be all that you need. A development of this scheme is to include a "stopper" of a few ohms in the gap X–X. This prevents dc ground current flow because its impedance is high relative to that of the correct ground path, but it effectively ties the input buffer to its parent ground at high frequencies and prevents it from floating if the interboard link is disconnected.

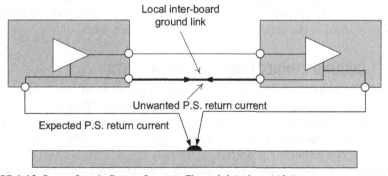

FIGURE 1.13 Power Supply Return Currents Through Interboard Links.

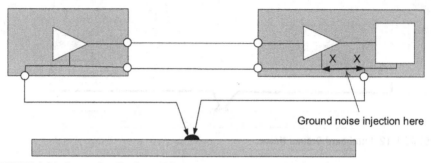

Ground noise injection here

FIGURE 1.14 Separating the Ground Returns.

Option 2: Use differential connections at the interface. The signal currents are now balanced and do not require a ground return; any ground noise is injected in common mode and is canceled out by the input buffer. This technique is common where high-speed or low-level signals have to be communicated some distance, but it is applicable at the interboard level as well. It is of course more expensive than typical single-ended interfaces since it needs dedicated buffer drivers and receivers.

1.1.9 STAR-POINT GROUNDING

One technique that can be used as a circuit discipline is to choose one point in the circuit and to take all ground returns to this point. This is then known as the "star point." Fig. 1.2 shows a limited use of this technique in connecting together chassis, mains earth, power supply ground, and 0-V returns to one point. It can also be used as a local subground point on printed circuit layouts.

When comparatively few connections need to be made, this is a useful and elegant trick, especially as it offers a common reference point for circuit measurements. It can be used as a reference for power supply voltage sensing, in conjunction with a similar star point for the output voltage (Fig. 1.2 again). It becomes progressively messier as more connections are brought to it and should not substitute for a thorough analysis of the anticipated ground current return paths.

1.1.10 GROUND CONNECTIONS BETWEEN UNITS

Much of the theory about grounding techniques tends to break down when confronted with the prospect of several interconnected units. This is because the designer often either has no control over the way in which units are installed or is forced by safety-related or other installation practices to cope with a situation that is hostile to good grounding practice.

The classic situation is where two mains-powered units are connected by one (or more) signal cable (Fig. 1.15). This is the easiest situation to explain and visualize;

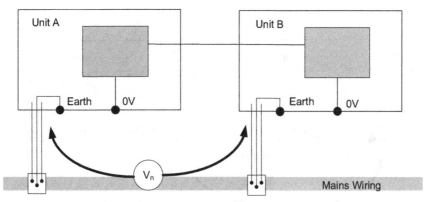

FIGURE 1.15 Interunit Ground Connection via the Mains.

actual setups may be complicated by having several units to contend with, or different and contradictory ground regimes, or by extra mechanical bonding arrangements.

This configuration is exactly analogous to that of Fig. 1.12. Ground noise, represented by V_n, is coupled through the mains earth conductors and is unpredictable and uncontrollable. If the two units are plugged into the same mains outlet, it may be very small, though never zero, as some noise is induced simply by the proximity of the live and neutral conductors in the equipment mains cable. But this configuration cannot be prescribed: it will be possible to use outlets some distance apart, or even on different distribution rings, in which case the ground connection path could be lengthy and could include several noise injection sources. Absolute values of injected noise can vary from less than a millivolt rms in very quiet locations to several volts, or even tens of volts, mentioned in Section 1.1.6. This noise effectively appears in series with the signal connection.

To tie the signal grounds in each unit together you would normally run a ground return line along with the signal in the same cable, but then

- noise currents can now flow in the signal ground, so it is essential that the impedance of the ground return (R_s) is much less than the noise source impedance (R_n)—usually but not invariably the case—otherwise the ground-injected noise will not be reduced;
- you have created a ground loop (Fig. 1.16, and compare this with Section 1.1.4), which by its nature is likely to be both large and variable in area, and to intersect various magnetic field sources, so that induced ground currents become a real hazard.

Breaking the Ground Link

If the susceptibility of the signal circuit is such that the expected environmental noise could affect it, then you have a number of possible design options.

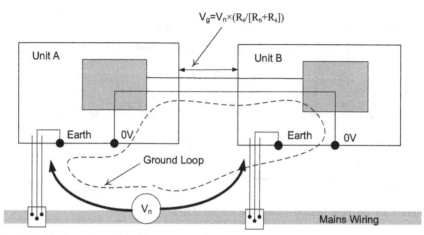

FIGURE 1.16 Ground Loop via Signal and Mains Earths.

- Float one or other unit (disconnect its mains ground connection), which breaks the ground loop at the mains lead. This is already done for you if it is battery powered, and in fact this is one good reason for using battery-powered instruments. On safety-class I (earthed) mains-powered equipment, doing this is not an option because it violates the safety protection.
- Transmit your signal information via a differential link, as recommended for interboard signals earlier. Although a ground return is not necessary for the signal, it is advisable to include one to guard against too large a voltage differential between the units. Noise signals are now injected in common mode relative to the wanted signal and so will be attenuated by the input circuit's common-mode rejection, up to the operating limit of the circuit, which is usually several volts.
- Electrically isolate the interface. This entails breaking the direct electrical connection altogether and transmitting the signal by other means, for instance a transformer, opto-coupler or fiber-optic link. This allows the units to communicate in the presence of several hundred volts or more of noise, depending on the voltage rating of the isolation; alternatively it is useful for communicating low-level ac signals in the presence of relatively moderate amounts of noise that cannot be eliminated by other means.

1.1.11 SHIELDING

Some mention must be made here of the techniques of shielding interunit cables, even though this is more properly the subject of Chapter 8. Shielded cable is used to protect signal wires from noise pickup, or to prevent power or signal wires from radiating noise. This apparently simple function is not so simple to apply in practice. The characteristics of shielded cable are discussed later (see Section 1.2.4); here we shall look at how to apply it.

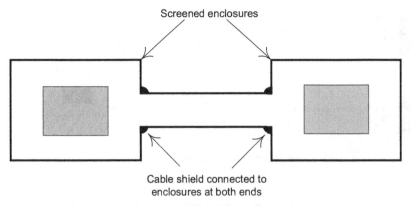

Screened enclosures

Cable shield connected to
enclosures at both ends

FIGURE 1.17 Radio Frequency (RF) Cable Shield Connections.

At which end of a cable do you connect the shield, and to what? There is no one correct answer, because it depends on the application. If the cable is used to connect two units, which are both contained within screened enclosures to keep out or keep in RF energy, then the cable shield has to be regarded as an extension of the enclosures and it must be connected to the screening at both ends via a low-inductance connection, preferably the connector screen itself (Fig. 1.17). This is a classic application of electromagnetic compatibility principles and is discussed more fully in Sections 8.5 and 8.7. Note that if both of the unit enclosures are themselves separately grounded then you have formed a ground loop (again). Because ground loops are a magnetic coupling hazard, and because magnetic coupling diminishes in importance at higher frequencies, this is often not a problem when the purpose of the screen is to reduce HF noise. The difficulty arises if you are screening against both high and low frequencies, because at low frequencies you should ground the shield at one end only, and in these cases you may have to take the expensive option of using double-shielded cable.

The shield should not be used to carry signal return currents unless it is at RF and you are using coaxial cable. Noise currents induced in it will add to the signal, nullifying the effect of the shield. Typically, you will use a shielded pair to carry high-impedance low-level input signals, which would be susceptible to capacitive pickup. (A cable shield will *not* be effective against magnetic pickup, for which the best solution is twisted pair.)

Which End to Ground for Low-Frequency Shielding?

If the input source is floating, then the shield can be grounded at the amplifier input. A source with a floating screen around it can have this screen connected to the cable shield. But, if the source screen itself is grounded, you will create a ground loop with the cable shield, which is undesirable: ground-loop current induced in the shield will couple into the signal conductors. One or other of the cable shield ends should be left floating, depending on the relative amount of unavoidable capacitive coupling to

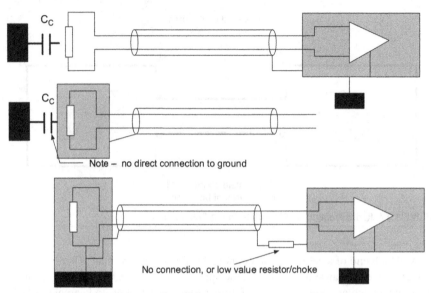

FIGURE 1.18 Cable Shield Connection Options.

ground (C_c) that exists at either end. If you have the choice, usually it is the source end (which may be a transducer or sensor) that has the lower coupling capacitance so this end should be floated.

If the source is single-ended and grounded, then the cable shield should be grounded at the source and either left floating at the (differential) input end or connected through a choke or low-value resistor to the amplifier ground. This will preserve dc and low-frequency continuity while blocking the flow of large induced high-frequency currents along the shield. The shield should not be grounded at the opposite end to the signal. Fig. 1.18 shows the options.

Electrostatic Screening

When you are using shielded cable to prevent electrostatic radiation from output or interunit lines, ground-loop induction is usually not a problem because the signals are not susceptible, and the cable shield is best connected to ground at both ends. The important point is that each conductor has a distributed (and measurable) capacitance to the shield, so that currents on the shield will flow as long as there are ac signals propagating within it (see Fig. 1.19). These shield currents must be provided with a low-impedance ground return path so that the shield voltages do not become substantial. The same applies in reverse when you consider coupling of noise induced on the shield into the conductors.

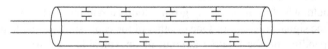

FIGURE 1.19 Conductor to Shield Coupling Capacitance.

Surface Transfer Impedance

At high frequencies, the notion of surface transfer impedance becomes useful as a measure of shielding effectiveness. This is the ratio of voltage developed between the inner and outer conductors of shielded cable due to interference current flowing in the shield, expressed in milliohms per unit length. It should not be confused with characteristic impedance, with which it has no connection. A typical single braid screen will be 10 mΩ/m or so below 1 MHz, rising at a rate of 20 dB/decade with increasing frequency. The common aluminum/mylar foil screens are around 20 dB worse. Unhappily, surface transfer impedance is rarely specified by cable manufacturers.

1.1.12 **THE SAFETY EARTH**

A brief word is in order about the need to ensure a mains earth connection, since it is obvious from the preceding discussion that this requirement is frequently at odds with antiinterference grounding practice. Most countries now have electrical standards which require that equipment powered from dangerous voltages should have a means of protecting the user from the consequences of component failure. The main hazard is deemed to be inadvertent connection of the live mains voltage to parts of the equipment with which the user could come into contact directly, such as a metal case or a ground terminal.

Imagine that the fault is such that it makes a short circuit between live and case, as shown in Fig. 1.20. These are normally isolated, and if no earth connection is made the equipment will continue to function normally—but the user will be threatened with a lethal shock hazard without knowing it. If the safety earth conductor is connected then the protective mains fuse will blow when the fault occurs, preventing the hazard and alerting the user to the fault.

For this reason a safety earth conductor is mandatory for all equipment that is designed to use this type of protection and does not rely on extra levels of insulation. The conductor must have an adequate cross section to carry any prospective fault

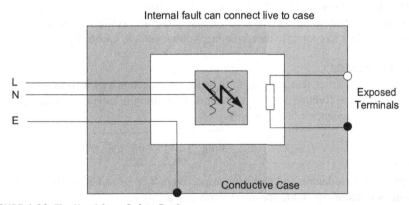

FIGURE 1.20 The Need for a Safety Earth.

current, and all accessible conductive parts must be electrically bonded to it. The general requirements for earth continuity are as follows:

- The earth path should remain intact until the circuit protection has operated;
- The impedance should not significantly or unnecessarily restrict the fault current.

As an example, EN 60065 requires a resistance of less than 0.5 Ω at 10 A for a minute. Design for safety is covered in greater detail in Section 9.1.

1.2 WIRING AND CABLES

This section will look briefly at the major types of wire and cable that can be found within typical electronic equipment. There are so many varieties that it comes as something of a surprise to find that most applications can be satisfied from a small part of the range. First, a couple of definitions: *wires* are single-circuit conductors, insulated or not; *cables* are groups of individual conductors, separately insulated and mechanically contained within an overall sheath.

1.2.1 WIRE TYPES

The simplest form of wire is tinned copper wire, available in various gauges depending on required current-carrying capacity. Component leads are almost invariably tinned copper, but the wire on its own is not used to a great extent in the electronics industry. Its main application was for links on PCBs, but the increasing use of double-sided and multilayer plated-through-hole boards makes them redundant. Tinned copper wire can also be used in rewirable fuse links. Insulated copper wire is used principally in wound components such as inductors and transformers. The insulating coating is a polyurethane compound that has self-fluxing properties when heated, which makes for ease of soldered connection, especially to thin wires.

Table 1.2 compares dimensions, current capacity, and other properties for various sizes of copper wire. In the United Kingdom, the wires are specified under BS EN 13602 for tinned copper and BS EN 60182 (IEC 60182-1) for enamel insulated and are sold in metric sizes. Two grades of insulation are available: Grade 1 being thinner; Grade 2 has roughly twice the breakdown voltage capability.

Wire Inductance

We mentioned earlier that any length of wire has inductance as well as resistance. The approximate formula for the inductance of a straight length of round section wire at high frequencies is shown in Eq. (1.14):

$$L = K \times l \times \left(2.3 \log_{10}\left(\frac{4l}{d}\right) - 1\right) \ \mu H \qquad (1.14)$$

where l and d are length and diameter respectively, $l \gg d$ and K is 0.0051 for dimensions in inches or 0.002 for dimensions in cm.

Table 1.2 Characteristics of Copper Wire

Wire Size (mm dia)	1.6	1.25	0.71	0.56	0.315	0.2
Approximate standard wire gauge (SWG)	16	18	22	24	30	35
Approximate American wire gauge (AWG)	14	16	21	23	28	32
Current rating (Amps)	22	12.2	3.5	2.5	0.9	0.33
Fusing current (Amps)	70	45	25	17	9	5
Resistance/meter at 20°C (W)	0.0085	0.014	0.043	0.069	0.22	0.54
Inductance of 1 m length (μH)	1.36	1.41	1.53	1.57	1.69	1.78

This equation is used to derive the inductance of a 1 m length (note that this is not quite the same as inductance per meter) in Table 1.2 and you can see that inductance is only marginally affected by wire diameter. Low values of inductance are not easily obtained by adding cross section and the reactive component of impedance dominates above a few kilohertz whatever the size of the conductor. A useful rule of thumb is that the inductance of a 1 in. length of ordinary equipment wire is around 20 nH and that of a 1 cm length is around 7 nH. This factor becomes important in high-speed digital and RF circuits where performance is limited by physical separation and also in circuits where the rate-of-change of current (di/dt) is high.

Equipment Wire

Equipment wire is classified mainly according to its insulation. This determines the voltage rating and the environmental properties of the wire, particularly its operating temperature range and its resistance to chemical and solvent attack. The standard type of wire, and the most widely available, is PVC insulated to BS4808, which has a maximum temperature rating of 85°C. As well as current ratings at 25°C you will find specifications at 70°C; these allow for a 15°C temperature rise, to the maximum rated temperature, at the specified current. Temperature ratings of 70°C for large conductor switchgear applications and 105°C to American and Canadian underwriters laboratories (UL) and Canadian Standards Association (CSA) standards are also available in PVC. PTFE is used for wider temperature ranges, up to 200°C, but is harder to work with. Other more specialized insulations include extraflexible PVC for test leads and silicone rubber for high temperature (150°C) and harsh environments. Many wires carry military, telecom, and safety authority approval and have to be specified on projects that are carried out for these customers.

Table 1.3 is included here as a guide to the electrical characteristics of various commonly available PVC equipment wires. Note that the published current ratings of each wire are related to permitted temperature rise. Copper has a positive temperature coefficient of resistivity of 0.00393 per °C, so that resistance rises with increasing current; using the room temperature resistance may be optimistic by several percent if the actual ambient temperature is high or if significant self-heating occurs.

Table 1.3 Characteristics of BS4808 Equipment Wire

Wire Size (No. of strands/mm dia)	1/0.6	7/0.2	16/0.2	24/0.2	32/0.2	63/0.2
Resistance (Ω/ 1000 m at 20°C)	64	88	38	25.5	19.1	9.7
Current rating at 70°C (A)	1.8	1.4	3.0	4.5	6.0	11.0
Current rating at 25°C (A)	3.0	2.0	4.0	6.0	10.0	18.0
Voltage drop/ meter at 25°C current	192 mV	176 mV	152 mV	153 mV	191 mV	175 mV
Voltage rating	1 KV	1 KV	1 KV	1.5 KV	1.5 KV	1.5 KV
Overall diameter (mm)	1.2	1.2	1.55	2.4	2.6	3.0
Near equivalent American wire gauge (not direct equivalent)	23	24	20	18	17	15

Wire-Wrap Wire

A further specialized type of wire is that used for wire-wrap construction. This is available primarily in two sizes, with two types of insulation: Kynar, trademark of Pennwalt, and Tefzel, trademark of DuPont. Tefzel is more expensive but has a higher temperature rating and is easier to strip. Table 1.4 lists the properties of the four types.

1.2.2 CABLE TYPES

Ignoring the more specialized types, cables can be divided loosely into three categories:

- power
- data and multicore
- RF

Table 1.4 Characteristics of Wire-Wrap Wire

	Kynar		Tefzel	
	30 AWG	26 AWG	30 AWG	26 AWG
Conductor diameter (mm)	0.25	0.4	0.25	0.4
Maximum service temperature (°C)	105	105	155	155
Resistance/m at 20°C (W)	0.345	0.136	0.345	0.136
Voltage rating (V)	–	–	375	375
Current rating at 50°C (A)	–	–	2.6	4.5

1.2.3 POWER CABLES

Because mains power cables are inherently meant to carry dangerous voltages they are subject to strict standards: in the United Kingdom, the principal one is BS6500. International ones are IEC 60227 for PVC insulated or IEC 60245 for rubber insulated. These standards have been harmonized throughout the CENELEC countries in Europe so that any equipment that uses a cable with a harmonized code number will be acceptable throughout Europe. BS6500 specifies a range of current ratings and allows a variety of sheath materials depending on application. The principal ones are rubber and PVC; rubber is about twice the price of PVC but is somewhat more flexible and therefore suitable for portable equipment, and can be obtained in a high-temperature heat and oil resisting, flame retardant (HOFR) grade. The current-carrying capacities and voltage drops for dc and single-phase ac, and supportable mass are shown in Table 1.5.

Unfortunately, American and Canadian mains cables also need to be approved, but the approval authorities are different (UL and CSA). Cables manufactured to the European harmonized standards do not meet UL/CSA standards and vice versa. So, if you intend to export your mains-powered equipment both to Europe and North America, you will need to supply it with two different cables. The easy way to do this is to use a CEE-22 6 A connector on the equipment and supply a different cable set depending on the market. This practice has been adopted by virtually all of the large-volume multinational equipment suppliers with the result that the CEE-22 mains inlet is universally accepted. There are also several suppliers of ready-made cable sets for the different countries!

The alternative, widely used for information technology and telecoms equipment, is to use a "wall-wart" plug top power supply and provide different ones for each market, so that the cable carries low-voltage dc and no approved mains cable is needed.

Table 1.5 Characteristics of BS6500 Mains Cables

Cross-Sectional Area (mm^2)	0.5	0.75	1.0	1.25	1.5	2.5
Current-carrying capacity (A)	3	6	10	13	16	25
Voltage drop per amp per meter (mV)	93	62	46	37	32	19
Maximum supportable mass (kg)	2	3	5	5	5	5
Correction Factor for Ambient Temperature						
60°C rubber and PVC cables	Temp. (°C)	35	40	45	50	55
	CF	0.92	0.82	0.71	0.58	0.41
85°C HOFR rubber cables	Temp. (°C)	35–50	55	60	65	70
	CF	1.0	0.96	0.83	0.67	0.47

From IEE wiring regulations 15th edition.

1.2.4 DATA AND MULTICORE CABLES

Multicore cables are used when you need to transport several signals between the same source and destination. They should never be used for mains power because of the hazards that could be created by a cable failure, nor should high-power and signal wires be run within the same cable because of the risks of interference. Conventional multicore is available with various numbers of conductors in 7/0.1, 7/0.2, and 16/0.2 mm, with or without an overall braided screen. As well as the usual characteristics of current and voltage ratings, which are less than the ratings for individual wires because the conductors are bunched together, inter-conductor capacitance is an important consideration, especially for calculating cross talk (to which we return shortly). It is not normally specified for standard multicore, although nominal conductor-to-screen capacitances of 150–200 pF/m are sometimes quoted. For a more complete specification you need to use data cable.

Data Communication Cables

Data cables are really a special case of multicore, but with the explosion in data communications they now deserve a special category of their own. Transmitting digital data presents special problems, notably

- the need to communicate several parallel channels at once, usually over short distances, which has given rise to ribbon cable;
- the need to communicate a few channels of high-speed serial data over long distances with a high data integrity, which has given rise to cables with multiple individually screened conductor pairs in an overall sheath that may or may not be screened.

Interconductor capacitances and characteristic impedances (which we will discuss when we come to transmission lines) are important for digital data transmission and are quoted for most of these types. Table 1.6 summarizes the characteristics of the most common of them.

Table 1.6 Characteristics of Data Transmission Cables

Cable Type	Ribbon		Round	
	Straight	Twisted Pair	Type A	Type B
Interconductor capacitance (pF/m)	50	72	40–115	41–98
Conductor-screen capacitance (pF/m)	–	–	66–213	72–180
Characteristic impedance (Ω)	105	105	–	50
Voltage rating (V)	300	300	300	30

Type A: multipair/multicore overall foil-screened cable.
Type B: multipair individually foil-screened cable.

Structured Data Cable

One particular cable application which forms an important aspect of data communications is so-called "structured" or "generic" cabling. This is general-purpose data-comms cable, which is installed into the structure of a building or campus to enable later implementation of a variety of telecom and other networks: voice, data, text, image, and video. In other words, the cable's actual application is not defined at the time of installation. To allow this, its characteristics, along with those of its connectors, performance requirements and the rules for acceptable routing configurations, are defined in ISO/IEC 11801 (the US TIA/EIA-568 covers the same ground).

Equipment designers may not be too interested in the specifications of this cable until they come to design a LAN or telecom port interface; then the cable becomes important. The TIA/EIA-568 (both ISO/IEC 11801 and EN 50173 have similar specifications) parameters for the preferred 100-Ω quad-pair cable are shown in Table 1.7. The standard allows for a series of categories with increasing bandwidth. Cat 5 and Cat 5e are popular and have been widely installed.

Other characteristics, particularly mechanical dimensions, cross talk performance (extended for Cat 5e and 6) and propagation delay skew are also defined.

Shielding and Microphony

Shielding of data and multicore falls into three categories:

*   Copper braid. This offers a good general-purpose electrical shield but cannot give 100% shield coverage (80%–95% is typical), and it increases the size and weight of the cable.

Table 1.7 Characteristics of 50 Ω Coaxial Cables

Cable Type	URM43	URM67	RG58C/U	RG174A/U	RG178B/U
Overall diameter (mm)	5	10.3	5	2.6	1.8
Conductor material	Sol 1/0.9	Str 7/0.77	Str 19/0.18	Str 7/0.16	Str 7/0.1
Dielectric material	Solid polythene/polyethylene				PTFE
Voltage rating[a]	2.6 kV pk	6.5 kV pk	3.5 kV pk	1.5 kV rms	1 kV rms
Attenuation dB/10 m at 100 MHz	1.3	0.68	1.6	2.9	4.4
At 1 GHz	4.5	2.5	6.6	10	14
Temperature range (°C)		−40 to +85			−55/+200
Cost per 100 m£[b]	58	174	83	36	129

[a] Voltage ratings may be specified differently between manufacturers.
[b] Prices are typical costs as in 2017.

- Tape or foil. The most common of these is aluminized mylar. A drain wire is run in contact with the metallization to provide a terminating contact and to reduce the inductance of the shield when it is helically wound. This provides a fairly mediocre degree of shielding but hardly affects the size, weight, and flexibility of the cable at all.
- Composite foil and braid. These provide excellent electrostatic shielding for demanding environments but are more expensive—about twice the price of foil types.

For small-signal applications, particularly low-noise audio work, another cable property is important—microphony due to triboelectric induction. Any insulator generates a static voltage when it is rubbed against a dissimilar material, and this effect results in a noise voltage between conductor and screen when the cable is moved or vibrated. Special low-noise cable is available, which minimizes this noise mechanism by including a layer of low-resistance dielectric material between braid and insulator to dissipate the static charge. When you are terminating this type of cable, make sure the low-resistance layer is stripped back to the braid, otherwise you run the risk of a near short circuit between inner and outer.

1.2.5 RADIO FREQUENCY CABLES

Cables for the transport of RF signals are almost invariably coaxial, apart from a few specialized applications such as HF aerial feeder which may use balanced lines. Coax's outstanding property is that the field due to the signal propagating along it is confined to the inside of the cable (Fig. 1.21), so that interaction with its external environment is kept to a minimum. A further useful property is that the characteristic impedance of coax is easily defined and maintained. This is important for RF applications as in these cases cable lengths frequently exceed the operating wavelength.

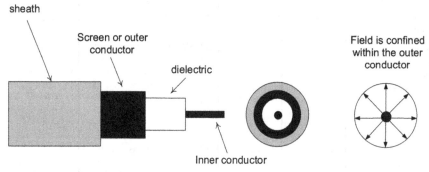

FIGURE 1.21 Coaxial Cable.

The generic properties of transmission lines—of which coax is a particular type—will be discussed in Section 1.3. The parameters that you will normally find in coax specifications are as follows:

- Characteristic impedance (Z_o): the universal standard is 50 Ω, since this results in a good balance between mechanical properties and ease of circuit application. 75 and 93 Ω are other standards that find application in video and data systems. Any other impedance must be regarded as special.
- Dielectric material. This affects just about every property of the cable, including Z_o, attenuation, voltage handling, physical properties, and temperature range. Solid polythene or polyethylene are the standard materials; cellular polyethylene, in which part of the dielectric insulation is provided by air gaps, offers lower weight and lower attenuation losses but is more prone to physical distortion than solid. These two have a temperature rating of 85°C. PTFE is available for higher temperature (200°C) and lower loss applications but is much more expensive.
- Conductor material. Copper is universal. Silver plating is sometimes used to enhance high-frequency conductivity through the skin effect, or copper can be plated onto steel strands for strength. Inner conductors can be single or stranded; stranded is preferred when the cable will be subjected to flexing. The outer conductor is normally copper braid, again for flexibility. The degree of braid coverage affects high-frequency attenuation and also the shielding effectiveness. Solid outer conductor is available for extreme applications that do not require flexing.
- Voltage rating. A thicker cable can be expected to have a higher voltage rating and a lower attenuation. You cannot easily relate the voltage rating to power-handling ability unless the cable is matched to its characteristic impedance. If the cable is not matched, voltage standing waves will exist, which will produce peaks at distinct locations along the cable higher than would be expected from the power/impedance relationship.
- Attenuation. Losses in the dielectric and conductors result in increasing attenuation with frequency and distance, so attenuation is quoted per 10 m at discrete frequencies, and you can interpolate to find the attenuation at your operating frequency. Cable losses can easily catch you out, especially if you are operating long cables over a wide bandwidth and forget to allow for several extra dB of loss at the top end.

Readily available coax cables are specified to two standards, the US MIL-C-17 for the RG/U (Radio Government, Universal) series and the UK BS2316 for the UR-M (Uniradio) series. The international standard is IEC 60096. Table 1.7 gives comparative data for a few common 50-Ω types.

One word of warning: never confuse screened audio cable with RF coax. The braids and dielectric materials are quite different, and audio cable's Z_o is undefined, and its attenuation at high frequencies is large. If you try to feed RF down it you will not get much at the other end! On the other hand, RF coax *can* be used to carry audio signals.

1.2.6 TWISTED PAIR

Special mention should be given to twisted pair because it is a particularly effective and simple way of reducing both magnetic and capacitive interference pickup. Twisting the wires tends to ensure a homogeneous distribution of capacitances. The capacitance to ground and also to extraneous sources is balanced. This means that the common-mode capacitive coupling is also balanced, which allows high common-mode rejection. Twisted and untwisted pairs are compared in Fig. 1.22, but note that if your problem is already common-mode capacitive coupling, twisting the wires will not help. For that, you need shielding.

Twisting is most useful in reducing low-frequency magnetic pickup because it reduces the magnetic loop area to almost zero. Each half-twist reverses the direction of induction so, assuming a uniform external field, two successive half-twists cancel the wires' interaction with the field. Effective loop pickup is now reduced to the small areas at each end of the pair, plus some residual interaction due to nonuniformity of the field and irregularity in the twisting. Assuming that the termination area is included in the field, the number of twists per unit length is unimportant: around 8–16 turns per foot (26–50 turns per meter) is usual. Fig. 1.23 shows measured

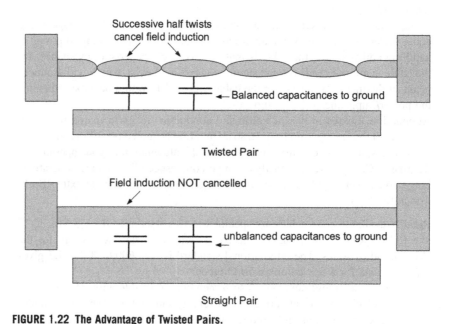

FIGURE 1.22 The Advantage of Twisted Pairs.

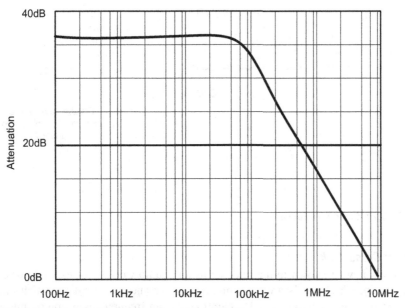

FIGURE 1.23 Magnetic Field Attenuation of Twisted Pair.

From Unscrambling the mysteries about twisted wire, R.B. Cowdell, IEEE EMC Symposium, 1979, p. 183.

magnetic field attenuation versus frequency for twisted 22 AWG wires compared with parallel 22 AWG wires spaced at 0.032″.

A further advantage of twisting pairs together is that it allows a fairly reproducible characteristic impedance. When combined with an overall shield to reduce common-mode capacitive pickup, the resulting cable is very suitable for high-speed data communication as it reduces both radiated noise and induced interference to a minimum.

1.2.7 CROSS TALK

When more than one signal is run within the same cable bundle for any distance, the mutual coupling between the wires allows a portion of one signal to be fed into another and vice versa. The phenomenon is known as cross talk. Strictly speaking, cross talk is not only a cable phenomenon but refers to any unwanted interaction between nominally uncoupled channels. The coupling can be predominantly either capacitive, inductive, or due to transmission-line phenomena.

The equivalent circuit for capacitive coupling at low-to-medium frequencies where the cable can be considered as a lumped component (in contrast to high frequencies where it must be considered as a transmission line) is as shown in Fig. 1.24.

In the worst case where the capacitive coupling impedance is much lower than the circuit impedance, the cross talk voltage is determined only by the ratio of circuit impedances.

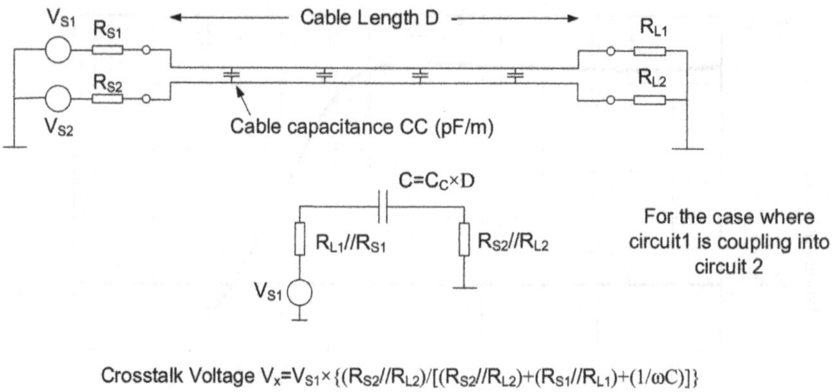

Crosstalk Voltage $V_x = V_{S1} \times \{(R_{S2}//R_{L2})/[(R_{S2}//R_{L2}) + (R_{S1}//R_{L1}) + (1/\omega C)]\}$

FIGURE 1.24 Cross Talk Equivalent Circuit.

Digital Cross Talk

Cross talk is well known in the telecomms and audio worlds, for example where separate speech channels are transmitted together and one breaks through onto another, or where stereo channel separation at high frequencies is compromised. Although digital data might seem at first sight immune from cross talk, in fact it is a serious threat to data integrity as well. The capacitive coupling is all but transparent to fast edges with the result that clocked data can be especially corrupted, as Fig. 1.25 shows. If the logic noise immunity is poor, severe false clocking can result. A couple of worked examples will demonstrate the nature of the problem.

Cross talk can be combated with a number of strategies, which follow from the above examples. These:

- reduce the circuit source and/or load impedances. Ideally, the offending circuit's source impedance should be high and the victim's should be low. Low impedances require more capacitance for a given amount of coupling.
- reduce the mutual coupling capacitance. Use a shorter cable, or select a cable with lower core-to-core capacitance per unit length. Note that for fast or high-frequency signals this will not solve anything because the impedance of the coupling capacitance is lower than the circuit impedances. If you use ribbon cable, sacrifice some space and tie a conductor to ground between each signal conductor; another alternative is ribbon cable with an integral ground plane. Best of all, use an individual screen for each circuit. The screen must be grounded or you gain nothing at all from this tactic!
- reduce the signal circuit bandwidth to the minimum required for the data rate or frequency response of the system. As can be seen from the above point, the coupling depends directly on the risetime of the offending signal. Slower risetimes mean less cross talk. If you do this by adding a capacitance in parallel with the input load resistor (across R_{L2} in Fig. 1.24), this will act as a

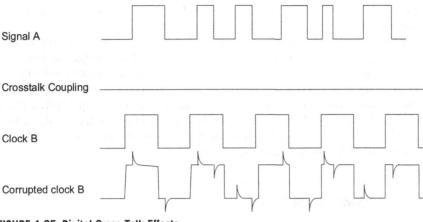

Signal A

Crosstalk Coupling

Clock B

Corrupted clock B

FIGURE 1.25 Digital Cross Talk Effects.

(a) Two audio circuits with 10-kΩ source and load impedances are run in 2 m of multicore cable with interconductor capacitances of 150 pF/m. What is the cross talk ratio at 10 kHz? The coupling capacitance C_C is 2 m of 150 pF/m = 300 pF. At 10 kHz, this has an impedance of 53 kΩ.

The source and load impedances in the cross talk circuit in each case are 10//10K = 5 kΩ. So the cross talk will be

5K/(5K + 5K + 53K) = 22 dB: unacceptable in just about any situation!

If the output drive impedance is reduced from 10 kΩ to 50 Ω then the cross talk becomes

49/(49 + 49 + 53K) = 60 dB

which is acceptable for many purposes, though probably not for hi-fi.

(b) Two EIA-232 (RS-232) serial data lines are run in 16 m of data cable (not individual twisted pair) which has a core/core capacitance of 108 pF/m. The transmitters and receivers conform to the EIA-232 spec of 300 Ω output impedance, 5 kΩ input impedance, ±10 V swing, and 30 V/μs risetime. What is the expected magnitude of interference spikes on one circuit due to the other?

Coupling capacitance here is 16 × 108 pF = 1728 pF.

The current that will be flowing after t seconds in an RC circuit fed from a ramping voltage with a constant dV/dt is defined in Eq. (1.15):

$$I = C\frac{dV}{dt}\left(1 - \exp\left[\frac{-t}{RC}\right]\right) \tag{1.15}$$

which for our case with dV/dt = 30 V/μs for 0.66 μs and a circuit resistance of 567 Ω is 25 mA. This translates to a peak voltage across the load resistance of (300//5K//5K) of

$$25 \times 10^{-3} \times 267 = 6.8 \text{ V}$$

This is one reason why EIA-232 is not suitable for long distances and high data rates!

potential divider with the core-to-core capacitance, as well as reducing the input impedance for high-frequency noise.

- use differential transmission. The bogey of cross talk is a major reason for the popularity of differential data standards such as EIA-422 (RS-422), and other more recent ones, at high data rates. Coupling capacitance is not necessarily reduced by using paired lines, but the cross talk is now injected in common mode and so benefits from the common-mode rejection of the input buffer. The limiting factor to the degree of rejection that can be obtained is the unbalance in coupling capacitance of each half of the pair. This is why twisted-pair cable is advised for differential data transmission.

1.3 TRANSMISSION LINES

Electronics is not a homogeneous discipline. It tends to divide into set areas: analog, digital, power, RF, and microwave. This is a pragmatic division because different mathematical tools are used for these different areas, and it is rare for any one designer to be proficient in all or even most of them. Unhappily for the designer, nature knows nothing of these civilized distinctions; all electrons follow the same physical laws regardless of who observes them and regardless of their speed.

When signal frequencies are low, it is possible to imagine that circuit operation is constrained by the laws of circuit theory: Thévenin, Kirchhoff et al. This is not actually true. Electrons do not read circuit diagrams, and they operate according to the rather grander and more universal laws of electromagnetic field theory, but the difference at low frequencies is so slight that circuit-theoretical predictions are indistinguishable from the real thing. Circuit theory serves electronic engineers well.

As the speed of circuit operation rises, though, it breaks down. It is not that electrons change their behavior at higher frequencies; there is no cut-off point beyond which everything is different. It is simply that the predictions of circuit theory diverge from those of electromagnetic field theory, and the latter, having the backing of nature, wins. One of the consequences of this victory is that perfectly ordinary lengths of wire and cable magically turn into transmission lines.

Transmission Line Effects

There is no straightforward answer to the question "when do I have to start considering transmission line properties?" The best response is, when the effects become important to you. One of the simplest electrical laws is that which relates frequency, wavelength, and the speed of light as shown in Eq. (1.16):

$$\lambda = 3 \times 10^8 / f \tag{1.16}$$

which is modified because of the reduction in velocity of propagation when a (lossless) dielectric medium is involved by the relative permittivity or dielectric constant of the medium:

$$\lambda_d = \lambda / \sqrt{e_r} \tag{1.17}$$

One rule of thumb is that a cable should be considered as a transmission line when the wavelength of the highest frequency carried is less than 10 times its length. You may be embarrassed by transmission-line effects at lengths of one fortieth the wavelength or less if you are working with precision high-speed signals, or you may not care until the length reaches a quarter wavelength—though by then you will certainly be getting some odd results.

Critical Lengths for Pulses

If as a digital engineer you work in terms of risetimes rather than frequency, then a roughly equivalent rule of thumb is that if the shortest risetime is less than three times the traveling time along the length of the cable, you should be thinking in terms of transmission lines. Thus for a risetime of 10 ns in coax with a velocity factor $\left(1/\sqrt{e_r}\right)$ of 0.66 the critical length will be two-thirds of a meter.

1.3.1 CHARACTERISTIC IMPEDANCE

Characteristic impedance (Z_o) is the most important parameter for any transmission line. It is a function of geometry as well as materials, and it is a dynamic value independent of line length; you cannot measure it with a multimeter. It is related to the conventional distributed circuit parameters of the cable or conductors by Eq. (1.18):

$$Z = \sqrt{\frac{R + j\omega L}{G + j\omega C}} \qquad (1.18)$$

where R is the series resistance per unit length (Ω/m); L is the series inductance (H/m); G is the shunt conductance (℧/m); C is the shunt capacitance (F/m).

L and C are related to the velocity factor by:

$$\text{velocity of propagation} = \frac{1}{\sqrt{LC}} = 3 \times 10^8 / \sqrt{e_r} \qquad (1.19)$$

For an ideal, lossless line $R = G = 0$ and Z_o reduces to $\sqrt{LC}$. Practical lines have some losses that attenuate the signal, and these are quantified as an attenuation factor for a specified length and frequency (Table 1.7 shows these for coaxial cables). Table 1.7 summarizes the approximate characteristic impedances for various geometries, along with velocity factors of some common dielectric materials. The value 377 (120π) crops up several times: it is a significant number in electromagnetism, being the *impedance of free space* (in ohms), which relates electric and magnetic fields in free-field conditions.

Driving a signal down a transmission line provides an important exception to the general rule of circuit theory (for voltage drives) that the driving source impedance should be low while the receiving load impedance should be high. When sent down a transmission line, the signal is only received undistorted if both source and load impedances are the same as the line's characteristic impedance. This is said to be the *matched* condition. It is easiest to consider the effects of matching and mismatching in two parts: in the time domain for digital applications and in the frequency domain for analog RF applications.

1.3.2 TIME DOMAIN

Imagine a step waveform being launched into a transmission line from a generator, which is matched to the line's characteristic impedance Z_o. We can view the waveform at each end of the line and, because of the finite velocity of propagation down the line, the two waveforms will be different. The results for three different cases of open, matched, and short line terminating impedance (these are the easily visualized special cases) are shown in Fig. 1.26. If you have a reasonably fast pulse generator, a wide bandwidth oscilloscope, and a length of coax cable you can perform this experiment on the bench yourself in 5 min.

A matched transmission line is actually a simple form of delay line, with delays of the order of tens of nanoseconds achievable from practical lengths. Discrete-component delay lines are smaller but work on the same principle, with the distributed L and C values being replaced by actual components.

In all cases the long-term result is as would be expected from conventional circuit theory: an open circuit results in V_p, a short circuit results in zero and anything in between results in the output being divided by the potential divider $Z_L/(Z_{out} + Z_L)$,

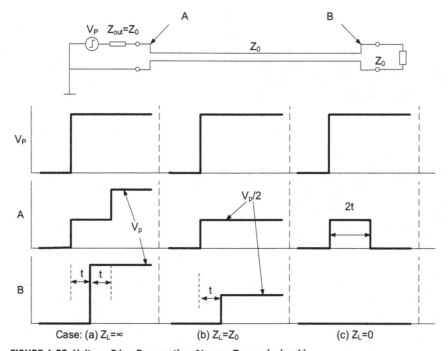

FIGURE 1.26 Voltage Edge Propagating Along a Transmission Line.

giving $V_p/2$ for the matched case. While the edge is in transit the driving waveform is different.

Forward and Reflected Waves

Transmission line theory explains the results in terms of a forward and a reflected wave, the two components summing at each end to satisfy the boundary conditions: zero current for an open circuit, zero voltage for a short. Thus in the short-circuit case, the forward wave of amplitude $V_p/2$ generates a reflected wave of amplitude $-V_p/2$ when it reaches the short, which returns to the driving end and sums with the already-existing $V_p/2$ to give zero. In the general case, the ratio of reflected to forward wave amplitude is given in Eq. (1.20):

$$\frac{V_r}{V_i} = \frac{Z - Z_o}{Z + Z_o} \qquad (1.20)$$

This explanation is most useful when you want to consider mismatches at both ends. Forward and reflected waves are then continually bounced off each mismatched end. Take as another example a drive impedance of $Z_o/2$ and an open-circuit load, which is a very crude approximation to an high density complementary metal oxide semiconductor (HCMOS) logic buffer driving an unterminated HCMOS input. This is shown in Fig. 1.27.

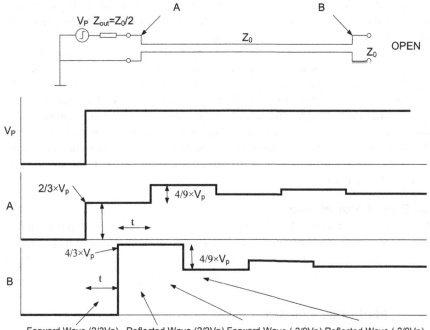

FIGURE 1.27 A Mismatched Transmission Line.

Ringing

The reflected wave from the open-circuit end now gets reflected in turn from the mismatched driver end with a lower amplitude, which is reflected back by the open circuit that gets reflected again from the driver with a lower amplitude. Eventually the reflections die away and equilibrium is reached. The waveforms at both ends show considerable "ringing." If you work with digital circuits, you will be familiar with ringing if you have ever observed your signals over a few inches of PCB track with a fast oscilloscope. The amplitude of the ringing depends entirely on the degree of mismatch between the various impedances, which are complex and for practical purposes essentially unknowable, and the period of the ringing depends on the transit time from driver to termination and hence on line length. A typical ringing frequency for a 0.6 mm wide track over a ground plane on 1.6 mm epoxy-glass PCB is 35 MHz divided by the line length in meters.

The Bergeron Diagram

An accurate determination of the amplitude of the reflections at both ends of a transmission line can be made using a Bergeron diagram. This shows the characteristic impedance of a transmission line as a series of load lines on the input and output characteristics of the line driver and receiver. Each load line originates from the point at which the previous load line intersects the appropriate input/output characteristic. To properly use the Bergeron diagram, you need to know the device characteristics both within and outside the supply rail voltage levels, since ringing carries the signal line voltage outside these points. Many manufacturers of high-speed logic ICs detail its use in their application notes.

Ringing in digital circuits is always undesirable since it leads to spurious switching, but it can be tolerated if the amplitudes involved are within the logic family's noise immunity band or if the transit times are faster than its response speed. In fact the idealized example in Fig. 1.27 shows a step edge, which is unrealistic, as practical risetimes will damp the response. The only way to avoid it completely is to consider every interconnection as a transmission line and to terminate each end with its correct characteristic impedance. Very fast circuits are designed in exactly this way; designers of slower circuits will only meet the problem in severe form when driving long cables.

The Uses of Mismatching

Mismatching is not always bad. For instance, a very fast, stable pulse generator can be built by feeding a fast risetime edge into a length of transmission line shorted at the far end (Fig. 1.28) and taking the output from the input to the line. A 1 m length of coax with velocity factor 0.66 will give a 10 ns pulse.

1.3.3 FREQUENCY DOMAIN

If you are more interested in RF signals than in digital edges, you want to know what a transmission line does in the frequency domain. Consider the transmission line of

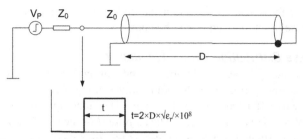

FIGURE 1.28 Pulse Generation With a Shorted Transmission Line.

Fig. 1.26 being fed from a continuous sine-wave generator of frequency f and matched to the line's Z_O. Again, the energy can be thought of as a wave propagating along the line until it reaches the load; if the load impedance is matched to Z_o then there is no reflection, and all the power is transferred to the load.

If the load is mismatched then a portion of the incident power is reflected back down the line, exactly like an applied pulse edge. A short or open circuit reflects all the power back. But the signal that is reflected is a continuous wave, not a pulse; so the voltage and current at any point along the line is the vector sum of the voltages and currents of the forward and reflected waves and depends on their relative amplitudes and phases. The voltage and current distribution down the length of the line forms a so-called "standing wave." The standing wave patterns for four conditions of line termination are shown in Fig. 1.29. You can verify this experimentally

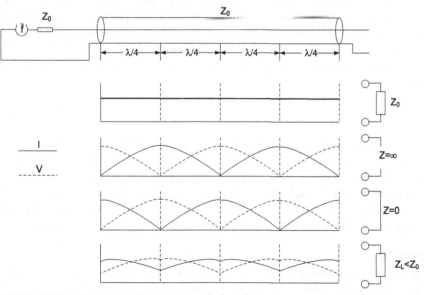

FIGURE 1.29 Standing Waves Along a Transmission Line.

with a length of fairly leaky coax and a "sniffer" probe, connected to an RF volt-meter, held close to and moved along the coax.

Standing Wave Distribution Versus Frequency

Note that the standing wave distribution depends on the wavelength of the applied signal and hence on its frequency. Standing waves at one frequency along a given length of line will differ from those at another. The standing wave pattern repeats itself at multiples of $\lambda/2$ along the line. The amplitude of the standing wave depends on the degree of mismatch, which is represented by the reflection coefficient Γ, the ratio of reflected current or voltage to incident current or voltage. Standing wave ratio (*s.w.r.*) is the ratio of maximum to minimum values of the standing wave and is given by Eq. (1.21):

$$s.w.r = \frac{1 + |G|}{1 - |\Gamma|} = \frac{R_L}{Z_o} \quad \text{for a purely resistive termination} \qquad (1.21)$$

Thus an *s.w.r.* of 1:1 describes a perfectly matched line; infinite *s.w.r.* describes a line terminated in a short or open circuit. The generator source impedance has no effect on the *s.w.r.* It depends only on the nature of the load at the far end.

Impedance Transformation

The variation of voltage and current along a mismatched transmission line is of great interest to the designers and operators of radio transmitters because it affects the efficiency of power transfer from the transmitter, through the feeder line to the antenna. It is also useful to high-frequency circuit designers as a means of making impedance transformations. Remembering that impedance is voltage divided by current, you can see from Fig. 1.29 that the impedance at any given point along the line varies considerably depending on the distance from the termination. For each quarter wavelength, the impedance varies from minimum to maximum. In fact, for a quarter-wave transmission line transformer, the impedance transformation is given by Eq. (1.22):

$$Z_{\text{in}} = \frac{Z_o^2}{Z_L} \qquad (1.22)$$

This useful property is of course frequency dependent; it only occurs at $\lambda/4$ and multiples thereof. If the frequency is changed then the line length departs from $\lambda/4$ and Z_{in} becomes reactive. A related property is that at even multiples of $\lambda/4$ (equivalent to saying any multiple of $\lambda/2$) the original load impedance is regained, whatever the value of Z_o. Thus a shorted line will have a virtually zero impedance $\lambda/2$ away from the short, which property can be used to create a distributed tuned circuit.

Lossy Lines

The preceding discussion assumes zero-loss lines, which in practice are unrealizable. For short line lengths the losses are usually insignificant—see Table 1.8 for

Table 1.8 Characteristics of TIA/EIA-568 (ISO/IEC 11801) 100-Ω Quad-Pair Cable

	Frequency, MHz	Cat 3	Cat 5	Cat 5e	Cat 6
Bandwidth		16 MHz	100 MHz		250 MHz
Characteristic impedance	0.1	75–150 Ω		N/A	
	≥1	100 ± 15 Ω			
Attenuation dB per 100 m	0.256	1.3	1.1		N/A
	1.0	2.6	2.1		2.0
	4.0	5.6	4.3		3.8
	10	9.8	6.6		6.0
	16	13.1	8.2		7.6
	31.25	N/A	11.8		10.7
	62.5		17.1		15.4
	100		22.0		19.8
	200		N/A		29.0
	250				32.8
Capacitance unbalance	1 kHz	3400 pF/km		330 pF/100m	
DC loop resistance		19.2 Ω/100 m, max unbalance 3%			
Return loss dB at 100 m cable length	1 – 10	12	23	$20 + 5 \log (f)$	
	10 – 20	$12 - 10 \log (f/10)$	23	25	
	20 – 100	N/A	$23 - 10 \log (f/20)$	$25 - 7 \log (f/20)$	
	200	N/A			18.0
	250				17.3

typical coax losses. Note that these are quoted for the matched condition. If the line is operated with standing waves, the loss is greater than if it is matched because increased voltages and currents are present and the average heat loss is greater for the same power output. The effect of attenuation in long lines is to cause an improvement in *s.w.r.* towards the generator, since the effect of a mismatch is attenuated in both directions. In the limit, a long cable can make a very good power attenuator!

If we reconsider the equation for the characteristic impedance of a transmission line in Eq. (1.23):

$$Z_o = \sqrt{\frac{R + j\omega L}{G + j\omega C}} \tag{1.23}$$

where R is the series resistance per unit length (Ω/m); L is the series inductance (H/m); G is the shunt conductance ($\mho$/m); C is the shunt capacitance (F/m).

We can rewrite this in a slightly different form, which makes it useful when considering the lossy versus lossless case in Eq. (1.24):

$$Z_o = \sqrt{\frac{L}{C \times \left[1 + j\left(\frac{G}{j\omega C - R/j\omega L}\right)\right]}} \tag{1.24}$$

In other words, as $R \gg 0$ and $G \gg 0$, then the characteristic impedance will tend towards the lossless case, but this equation shows how much effect potentially the losses may have on the impedance of the line in practice.

There is also the consideration that in general the characteristic impedance Z_o is clearly complex (by implication introducing a phase shift), but if $G/C = R/L$, then the impedance will be real.

Given this condition, we can then say that the line has a characteristic resistance, R_0, and that this is equal to the characteristic impedance value Z_o.

We can use the expressions for the calculation of characteristic impedance given in Table 1.9, and the dielectric constants given in Table 1.10, to either calculate the impedance from the relevant geometry or design a specific geometry to match to the required impedance.

A 24 standard wire gauge (swg) copper-cored cable is to be designed to have a characteristic impedance of 50 Ω. What should the inside diameter of the outside conductor be to achieve this? Assume that the coaxial cable is filled with PTFE.

From Table 1.10 we can see that the relative permittivity e_r is 2.1.

The diameter of the conductor in a 24 swg cable from Table 1.2 is 0.56 mm.

The characteristic impedance of the cable is required to be 50 W and we know from Table 1.9 that the equation for the characteristic impedance is given by the expression in Eq. (1.25):

$$Z_o = 60/\sqrt{e_r} \times \ln(D/d) \tag{1.25}$$

Table 1.9 Characteristic Impedance and Geometry

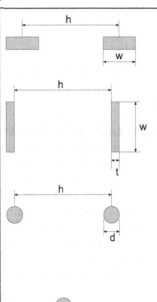

1. **Side-by-side parallel strip**

$$Z_o = 120/\sqrt{e_r} \times \ln\left\{h/w + \sqrt{\left[(h/w)^2 - 1\right]}\right\}$$

2. **Face-to-face parallel strip**

$$Z_o = 377/\sqrt{e_r} \times h/w \quad \text{if } h > 3t, \; w \gg h$$
$$120/\sqrt{e_r} \times \ln 4h/w \quad \text{if } h \gg w$$

3. **Parallel wire**

$$Z_o = 120/\sqrt{e_r} \times \ln\left\{h/d + \sqrt{\left[(h/d)^2 - 1\right]}\right\}$$
$$120/\sqrt{e_r} \times \ln 2h/d \quad \text{if } d \ll h$$

(Z_o of typical PVC-insulated pairs and twisted pairs is around 100 Ω)

4. **Wire parallel to infinite plate**

$$Z_o = 60/\sqrt{e_r} \times \ln\left\{2h/d + \sqrt{\left[(2h/d)^2 - 1\right]}\right\}$$
$$60/\sqrt{e_r} \times \ln 4h/d \quad \text{if } d \ll h$$

5. **Strip parallel to infinite plate**

$$Z_o = 377/\sqrt{e_r} \times h/w \quad \text{if } w > 3h$$
$$60/\sqrt{e_r} \times \ln 8h/w \quad \text{if } h > 3w$$

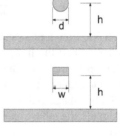

6. **Coaxial**

$$Z_o = 60/\sqrt{e_r} \times \ln(D/d)$$

Table 1.10 Dielectric Constants

Dielectric Constants of Various Materials	e_r	Velocity Factor $(1/\sqrt{e_r})$
Air	1.0	1.0
Polythene/polyethylene	2.3	0.66
PTFE	2.1	0.69
Silicone rubber	3.1	0.57
FR4 fiberglass PCB	4.5 (typ)	0.47
PVC	5.0	0.45

So, we can rearrange this into a form that we can then calculate what the inside diameter of the outside conductor (D) needs to be:

$$D = d \times \exp\left(\frac{Z_o \sqrt{e_r}}{60}\right) \tag{1.26}$$

which for this case results in:

$$D = 1.873 \ \text{mm}$$

Understanding the Transmission Line Impedance Graphically

We can also interpret the transmission line in terms of the relationship between the incident and reflected signals on the line plus a termination. As we have discussed previously, the incident voltage applied to a transmission line will result in standing waves depending on the termination impedance in conjunction with the impedance of the transmission line itself, and it is useful to have this ability to connect the voltage standing wave ratio (VSWR) with the reflection, in particular the reflection coefficient, as this is something that we can measure relatively easily with a probe on the line.

The reflection coefficient can be calculated as follows:

$$\rho_v = \frac{|\text{reflected voltage}|}{|\text{incident voltage}|} = \frac{\text{VSWR} - 1}{\text{VSWR} + 1} \tag{1.27}$$

The VSWR from a practical perspective is the ratio of the maximum to the minimum voltage along the line.

In addition to the magnitude of the reflection, there is also a phase associated with the transmission line, and this can be measured by using a probe to locate the minimum voltage point on the line, and from this and the wavelength of the applied signal, the phase can be calculated using the following expression in Eq. (1.28):

$$\theta_v = 720 \times \left(\frac{x}{\left(\lambda - \frac{1}{4}\right)}\right) \tag{1.28}$$

where x is the distance of the minimum from the *load* and the wavelength λ is in meters.

With the magnitude and phase of the reflection coefficient, we can plot the resulting complex value and from it establish whether the load is matched, infinite, capacitive, resistive, or inductive. A sample reflection coefficient chart is shown in

Fig. 1.30. This is very like a "Smith" chart which has normalized impedance values superimposed so that the values of load can be read off the chart, and then a "stub" can be designed to match the impedance of the load to the line.

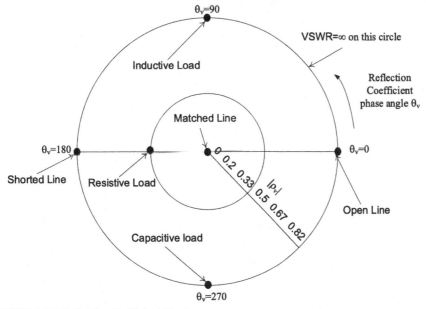

FIGURE 1.30 Reflection Coefficient Chart.

Printed Circuits

The Circuit Designer's Companion. http://dx.doi.org/10.1016/B978-0-08-101764-7.00002-5

Virtually every electronic circuit built nowadays uses a printed circuit board (pcb) as both its interconnecting medium and mechanical mounting substrate. The pcb is custom-designed for the circuit it carries, and its selection is an important part of the circuit designer's task, though this is frequently delegated to the (perceived) inferior function of pc layout draftsman. The design of the pcb has a strong effect on the mechanical and electrical performance of the final product. This chapter looks at the main factors that you should consider when working with your new pcb design.

2.1 **BOARD TYPES**

An unprocessed board is a laminate of a conductive material and an insulating dielectric substrate. The different materials that are used, and the different ways of laminating and interconnecting between conductive layers, decide the type.

2.1.1 **MATERIALS**

The conductive layer is almost invariably copper foil, adhesive bonded under heat and pressure to the substrate. The copper cladding thickness is usually specified (for historical reasons) by its weight per square foot, the most common being 1 or 2 oz, other thicknesses being 0.25, 0.5, 3, and 4 oz. Thickness of 1-oz copper is typically 0.035 ± 0.002 mm, other weights being thicker or thinner pro rata. The main deciding factor in choosing copper weight is its resistivity. Fig. 2.3 gives resistance versus track width for the different weights.

The most common laminates are epoxy glass and phenolic paper. Phenolic paper (or synthetic resin—bonded paper) is cheaper and can be punched readily, so its major application is in high-volume domestic and other noncritical sectors. It is electrically inferior to epoxy glass, is mechanically brittle, has a poor temperature range, absorbs moisture readily, and is unsuitable for plated-through-hole (PTH) construction. You will only find it used for very cost-sensitive, low-performance applications. Otherwise, epoxy glass is universal except for specialist applications such as RF circuits.

Epoxy Glass

Epoxy resin with woven glass cloth reinforcement is used for PTH and multi-layer boards. It can also be used for simpler constructions if its better mechanical and electrical properties are needed. It offers much better dimensional stability and is stronger than phenolic paper, but this does mean that high-volume boards cost more because it must be drilled rather than punched. For even denser and more dimensionally critical boards, epoxy aramid is another suitable material. Table 2.1 gives a selection of material specifications for some frequently used types of board material. The common designation of "FR4" refers to the American National Electrical Manufacturers' Association (NEMA) specification for flame-retardant epoxy-woven glass board, which is offered by most laminate manufacturers; there are many flavors of FR4, having better or worse specifications but all of them meeting the NEMA requirements. Flame-retardant grades are available for all common materials and are shown by the manufacturers' ID being marked in red.

For flexible boards the base films are polyester or polyimide. Polyester is cheap but cannot easily be soldered because of its low softening temperature so is used primarily for flexible "tails." Polyimide is more expensive but components can be mounted on it.

Table 2.1 Printed Circuit Board Laminate Material Properties

Material	Surface Resistance (MΩ)	Dielectric ε_r	Dielectric tan δ	Dielectric Strength (kV/mil)	Tempco x–y (ppm/°C)	Maximum Temperature (°C)
Standard FR4	Minimum 1.10^4	Maximum 5.4, typical 4.6–4.9	Maximum 0.035	1.0 min	13–16	110–150
FR408 (high quality)	1.10^6	3.8	0.01	1.4	13	180
Epoxy aramid (close tolerance)	5.10^6	3.8	0.022	1.6	10	180
Polyimide (Kapton)	–	3.4	0.01	3.8	20	300
Polyester (Mylar)	–	3.0	0.018	3.4	27	105

2.1.2 **TYPE OF CONSTRUCTION**

Most circuit requirements can be accommodated on one of the following board types, which are listed roughly by increasing cost.

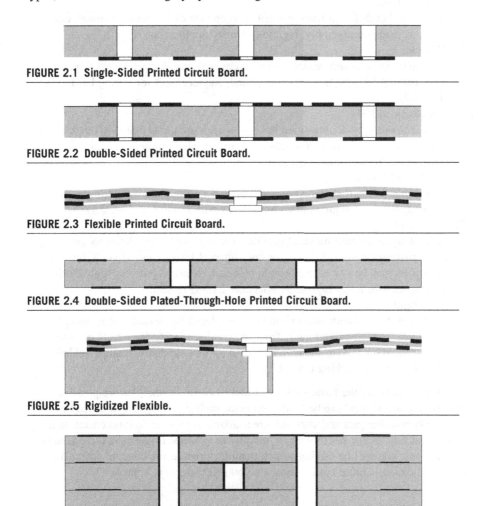

FIGURE 2.1 Single-Sided Printed Circuit Board.

FIGURE 2.2 Double-Sided Printed Circuit Board.

FIGURE 2.3 Flexible Printed Circuit Board.

FIGURE 2.4 Double-Sided Plated-Through-Hole Printed Circuit Board.

FIGURE 2.5 Rigidized Flexible.

FIGURE 2.6 Multilayer Printed Circuit Board.

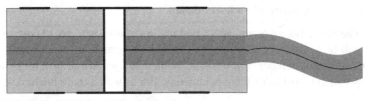

FIGURE 2.7 Flexi-rigid Printed Circuit Board.

1. Single-sided
Cheap and cheerful for simple, low-performance and/or high-volume circuits (Fig. 2.1).
2. Double-sided
As above, but different track pattern on each side of the board, through connections made on assembly, e.g., low-density applications (Fig. 2.2).
3. Flexible
Base material thin and flexible, may be covered with further layers to protect track pattern. May be through-hole plated. Replacement for wiring harnesses (Fig. 2.3).
4. Double-sided PTH
Similar to ordinary double-sided but hole barrels are metal plated to interconnect both sides, so a different production technique. Medium density general industrial etc. applications (Fig. 2.4).
5. Rigidized Flexible
As flexible, but part stiffened by rigid plate for component mounting. Used where application requires few components with flexibility (Fig. 2.5).
6. Multilayer
Several layers of base material laminated into a single unit. Variants may have holes passing through all layers or only internal layers (buried vias). Internal layer pairs may be power and ground planes. Expensive, but very high densities achievable with many layers. See Section 2.1.5 (Fig. 2.6).
7. Flexi-rigid
Multilayer boards with some rigid layers replaced by flexible areas (usually polyimide), which extend away from the rigid section to form tails or hinges. Several rigid areas may be interconnected by flexible areas to allow folded shapes for dense packing (Fig. 2.7).

Many variants of the basic multilayer and flexi-rigid principles are possible, for instance, up to 24 layers can be fabricated in the multilayer construction (see Section 2.1.5). Other techniques and materials are available for specialized functions such as switch contacts, motor assemblies, or microwave systems. If you have these kinds of applications, you will be talking to pcb manufacturers at the concept stage. But, if your application is more typical, how do you select the best approach?

2.1.3 CHOICE OF TYPE

As in any design problem, many factors need to be balanced to achieve the best final result. The most important are cost, packing density, and electrical performance, with other factors appearing to a lesser extent.

- *Cost.* The above list gives an idea of ranking in terms of bare board cost, though there will be a degree of overlap. The actual cost formula involves board quantity ordered, number of processes, number and variety of drill holes, and the raw material cost as its main parameters. But as well as bare board cost you should also consider the possible effect on overall unit costs; the choice of board type can affect assembly, test, repair, and rework. For instance, if you expect to rework a significant number of units, this would rule out phenolic paper because of its poorer copper-laminate adhesion. At the other extreme, it could work against multilayer boards because of the danger of destroying an otherwise good board by damaging one through hole.

- *Space limitations.* If your board size is fixed and so is your circuit package count, then you have automatically determined packing density and to a great extent the optimum board type for lowest cost. Non-PTH boards give the lowest packing density. Double-sided PTH can offer between 4 and 7 cm^2 per 16-pin dual-in-line (DIL) through hole package depending on track spacing and dimensions, while multilayers can approach the practical limit of 2 cm^2 per pack. Multilayers are also the only way to fully realize the space advantages of flatpack or surface-mount components. Large, discrete-leaded components (resistors, capacitors, transformers, etc.) reduce the advantage of multilayers because they effectively offer more area for tracking on the surface. If there is no limitation on board size, then using a larger, cheaper, low-density board must be balanced against the cost of its larger housing.

- *Electrical characteristics.* Phenolic paper may not have sufficiently high bulk resistance, voltage breakdown, or low dielectric loss; or the laminate thickness may be determined by required track characteristics, although usually it's easier to work the other way round; or thicker copper may be required for low resistivity. If surface leakage is likely to be a problem, conformal coating is one solution (see Section 2.4.1). For proper power/ground plane distribution, a minimum four-layer multilayer construction is normal; for lower densities the ground plane can be put on one side of a double-sided board.

- *Mechanical characteristics.* Weight, stiffness, and strength may all be important. If you need good resistance to vibration or bowing, either a thicker laminate or stiffening bars may be required. Usually, strength is not critical unless the board carries very heavy components such as a large transformer, in which case epoxy glass is essential. Coefficient of thermal expansion and maximum temperature rating may need checking if you have a wide-temperature range application.

- *Availability.* Just about any pcb manufacturer should be able to cope with ordinary single-sided and double-sided and PTH boards, though prices can vary widely. As you progress toward the more exotic flexible and multilayer constructions your options will diminish and you could become locked in to a single source or face unacceptable delivery times. Also, designing complex multilayer boards requires an advanced level of skill, even with improving CAD systems, and your design resources may not cover this when faced with short timescales.

- *Reliability and maintainability.* These factors normally favor uncomplicated construction with high-quality materials, for which double-sided PTH on epoxy glass wins out easily.

- *Rigid versus flexi.* The flexi-rigid construction may offer lower assembly costs and better packing density and reliability by eliminating interboard wiring and connectors. Against this it is more expensive and more prone to difficulties in supply and may run counter to a repair philosophy based on modular replacement.

2.1.4 **CHOICE OF SIZE**

If you have a free hand in selecting the size of the board or boards in your system, then another set of factors need to be considered. It may be best to go for a standard size such as Eurocard (100 × 160 mm) or double-Eurocard (233.4 × 160 mm) especially if the boards will be fitted into a modular rack system. Modularizing the total system like this may not be optimum, though; for smaller systems a single large board will do away with the cost of interconnecting components and will be cheaper to produce than several smaller boards of equivalent area. On the other hand, board material costs depend to some extent on wastage in cutting from the stock laminate, and larger boards are likely to waste more unless their dimensions are matched to the stock. Also, there is a practical limit to the size of very large boards, fixed by considerations of stiffness, dimensional tolerancing and handling, not to mention the capability of the board manufacturer, photoplotter, and layout generation. Your own assembly department will also have limits on the board size they can handle. Optimum size for large boards is generally around 30–50 cm on the longer edge.

With the advent of modern production techniques and the concept of multiple panels, many companies now offer a cheaper service where many boards are produced in a single process, thus sharing the tooling costs. This "pooling" service can make it cost-effective to go for much smaller board sizes, even when prototyping or for small quantities—more on this in the section on panelization.

Subdivision Boundaries

If you are going to split the system into several smaller boards, then position the splits to minimize the number of interconnections. Often this corresponds with splitting it into functional subassemblies, and this helps because it is easiest to test each assembled board as a fully functioning subunit. Test cost per overall system will decrease as the board size increases, but the expense of the required automatic test equipment will rise. Allow a safety factor in your calculations of the board area needed for each subsystem: subsequent circuit modifications nearly always increase component count, and very rarely reduce it.

Panelization

It is unusual for smaller boards to be built one per stock laminate. In a production run, the main cost is in the processing stages, and if, say, half a dozen boards can be made at a time in one piece the processing and handling applies simultaneously to all of them as if they were one board. The artwork for one board is simply repeated in steps across the whole panel (Fig. 2.8). At the end of the processing—quite probably even after delivery, population and assembly of components—they can be separated. They do not even have to be the same design, as long as they have identical

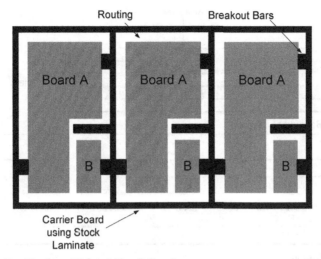

FIGURE 2.8 Panelization of Printed Circuit Boards.

layer stack-ups. You may, for instance, want to put all the boards for one project into one panel, since the numbers required for each will be identical.

The size of the board can then be matched to a multiple of a given stock size, allowing for routing and/or score lines to give separation. It can also be cost-effective if you have a board with a large waste area, for instance, a thin L-shape, to put other boards into the waste region and make use of it. Although panelization reduces your options in some respects, because all the boards on a given panel must have the same layer count and thickness, it is worth accepting these limitations for the sake of greater cost-effectiveness.

2.1.5 HOW A MULTILAYER BOARD IS MADE

The stages of processing for a multilayer board are well established (Fig. 2.9–2.16 shows the process of making a six-layer board as an example). The finished board

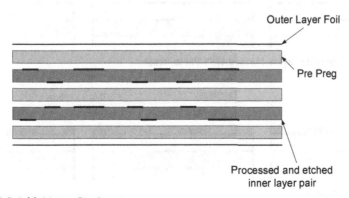

FIGURE 2.9 Initial Layer Stack-up.

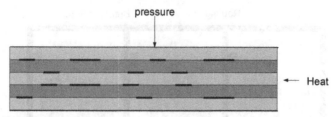

FIGURE 2.10 Bonding of Layers.

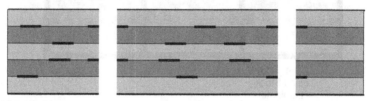

FIGURE 2.11 Drilled.

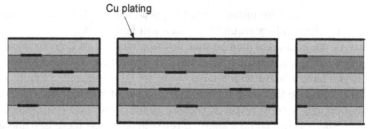

FIGURE 2.12 Holes and Outside Surface Plating.

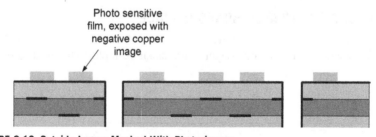

FIGURE 2.13 Outside Layers Masked With Photo-image.

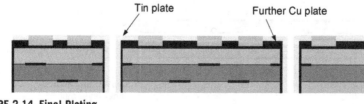

FIGURE 2.14 Final Plating.

Etch

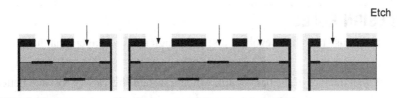

FIGURE 2.15 Photo-Image Film Removed and Unplated Areas Etched.

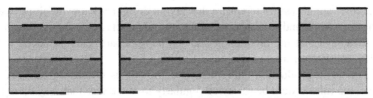

FIGURE 2.16 Tin-Stripped, Final Cu Pattern Ready for Surface Finish and Solder Mask.

consists of a sandwich (Fig. 2.9) of double-sided copper inner layer pairs—each one processed and etched separately with its track patterns—which are separated by layers of "prepreg," plain epoxy-glass material without copper laminate. The outer copper layers are applied as foil, initially unprocessed. Because of this method of assembly, multilayer boards always consist of an even number of copper layers—four, six, eight, and so on.

The individual sets of inner layers, prepregs, and outer foil are bonded together (Fig. 2.10) under heat and pressure. The alignment tolerances between the layers have to be established at this step. The assembly is then drilled according to the required pattern (Fig. 2.11). The holes may be used for vias, in which case they will only interconnect different layers, or for component lead wires, or for other purposes such as mounting the final assembly. After this, the whole assembly is put through a copper plating process (Fig. 2.12), which adds copper to all exposed surfaces: through the barrel of each hole (hence the designation "plated-through-hole") and also on the surface of the copper foil on each side. The vias are created at this step by the contact of the hole plating with the edges of the appropriate pads on the inner layers.

The two outside layers are then processed by applying a photosensitive film, and then exposing and developing this (Fig. 2.13) to leave a negative image of the required outer track and pad pattern. Further plating (Fig. 2.14) adds more copper and a protective tin coat to the exposed areas; the photosensitive film is removed, and the board is etched (Fig. 2.15) to take off the copper that was underneath it. The tin is then stripped away leaving the complete three-dimensional copper pattern outside and within the board (Fig. 2.16), ready for the application of solder mask and surface finishing. The thickness of the outer copper layers is the sum of the total plating thickness and the initial foil thickness.

2.2 **DESIGN RULES**

Most firms that use pcbs have evolved a set of design rules for layout that have two aims:

- ease of manufacture of the bare board, which translates into lower cost and better reliability;
- ease of assembly, test, inspection, and repair of the finished unit, which translates as above.

These rules are necessary to ensure that layout designers know the limits within which they can work, and so that some uniformity of purchasing policy is possible. It has the advantage that the production and servicing departments are faced with a reasonably consistent series of designs emanating from the design department, so that investment in production equipment and training is efficiently used. At the same time, the rules should be reviewed regularly to make sure that they do not unnecessarily restrict design freedom in the light of board production technology advances. Any company, for instance, whose design rules still specify minimum track widths of 0.3 mm is not going to achieve the best available packing densities. The design rules should not be enforced so rigorously that they actually prevent the optimum design of a new product.

BS6221 Part 3, "Guide for the design and use of printed wiring boards," presents a good overview of recommended design practices in pc layout and can be used to form the basis of in-house design rules.

Factors in board design that should be considered in the design rules are as follows:

- track width and spacing
- hole and pad diameter
- track routing
- ground distribution
- solder mask, component identification, and surface finish
- terminations and connections

Typical capabilities for pcb producers at the time of writing are shown in Table 2.2. Having decided on a particular supplier, the first thing you should be clear about is their capabilities in these key areas.

2.2.1 **TRACK WIDTH AND SPACING**

The minimum width and spacing determine to a large extent the achievable packing density of a board layout. Minimum width for tracks that do not carry large currents is set by the controllability of the etching process and layer registration for multilayers, which may vary between board manufacturers and will also, at the margins, affect the cost of the board. You will need to check with your board supplier what minimum width he or she is capable of processing. Table 2.2 suggests typical capabilities in this respect at the time of writing, for 1-oz thick copper; thicker copper

Table 2.2 Typical Best Capabilities of Printed Circuit Board Manufacturers

Board thickness range	0.35–3.5 mm
Maximum number of layers	14–24
Minimum track and gap width (1-oz copper)	0.1 mm possible, 0.15 mm preferred
Minimum PTH diameter	0.2 mm possible, 0.3 mm preferred
Maximum PTH aspect ratio	12:1 possible, 6:1 preferred
Registration: Drill to pad	0.03 mm
Registration: Layer to layer, including solder mask	0.075 mm

PTH, *plated-through-hole.*

will demand greater widths because of undercutting in the etching. The narrowest widths may be acceptable in isolated instances, such as between IC pads, but are harder to maintain over long distances. Fig. 2.17 shows the trade-off in track width and spacing versus number of tracks that can be squeezed between IC pads.

Conductor Resistance
Track width also, rather more obviously, affects resistance and hence voltage drop for a given current. Eq. (2.1) shows the theoretical resistance for various thicknesses of copper track per centimeter. It is derived from Eq. (2.1):

$$R = \frac{\rho l}{A} \tag{2.1}$$

where ρ is the conductor resistivity; l is the length; A is the track cross-sectional area.

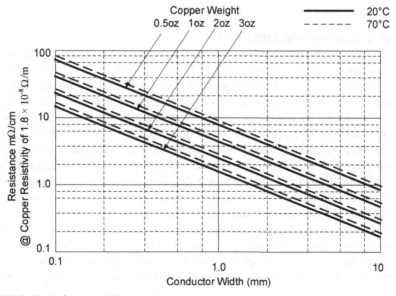

FIGURE 2.17 Resistance of Copper Tracks.

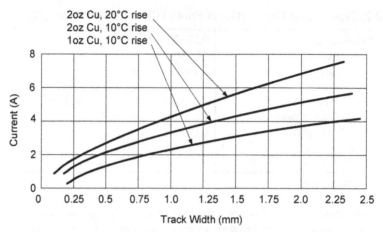

FIGURE 2.18 Safe Currents in Printed Circuit Board Tracks.

Abstracted from BS6221: Part 3: 1984.

These figures should only be taken as a guide because the actual manufactured tolerances, including base copper, plating and tin—lead thickness, will be very wide, amounting to a two-to-one variation in final value. The temperature coefficient of copper means that resistance will vary by several percent over typical ambient temperature ranges, and with self-heating. PTHs of greater than 0.8 mm diameter present less than a milliohm resistance.

The maximum current-carrying ability is determined by self-heating of the tracks. Fig. 2.18 shows the safe current versus track width for a given temperature rise.

Voltage Breakdown and Cross Talk

Track spacing is also determined by production capability and electrical considerations. Minimum spacings similar to track widths as shown in Table 2.2 are achievable by most pcb manufacturers. Cross talk and voltage breakdown are the electrical characteristics, which affect spacing. For a benign environment—dry and free from conductive particles—a spacing of 1 mm per 200 V, allowing for manufacturing tolerances, is adequate for preventing breakdown. BS6221 part three gives greater detail. When mains voltages are present, wider spacing is normally set by safety approval requirements. Spacings less than 0.5 mm risk solder bridging during wave soldering, depending on the transport direction, if solder resist is not used.

Cross talk (see Section 1.2.7) is likely to be the limiting factor on low-voltage digital or high-speed analogue boards. The mechanism is similar to that for cables; calculating track-to-track capacitance is best performed by electromagnetic field solvers for individual tracks working on the actual board layout. The simplest rule of thumb is that a track spacing greater than 1 mm will result in cross talk voltages less than 10% of signal voltages for most board configurations. Electrically short connections can be spaced much closer than this without undue concern. Cross talk can be reduced by routing ground conductors between pairs of signal lines considered susceptible.

Constant Impedance

For high frequency, longer tracks, including those that carry fast digital signals, it is necessary to design the conductors as a transmission line. The basic idea behind transmission line propagation is covered in Section 1.2.8; the important principle is that a signal conductor and its return path are considered as a whole and designed for a specific characteristic impedance (Z_0). Z_0 is a function of the geometry of the line and the relative permittivity of the material, which encloses it. Clearly these two factors are specific to each pcb design and construction. This means that, if you are going to have to rely on transmission lines in the circuit, you will have to designate certain layers as "constant impedance" layers.

It is usual, though not universal, to use microstrip construction (Table 1.9) in which a signal track is routed against a ground plane, which is the return path, or (on the inner layers) stripline construction in which the track is sandwiched between two ground planes. The principal geometrical factors in determining Z_0 are then the track width and its separation distance from the ground plane. Track width tolerance is controlled by the etching process, and so a wider track (which means a lower Z_0) gives a closer tolerance. The layer separation depends on the pressure applied during layer bonding and on the thickness tolerance of the prepreg or laminate materials. A thinner separation again gives a lower Z_0.

For transmission lines on the inner layers, Z_0 also depends on the square root of the relative permittivity ε_r of the epoxy-glass material. If the line is carried on the surface layer, the dependence is more complicated because part of the dielectric (above the line) is air, while below the line it is epoxy glass. How tightly ε_r is controlled depends on the quality (and cost) of the base material. General-purpose FR4 does not have a tight specification on this parameter, although more premium materials are available specifically for the purpose of building constant impedance boards. Table 2.1 gives some details of typical materials.

2.2.2 HOLE AND PAD SIZE

The diameter for component mounting holes should be reasonably closely matched to the component lead diameter for good soldering, an allowance of 0.15–0.3 mm greater than the lead diameter giving the best results. Automatic insertion machines may require larger allowances. There is a trade-off between the number of differently sized holes and board cost, so it is not advisable to specify different hole diameters for each different lead diameter. Typically you will use 0.8 mm diameter for DIL package leads and most small components, 1.0 mm for larger components and other sizes as required. Remember to specify hole diameters *after* plating on PTH boards. Also, remember to double check component lead diameters against individual holes, when you are writing up the board specification—or if your CAD system marks up hole diameters for you, make sure its component library is up-to-date. Some capacitors and power rectifier diodes in particular have larger leads than you may think! It is embarrassing to have to tell your production department to drill out holes on a double-sided PTH board, and then solder both sides because you got a lead diameter wrong. It is even more embarrassing if it's a multilayer board because you cannot drill out holes that connect to internal layers.

Vias

Via holes—PTHs that join tracks on different layers—can be any reasonable diameter, subject to current rating. The smallest useable diameter is linked to board thickness, and generally a ratio of thickness to diameter (aspect ratio) of up to 6:1 will not present too many problems in plating. But, unless you are forced into small diameter vias by constraints of packing density, you should keep them either equivalent to the smallest component hole diameter so that drill sizes are kept to a minimum or make them one size smaller (e.g., 0.6 mm) so that the likelihood of false insertion of component leads is low.

Through Hole Pads

Pads can be oval or round. Oval pads for DIL packages on a 0.1″ pitch are a hangover from early days in pcb technology when a large pad area was needed to assure a good soldered joint and good adhesion of the pad to the board. This is still advisable for non-PTH boards, and a typical round pad for a 0.8 mm diameter hole would be around 2 mm diameter, which leaves no room for tracking between pins. An oval allows one track between each pad. Some care is needed with oval pads as the hole-to-pad tolerance in the width dimension can be tight.

PTH technology reinforces the pad-to-board bond on both sides by the plating in the barrel of the hole, and the solder flows into the annulus between lead and hole plating, so that large or oval pads are unnecessary on PTH boards. The pad need only contain the drilled hole before plating, with allowance for all manufacturing tolerances. For a 0.8 mm diameter hole the pad diameter can be between 1.3 and 1.5 mm (Fig. 2.19).

Pads for larger holes on non-PTH boards should exceed the hole diameter by at least 1 mm to obtain good adhesion. The ratio of pad-to-hole diameter should be around 2 for epoxy glass and 2.5—3 for phenolic paper boards.

Surface Mount Pads

Pad sizes for SM components are determined by the individual component terminals and by the soldering technology used: wave or reflow (see Section 2.3.1). You have virtually no choice over these dimensions once you have chosen the assembly method and the component. CAD systems will all have libraries of the component and associated pad dimensions and will automatically layout the artwork for the correct sizes—provided, of course, that their libraries are correct. You may need to check this on occasion, and when you introduce a new component to the library you will need to be sure that you have incorporated the proper pad layout; component suppliers will normally include recommended pad dimensions in their datasheets.

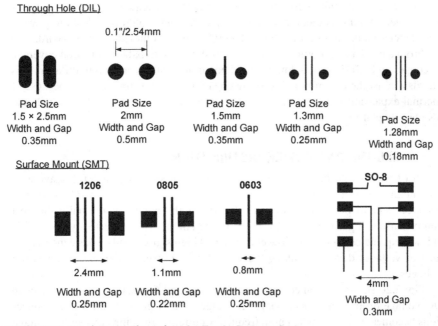

FIGURE 2.19 Spacing and Dimensions for Tracks Between Pads.

2.2.3 TRACK ROUTING

The first rule of good routing is to minimize track length. Short tracks are less susceptible to interference and cross talk, have lower parasitic reactances, and radiate less. Routing should proceed interactively with package placement to achieve this. It is often possible to improve routing prospects when using multifunction packages (typically gates or op-amps) by swapping pins, so unless there are overriding circuit considerations you should not fix pinouts on these packages until the layout has been finalized. A good CAD package with an extensive and intelligent component library will do this automatically. Similarly, you may have decided on grounds of package economy to use up all the functions in a package, but this could be at the expense of forcing long tracks to one or other function. An optimum board layout may require a few more packages than the minimum.

Many CAD autorouting software packages will lay all tracks of one orientation on one layer of the board and all of the other orientation on another layer (X−Y routing). This works and is fast, especially for low-performance digital boards, but hardly ever produces an optimum layout in terms of minimum track length and number of via holes and can be disastrous on analogue boards. Normally you should anticipate expending some skilled design effort at the layout stage in cleaning up the CAD output to produce a cheaper board with better electrical performance.

45 degrees-angled bends are preferable to right angles, as they allow a slight increase in tracking density. Right angles and acute angles in tracks are best avoided as

they provide opportunities for etchant traps and the subsequent risk of track corrosion. When two tracks meet at an acute angle the join should be filleted to prevent this. Tracks should not be run closer than 0.5 mm from the edge of the board.

From a mechanical point of view, the aim should be to balance the total coverage of copper on both sides of a double-sided board or on all layers of a multilayer. This guards against the risk of board warping due to differential strains, either because of thermal expansion in use or because of stress relief when the board is etched, and also assists plating.

2.2.4 GROUND AND POWER DISTRIBUTION

Much of what could be said here has already been said in Chapter 1. Layout of ground interconnections is as important on pcbs as it is between them. Common impedances should be avoided in critical parts of the circuit, but usually the best compromise is a ground "bus" for much of the circuit. This is perfectly acceptable at low frequencies, low gains, low currents, and high signal levels, where the magnitude of voltages developed along the ground rail are well below circuit operating voltages.

Fig. 2.20 shows resistance versus track width, and this can be used to calculate DC voltage drops where necessary. The inductance of ground and power connections becomes important when high frequency currents are being carried; the voltage drops now depend on rate of change of current.

Ground Rail Inductance

pc track inductance is primarily a function of length rather than width. The formula given in Section 1.2.1 shows that for wires, inductance depends directly on length but only logarithmically on width. pcb tracks in isolation follow the same law, but the inductance of a track on its own is often misleading as it will be modified by proximity to other tracks carrying return current. Fig. 2.20 shows the principle of magnetic field cancellation of two close tracks carrying equal but opposite currents.

This shows that the most effective way of reducing total power and ground inductance is to run the signal and return paths very close together. This maximizes their mutual inductance, and since the currents now oppose each other, the total inductance is *reduced* by the mutual inductance rather than increased. One way to achieve

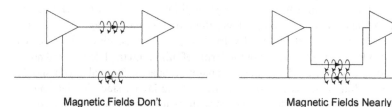

Magnetic Fields Don't
Cancel ≡ High Inductance

Magnetic Fields Nearly
Cancel ≡ Low Inductance

FIGURE 2.20 Signal and Return Currents.

this is to run power and ground rails with identical geometry on opposite sides of the board. But, if a compromise is necessary, it is better to use the board space to produce the best ground system possible and control power rail noise by decoupling (see Section 6.1.4).

Gridded Ground Layout

Good low-inductance power and ground rails can be assured on double-sided and multilayer boards by either of two techniques: grid layout or an overall or partial ground plane. The former is really an approximation of the latter. For many purposes, particularly when the layout is a regular pattern of ICs, and if the power and ground lines are well decoupled, a grid can approach the performance of a ground plane. Certainly a well-designed grid will be more effective than a poorly designed ground plane. A grid allows you to minimize sections of common impedance and ensures that lengthy ground separations are interconnected by several current paths. It is not advisable for sensitive analogue layouts when you need control of ground return current paths. In any case, an ideal grid will be too restricting for the signal paths and some compromise will be needed.

The Ground Plane

A ground plane scores when, as in analogue circuits or digital circuits with a mixture of package sizes, ground connections are made in a random rather than regular fashion across the layout. A ground plane is not necessarily, or even usually, created simply by filling all empty space with copper and connecting it to ground. Because its principal purpose is to allow the flow of return current (Fig. 2.21), there should be a minimum of interruptions to it, and for this reason ground planes are more successful on a multilayer board where an entire layer can be devoted, one to the power plane and one to the ground. This has the additional advantage of offering a low impedance between power and ground at high frequencies because of the distributed interplane capacitance.

Individual holes make no difference to the ground plane, but large slots (Fig. 2.22) do. When the ground plane is interrupted by other tracks or holes the normal low-inductance current flow is diverted round the obstacle and the inductance is effectively increased. Interruptions should only be tolerated if they do not

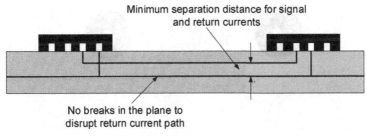

Minimum separation distance for signal and return currents

No breaks in the plane to disrupt return current path

FIGURE 2.21 The Purpose of the Ground Plane.

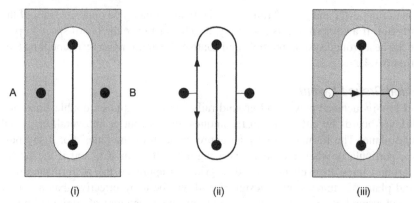

FIGURE 2.22 Current Flows From A to B in (i) and (ii) Are Equivalent and (iii) Is Preferable.

cut across lines of high *di/dt* flow: that is, underneath tracks carrying high switching currents or fast logic edges. Even a very narrow track interconnecting two segments of ground plane is better than none. At high frequencies, and this includes digital logic edge transitions, current tends to follow the path that encloses the least magnetic flux, and this means that the ground plane return current will prefer to concentrate under its corresponding signal track.

Some board manufacturers do not recommend leaving large areas of copper because it may lead to board warping or crazing of the solder resist. If this is likely to be a problem, you can replace solid ground plane by a cross-hatched pattern without seriously degrading its effectiveness. To make a soldered connection to the ground plane—or any other large area of copper—on the surface of the board, you should "break out" the pad from the area and connect via one or more short lengths of narrow track (Fig. 2.23). This prevents the plane acting as a heat sink

FIGURE 2.23 Connecting Pads to the Ground Plane.

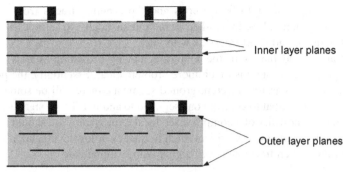

FIGURE 2.24 Inner Layer Versus Outer Layer Planes.

during soldering and makes for more reliable joints. This is not an issue for internal ground planes since the PTH increases the thermal resistance to the plane.

Inside or Outside Layers

A common question is whether the power and ground planes should be put on the outer layers of a four-layer board or the inner layers. The situation is illustrated in Fig. 2.24. From the point of view of control of the current return path, the two approaches are largely equivalent: the return current in the plane is one layer away from its adjacent signal current whichever choice is made.

Putting the planes on the outside layers has the merit of providing an electric field screen for the tracks within. This might seem to be advantageous, but in fact as soon as you put components on the board then the screening is compromised; and the denser the component packing then the less use the screen is going to be, since there will be more area (the components, their leadouts, and their pads) outside the screen.

If two planes on the inner layers provide power and ground distribution, then putting them close together has the great advantage of giving a distributed, low inductance capacitance between them, which helps greatly with high frequency decoupling. It needs to be augmented with standard decoupling capacitor techniques, but the closer the planes are the better is the board's high frequency performance. Putting planes on the outside throws away this advantage entirely.

So a general rule would be if your board has few components but a lot of high frequency tracks, for instance, a system backplane, then by all means screen them with planes on the outside. But if the board has densely packed components that need good HF decoupling—which is probably 90% of designs—then the planes should go on the inside.

Multiple Ground Planes

There is no objection to, and a great deal of advantage from, having several layers devoted to notionally identical ground planes in a multilayer (say eight or above) construction. This allows each set of track layers to be placed next to a ground plane layer and minimizes the separation distance for each track layer through the board. It is necessary though either to be rigorous about preventing tracks from jumping layers from one ground plane to another, or more realistically, to "stitch" the planes together at frequent and short intervals with vias.

Separating ground planes in the x−y direction by providing "moats" around different segments of ground is much more problematical. Essentially, the problem is that as soon as you assign different ground segments, there will be some signals that have to cross the moat to get from one segment to another. These signals are then exposed to the full hostility of a compromised local return path. You should never do this if the signals are in any way critical (high frequency or low level) and, if they are not and you must, then treat each such signal with extreme care.

2.2.5 COPPER PLATING AND FINISHING

The surface of the conductors on the outside layers of a pcb needs to be finished, to allow component soldering, or connections, or protection against oxidation. The most common finishing for surface mount boards is hot air solder leveling, which applies a thin layer of solder and then blasts it with a hot air knife to make sure the surface will be flat enough to take the solder paste application for the chip components. But for other applications, it is also possible to specify gold, silver, or nickel plating, or carbon ink. Silver may be used for RF circuits to reduce the circuit losses; gold and nickel would typically be used for connector contacts. Carbon ink is a cheap and simple finish for keypad contacts, where the resistivity of the contact is unimportant, and it can also be used for low-specification resistors.

Each plating process is a separate operation and naturally increases the cost of the bare board, even before you add the cost of the plating material itself. The thickness of the plating can be varied from fractions of a μm to more than 10 μm and will depend on the performance required of the surface, for instance, whether there will be many reconnection operations or just a few. It is possible to mask areas of the board to prevent plating of that area, and it is also possible to add "peelable" masks over particular areas of plating to prevent their contamination by a wave soldering operation. For instance, gold-plated connector areas would end up coated with solder if they were passed unprotected through a solder bath, so the mask remains in place and is removed once the whole assembly process has been completed.

2.2.6 SOLDER RESIST

Also known as solder mask, the solder resist is a thin, tough coating of insulating material applied to the board after all copper processing has been completed. Holes are left in the resist where pads are to be soldered. It serves to prevent the risk of short circuits between tracks and pads during soldering and subsequently, and it is also sometimes used as an anticorrosion coating and to provide a dark, uniform background for the component identification legend. It can be a screen-printed and oven-cured epoxy resin, a photographically exposed and developed dry film or a photo-cured liquid film.

Screen-Printed Resists

Screen-printed epoxy resin is a well-established method and is inexpensive, but achievable accuracies are poor compared to modern etching accuracies. The consequence of this is that there has to be an allowance for misregistration and resist bleeding of about 0.3—0.4 mm between the edges of pads and the edge of the solder resist pattern. It is easy to generate the artwork for this—simply repeat the pad pattern with oversize pads and generate a negative photographic image—but it cuts into the spacing between pads, and if fine tracks are run between pads, they may not be completely covered by the resist. Fig. 2.25 shows this effect. This nullifies the supposed purpose of the resist, to prevent bridges between pads and adjacent tracks! Also, screen-printed resists over large areas of copper that has been finished with tin—lead plating may crack when the board is wave soldered, as the plating melts and reflows. This is unsightly but not normally dangerous, as long as you do not rely on the resist as the only corrosion barrier.

Photo-Imaged Film

Photo film resists are capable of much higher registration accuracy and resolution (typically better than 0.1 mm) and are therefore preferred for high-density boards. They have their own problems, apart from expense, the main one being that dry films suffer lack of adhesion to poorly prepared board surfaces; liquid films have become the usual method as a result.

A solder resist should not always be regarded as essential (although for dense, wave-soldered surface mount boards, it is). It can be useful in reducing the risk of board failure through surface contamination or solder bridges but is not infallible. There is a danger that it is specified without thought or used as a crutch to overcome bad soldering practices. A well-designed board in a good production environment may be able to do without it.

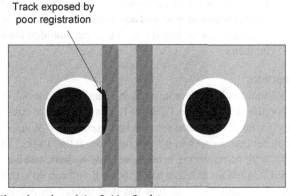

Track exposed by
poor registration

FIGURE 2.25 Misregistration of the Solder Resist.

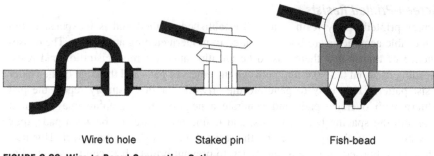

Wire to hole   Staked pin   Fish-bead

FIGURE 2.26 Wire to Board Connection Options.

2.2.7 **TERMINATIONS AND CONNECTIONS**

Any pcb that is part of a system must have connections to it. In the simplest case this is a wire soldered to a pad. If the board is plated-through, then this approach is acceptable, as the combined strength of solder in the plated hole plus the pad lands on both sides will be enough to cope with any normal wire flexing. Wires should not be soldered straight to non-PTH boards because the wire strains will be taken by the pad-to-board bond, which will quickly fail. Greater mechanical strength is needed.

This can be offered by feeding the wire back through a second hole in the board to give a measure of strain relief. Alternatively, use staked pins or "fish-beads." A pin press-fitted into a hole transmits all mechanical strains directly into the board laminate, so the reliability of the pad-to-track transition is unaffected. Wire dressing to the pin is slightly more labor-intensive, but this is not usually a disadvantage. Fish-beads are easier to use but more expensive than pins; and because they are not staked to the board, they can transmit some strain to the pad. Both types are eminently suitable for individual test points. Fig. 2.26 shows the options.

Direct wire connections should not be dismissed out of hand because they are, after all, cheap. But as soon as several connections must be taken to the board, or repetitive disconnection is required, you will automatically expect to use a multiway pcb connector. This can take one of two forms, a two-part molded male/female system or an "edge" connector.

Two-Part Connectors
There are many standard pcb connectors available, and it is impossible to do justice to all of them. Popular ones are the DIN-41612 range for Eurocard-sized module boards, the many variants of square-pin stacked connectors pioneered by Molex and available from several sources in pin pitches from $0.05''$ to $0.2''$, the insulation displacement types with pc-mounting headers and free sockets, and the subminiature "D" range to MIL-C-24308 for external data links. All you can do is to compare data sheets from several manufacturers to find the best overall fit for your particular requirement. In connectors it is especially true that you get what you pay for in terms of quality.

Contact resistance will be important if the connector carries appreciable current, as will be the case for power supply rails. Even more important is whether a low contact resistance will be maintained over time in the face of corrosion and repeated mating cycles. This depends mainly on the thickness of gold plating on the mating surfaces. A wise precaution is to dedicate several ways of a multiway connector in parallel for each power and ground rail, to guard against the effects of increasing contact resistance and faulty pins.

Insertion and withdrawal force specifications tend to be overlooked, but serious damage to a board can result if too much force has to be used to plug it in. Conversely, if the withdrawal force is low and/or the plug–socket pair is not latched, there is the risk of the connectors falling apart or vibrating loose. As with single wires, it is better to provide a separate route for diverting the mechanical strain of the interconnection into the board laminate rather than relying on the solder joints to individual pins. Connector moldings that allow for nut-and-bolt fixing are best in this respect. Incidentally, remember to specify on the production drawing that fixings should be tightened *before* the pins are soldered—otherwise strain will be put on all the soldered joints as soon as the nuts are torqued up, resulting in unreliable connections later.

Edge Connectors

Edge connector systems are popular with many designers because they are relatively cheap. In this case, the connector "finger" pattern is part of the pcb layout at the board edge, and the board itself plugs into a single-piece female receptacle on the mating part. The board should have a machined or punched slot at some point in the pattern, which mates with a blanking key in the receptacle, to ensure that the assembly is plugged in the right way round and to align the individual connectors. The ends of the receptacle can then be left open. This is safer than aligning the board edge with the end of a closed receptacle, as it is more accurate and less susceptible to damage.

The fingers on the board must be protected from corrosion by gold plating. The plating should cover the sides as well as the tops of the fingers, otherwise long-term corrosion from the edges will be a problem. Pay attention to dimensional tolerancing of the pcb, both on board thickness to ensure correct contact pressure and on machining to ensure accurate mating. A useful trick if you have spare edge connector contacts is to run them to a dummy pad in some unused space inboard. They will prove invaluable when you discover a need for more connections during prototype testing!

One final point on pcb connectors is to make sure that your chosen connector type and board technology are compatible. Very high-density connectors are available, which have multiple rows of pins at less than 0.1″ spacing. These require very fine tracking to get between the outer rows to the inner ones. Also, they are a nightmare to assemble to the board, and you may find that your production department wants larger holes than you bargained for, which makes the tracks even finer. Do not go for high-density connectors unless you really have to: stick with the chunky ones.

2.3 BOARD ASSEMBLY: SURFACE MOUNT AND THROUGH HOLE

There are two ways of mounting components to the board:

- surface mount, in which the surface mount devices (SMDs) have termination areas rather than leads and are held in place only by solder between pads on the board and their terminations
- through hole, in which the components have lead wires, which are taken through holes in the board. Such components are larger, and the net result is a considerably lower component density per unit area.

These are compared in Fig. 2.27. Because of its advantages, surface mount technology (SMT) makes up around 90% of board assemblies, but through hole still accounts for the remainder and is unlikely to disappear completely.

The advantages and disadvantages of SM construction versus through hole can be summarized as follows:

Advantages

- *Size.* Very much higher packing densities can be achieved. Components can be mounted on both sides if necessary. Enables applications that would be impossible with through hole.
- *Automation.* SM component placement and processing lends itself to fully automated assembly and so it is well suited to high-volume production. Unit assembly costs can be lowered.
- *Electrical performance.* Reduced size leads to higher circuit speeds and/or lower interference susceptibility. Higher performance circuits can be built in smaller packages—a prime marketing requirement.

Disadvantages

- *Investment.* To properly realize the potential of automated production, a sizeable capital investment in machinery is needed. This is normally measured in hundreds of thousands of pounds.
- *Experience.* A company cannot jump into SM production overnight. At every stage in the design and production process, special skills and techniques have to be learned.

Surface Mount Components   Equivalent through hole components

FIGURE 2.27 The Difference Between Surface Mount and Through Hole.

- *Components.* Most component types are available as SMDs, but still there are specialized types, which are only obtainable as through-hole mounting. Higher-power and large components can never be surface mounted.
- *Criticality of mechanical parameters.* Traditionally, any electrically equivalent component would do if it fitted the footprint. With SM, mechanical equivalence is as critical as electrical equivalence because the placement and soldering processes are that much less forgiving. This leads either to unreliability or to difficulties with component sourcing. Also, solderability and component shelf life become dominant issues.
- *Test, repair, and rework.* It is possible to test and rework faulty SM boards, with tweezers, a hot air gun, and a magnifying glass. It is a lot easier to test and rework through hole boards. Be prepared for extra expense and training at the back end of the production line.

The investment that is needed for a company that is contemplating doing its own SM production extends beyond simply acquiring the production equipment. There is probably an equivalent investment in CAD design tools, storage and procurement systems (to cut down component shelf life), test equipment, and rework stations, plus a hefty amount of time for retraining. Many companies will therefore prefer to use the services of a subcontract assembly house, despite a loss in their own profit margin, if their product volume cannot justify this investment. An important benefit of this approach is that it allows a firm to experiment with SMT, and gain some market and product experience with it, before full commitment. The other side of the coin is that the production staff do not gain any significant experience.

The stages of design, assembly, and test are very much more tightly coupled in SM than they need to be in conventional manufacture. The successful production and testing of an assembly is critically related to the pcb layout and design rules employed.

2.3.1 SURFACE MOUNT DESIGN RULES

Pad dimensions and spacing are relative to the component body and leadouts and depend on the soldering method that will be used with the board. Components can either be placed over a dot of adhesive to hold them in place and then run through a wave soldering operation or they can be placed on a board, which has had solder paste screen printed onto the pads, in which case the tackiness of the solder paste holds them lightly in position until it is reflowed, either by an infrared oven or a hot vapor bath. Fig. 2.28–2.31 shows the different stages for each type of production.

Solder Process

If the board will be wave soldered, then the IC packages should be oriented along the direction of board travel, across the wave, and a minimum spacing should be observed. This optimizes solder pickup and joint quality. Pad dimensions need to be fairly large to take up placement tolerances, since the absolute position of the

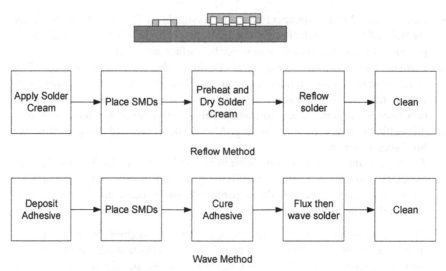

Reflow Method

Wave Method

FIGURE 2.28 Single-Sided Surface Mount Device Assembly.

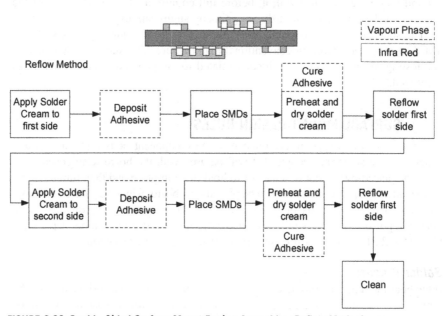

FIGURE 2.29 Double-Sided Surface Mount Device Assembly—Reflow Method.

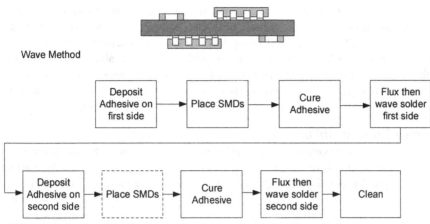

Wave Method

Deposit Adhesive on first side → Place SMDs → Cure Adhesive → Flux then wave solder first side

Deposit Adhesive on second side → Place SMDs → Cure Adhesive → Flux then wave solder second side → Clean

FIGURE 2.30 Double-Sided Surface Mount Device Assembly—Wave Method.

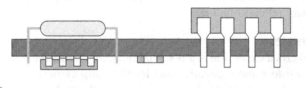

Wave Method

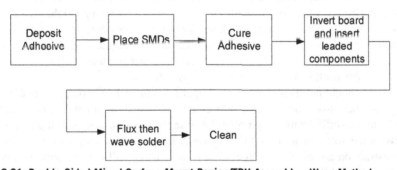

Deposit Adhesive → Place SMDs → Cure Adhesive → Invert board and insert leaded components

Flux then wave solder → Clean

FIGURE 2.31 Double-Sided Mixed Surface Mount Device/TPH Assembly—Wave Method.

package cannot be altered once the adhesive is cured. The advantage of wave soldering is that it can simultaneously process surface mount and conventional components, if they are mounted on opposite sides. The maximum height of the SM components will be determined by the risk of being detached from the board when passing through the wave.

Vapor phase or infrared soldering allows tighter packing and smaller pads, and the orientation is not critical. When the solder paste reflows, surface tension will draw misaligned components into line with their pads, so that placement tolerance is less critical. Shadowing can be a problem with components of varying height placed close together, as can variable heat absorption if there are large and small packages with different heat reflection coefficients. Board layout must start from the knowledge of which soldering method will be used and also what tolerances will be encountered in the placement process and in the components themselves. If wave soldering pad dimensions are used for reflow soldering, solder joint quality will suffer. Most layout designers use different pad dimensions for the two systems.

Printed Circuit Board Quality

pcb finish is more important than it is with through hole components. The overriding requirement is flatness of the surface, since component sizes are that much smaller and since good soldered joints depend on close contact of the leadouts with the pads. A solder resist is essential to control the soldering process. Photo-imageable film resists are to be preferred to screen printed (see Section 2.2.6) since their thickness is well controlled also because the tolerances on solder mask windows are that much tighter. There must be no bleeding of the resist onto the pads. Hot air leveling of the tin/lead finish on the solder pads is usual to prevent the bumps that form on the surface of ordinary reflowed tin/lead.

Thermal Stresses

Differential thermal expansion is a potential reliability threat for some SM components. Chip resistors and capacitors, and leadless chip carriers (LCCs), are made with a ceramic base material whose coefficient of thermal expansion is not well matched to epoxy fiberglass. Originally these components were developed for hybrid circuits, which use ceramic substrates, and for which good thermal matching is possible. At the same time, leadouts for these components are deposited directly on the ceramic so that there is no compliance between the leadout (at the soldered joint) and the case. As a result, strains set up under thermal cycling can crack the component itself or the track to which it is soldered.

You should not therefore use the larger ceramic or LCC components directly on epoxy fiberglass board. Leaded SM components such as small outline or flat-pack ICs do not suffer from this problem because there is a section of compliant lead between the soldered joint and the package. Plastic-leaded chip carriers with J-lead construction are useable for the same reason, and small chip ceramic components are also useable because of their small size.

Cleaning and Testing

Cleaning an SM board is trickier than for a through hole assembly because there is less of a gap underneath the packages. Flux contamination can get into the gaps but it is harder to flush out with conventional cleaning processes. There is considerable effort being put into developing solder fluxes that do not need to be cleaned off afterward.

Testability is an important consideration. It is bad practice to position test probes directly over component leadouts. Apart from the risk of component damage, the pressure of the probe could cause a faulty joint to appear sound. All test nodes should be brought out to separate test pads, which have no component connections to them, and which should be on the opposite side of the board to the components. There is an extra board space overhead for these pads, but they need be no more than 1 mm in diameter. Testing a double-sided densely packed SM assembly is a nightmare (see Section 9.3.3).

2.3.2 PACKAGE PLACEMENT

There is a mix of electrical and mechanical factors to consider when placing components and ICs. Normally the foremost is to ensure short, direct tracks between components, and it is always worth interactively optimizing the placement and track routing to achieve this. You may also face thermal constraints, for instance, precision components should not be next to ones that dissipate power, or you may have to worry about heat removal.

Over and above individual requirements there is also the general requirement of producibility. There should be continuous feedback between you as circuit designer and the production and service departments to make sure that products are easy, and therefore cheap, to produce and repair. Component and package placement rules should evolve with the capabilities of the production department. Some examples of producibility requirements follow:

- pick-and-place and auto insertion machines work best when components and packages are all facing the same way and are positioned on a well-defined grid
- small tubular components (resistors, capacitors, and diodes) should conform to a single lead pitch to minimize the required tooling heads. It does not matter what it is ($0.4''$ and $0.5''$ are popular) as long as it's constant
- inspection is easier if all ICs are placed in the same orientation, i.e., with each pin one facing the same corner of the board, and similarly all polarized components are facing the same way
- spacing between components should take into account the need to get test probes and auto insertion guides around each component
- spacing of components from the board edges depends on handling and wave soldering machinery, which may require a clear area (sometimes called the "stacking edge") on one or two edges
- if the board is to be wave soldered, rows of adjacent pins are best oriented across the direction of flow, parallel to the wave, to reduce the risk of solder bridges between pins or pads.

2.3.3 COMPONENT IDENTIFICATION

Most pcbs will have a legend, usually yellow or white, screen printed on the component side to describe the position of the components. This can be useful if assembly of the finished unit relies on manual insertion, but its major purpose is for the test and service departments to assist them in seeing how the board relates to its circuit diagram. With low-to-medium density boards you can normally find space beside each component to indicate its number, but as boards become more densely packed this becomes increasingly difficult. Printing a component's ID underneath it is of no use to the service engineer, and if component placement is automatic, it's no use to the production department either. Particularly if the board consists almost entirely of small outline or DIL ICs, these can be identified by a grid reference system with the grid coordinates included on the component side track artwork. Therefore you should consider whether the extra expense of printing the legend onto a high-density board is justified.

Assuming it is, there are a few points to bear in mind when creating the legend master. If possible, print onto flat surfaces, not over edges of tracks or pads; the uneven surface tends to blur the print quality. Never print ink over or near a hole (allowing for tolerances). Even if it's not a solderable hole, the unprinted ink will build up on the screen and after several passes will leave a blot, ruining the readability. If there are large numbers of vias, these should be filled in or "tented" to prevent the holes remaining and rendering areas of the board useless for legend printing.

Polarity Indication

There are several ways of indicating component polarity on the legend. Really the only criterion for these should be legibility *with the component in place*, so that they are as useful to inspectors and test engineers as they are to the assembly operators. Once a particular method has gained acceptance, it should be adhered to; using different polarity indicators on different boards (or even on the same board) is a sure way of confusing production staff and gaining faulty boards.

Guarding

The next level of defense is guarding. This technique accepts that there will be some degree of leakage to the high impedance node, but minimizes the current flow to it by surrounding it with a conductive trace that is connected to a low impedance point at the same potential. Like the similar circuit technique of bootstrapping, if the voltage difference between two nodes is forced to be very low, the apparent resistance between them is magnified. The electrical connections and pc layouts for the guards for the basic op-amp input configurations are shown in Fig. 2.32. The guard effectively absorbs the leakage from other tracks, reducing that reaching the high impedance point.

There should be a guard on both sides of a double-sided board. Although the guard virtually eliminates surface leakage, it has less effect on bulk leakage through the board; fortunately this is orders of magnitude higher than leakage due to surface contamination. The width of the guard track is unimportant as far as surface resistance goes, but a wider guard will improve the effect on bulk resistance.

Guarding is a very useful technique but obviously requires some extra thought in circuit design. Generally, you should consider it in the early design stages whenever you are working with impedance levels, which are susceptible to surface resistance variations. You may then be in the happy position of never noticing the problem.

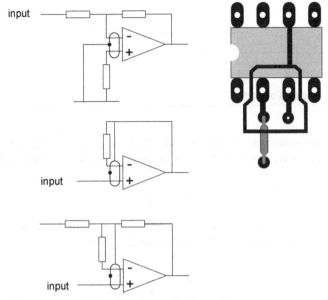

FIGURE 2.32 Input Guarding for Various Opamp Configurations.

2.3.4 UNDERSTANDING THERMAL BEHAVIOR

There are three main methods by which heat may be transmitted through a medium (such as air or pcb material):

- Convection
- Conduction
- Radiation

Of these, convection and conduction are of the most interest from a design point of view, as they can be a controllable method of removing heat from the components in the system; however, radiation, however, is less controllable and is potentially a problem. The heat generated by a component(s) or intrinsic to an environment can readily radiate to other components causing potential problems.

Thermal Conduction
Conduction is the mechanism for transfer of energy between materials at different temperatures.

The heat flow rate Q through a material (Fig. 2.33) can be calculated across a material of infinitesimally small thickness dx, with a temperature difference across it of dT using the expression:

$$Q = kA \frac{dT}{dx}$$

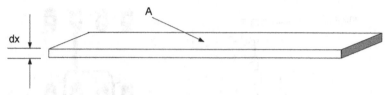

FIGURE 2.33 Thermal Conduction.

If the steady-state condition has been reached, such that the temperature on one side of the block is T_H and T_L on the other, Q can be expressed simply as in Eq. (2.2):

$$Q = kA \frac{(T_H - T_L)}{L} \qquad (2.2)$$

where k is the thermal conductivity, and L is the length of the material.

The thermal resistance can be defined using Eq. (2.3):

$$R_\theta = \frac{L}{kA} \qquad (2.3)$$

Thermal Convection

In contrast to conduction where the heat flow is through a static solid material, convection cooling occurs through a gas or fluid medium. The most common form of convection cooling is air cooling via a heat sink. In convection cooling systems, it is the movement of the medium (air or fluid) that carried the heat from the hot areas in the system to the cold.

The reason why heat sinks have fins is to present as large a surface area as possible to the medium to transport the heat away from the device. The thermal resistance can be drastically reduced in a convection cooling approach by using a forced air cooling system. If the system does not have forced air, it is called *natural* convection cooling; however, if the system has forced air it is called *forced* convection cooling.

The improvement that forced convection can bring is significant. At sea level the following expression can be used to estimate the air flow (in cfm—cubic feet per minute or lpm—liters per minute) required to dissipate a power loss at a specific temperature difference

$$\text{AirFlow (cfm)} = \frac{1.76^* W_{\text{loss}}}{\Delta T} \qquad (2.4)$$

$$\text{AirFlow (lpm)} = \frac{28.32^* 1.76^* W_{\text{loss}}}{\Delta T} \qquad (2.5)$$

Altitude also reduces the effectiveness of convection cooling as the air pressure decreases.

Table 2.3 Examples of Surface Emissivity

Surface	Emissivity
Aluminum (Polished)	0.04
Aluminum (Painted any color)	0.9
Aluminum (rough)	0.056
Aluminum (matt Anodized)	0.8
Copper (Rolled Bright)	0.03
Steel (plain)	0.5
Steel (Painted any color)	0.8

Thermal Radiation

Radiation cooling is a problem in power supplies as it is difficult to establish a clear path for extraction of excess heat using this method. Instead, the heat radiates in all directions and may cause problems with components nearby. The radiated power loss is proportional to the difference of temperatures to the power of 4 (Stefan–Boltzmann Law) in Eq. (2.6):

$$P_{\text{radiation}} = 5.67 \times 10^{-8} \times E \times \left(T^4 - T^4_{\text{ambient}}\right) \qquad (2.6)$$

where E is the emissivity of the material, with some example values of surface emissivity given in Table 2.3.

Thermal Capacity

The thermal capacity (or specific heat) is a term that causes apparent discrepancies with the thermal behavior of materials, and the thermal capacity is analogous to a capacitor in the electrical domain and influences how fast a material heats up. As an example, if a heat exchanger has a high thermal capacity, it will take some time for the final temperature to be reached, but it will be the same final temperature as a heat exchanger with the same thermal resistance but lower thermal capacity (Fig. 2.34).

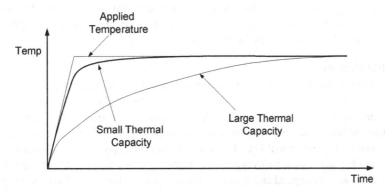

FIGURE 2.34 Illustration of Thermal Capacity.

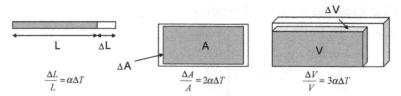

FIGURE 2.35 Thermal Expansion in One, Two, and Three Dimensions.

Material	Coefficient of Expansion α (per °C)
Aluminium	$24*10^{-6}$
Brass	$19*10\text{-}6$
Steel	$13*10\text{-}6$
Copper	$17*10\text{-}6$
Gold	$14*10\text{-}6$
Silver	$18*10\text{-}6$

FIGURE 2.36 Coefficients of Thermal Expansion for Various Materials.

Thermal Expansion

As the temperature increases, an object will also expand. This is quantified using the thermal coefficient of expansion. The material can expand in one, two, or three dimensions as shown in Fig. 2.35, with coefficients of thermal expansion provided for some common materials in Fig. 2.36.

Thermal Shock

Materials expand and contract naturally depending on the thermal conditions prevalent. Thermal shock is defined as the failure or fracture of a materials or structure due to stresses induced by thermal expansion. One aspect of this is manifest if two materials with different expansion coefficients are bonded together and the electronics cause a significant heat rise.

In this instance, the materials may be cracked if the rates of expansion are different enough for the resulting stresses to cause a structural failure. The stress in a material can be estimated by calculating the expansion and then applying Hooke's Law ($F = -kx$) to obtain the stress.

Thermal Cycling

As the changes in temperature will cause mechanical stress in materials, it is useful to consider various types of temperature cycling. There may be significant stresses in the die and substrate of power devices due to localized temperature variations.

There are also variations in environmental temperatures as the weather changes or in the operating environment (PC, Car Engine, Aircraft, Spacecraft). Thermal stress will occur also as the materials cool down, so the stresses are applied to the materials as they continuously expand and contract. The thermal cycling may also have a significant effect on other mechanically related aspects of the system such as solder joints.

Solder Cracking

Solder cracking can occur internally or externally.

- Internal solder cracks occur where the solder itself is cracked. This is especially insidious as it may be very difficult to spot or pinpoint the fault until a failure occurs.
- External solder cracks occur between the solder and the surface to which the solder is bonded.

The Impact of Thermal Stress on Reliability

The reliability of components will be adversely affected by any form of electrical and/or thermal stress. A typical approach to calculating the mean time before failure, usually specified in hours, is to take the base failure rate and add electrical and thermal stress "accelerators" to estimate the true reliability under fault conditions:

$$\lambda_m = \sum_{i=1}^{n} A_{ti} S_i \lambda_i \tag{2.7}$$

where A_{ti} = thermal accelerator; S_i = electrical accelerator; λ_i = base failure rate; n = number of modules.

The key term in this context is the thermal accelerator (A), which can be estimated using Eq. (2.8):

$$A_{ti} = e^{\left[\frac{e_a}{k}\left(\frac{1}{T_{ref}}-\frac{1}{T_{op}}\right)\right]} \tag{2.8}$$

where k = Boltzmann's constant (8.6e-5); e_a = eV; T_{ref} = reference temperature (K), T_{op} = operational temperature (K)

The change in failure rate can be seen graphically as the temperature increases in Fig. 2.37. In this case the failure rate is a ratio with respect to the base failure rate at 25°C.

Failure Rate (Ratio)

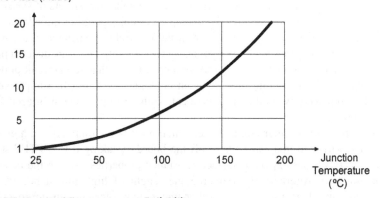

FIGURE 2.37 Impact of Temperature on Reliability.

2.4 SURFACE PROTECTION

The insulation resistance between adjacent conductors on a bare pcb surface depends on the conductor configuration, the surface resistance of the base material, processing of the board, and environmental conditions, particularly temperature, humidity, and contamination. For a new board with no surface contamination the expected insulation resistance between two parallel conductors can be derived from:

$$R_i = 160 \times R_m \times (w/l) \qquad (2.9)$$

where R_m is the material's surface resistance specification at a given temperature; (see Table 2.1); w is the track spacing; l is the length of the parallel conductors.

Variations in Surface Resistance

Generally this value is of the order of thousands of megohms and could be safely ignored for most circuits. Unfortunately it is not likely to be the actual value you would measure. This is because plating and soldering processes, dust and other surface contamination, moisture absorption, and temperature variation will all have the effect of lowering the insulation resistance. Variations between 10 and 1000 times less can be observed under normal operating conditions, and a severe environment can reduce it much more.

When you are working with high impedance or precision circuits you may not be able to ignore the change in surface resistance. Until you realize what the problem is, its effects can be mysterious and hard to pin down: they include variability of circuit parameters with time of day, handling of the board, weather (relative humidity), location and orientation of the assembly, and other such factors that normally you would expect to be irrelevant. Typical variations are change of bias point on high input impedance amplifiers and unreliable timing on long time constant integrators.

Circuit Design Versus Surface Resistance

There are a number of strategies you can use to combat the effects of surface resistance. The first and most obvious is to keep all circuit operating impedances as low as possible, so that the influence of an unstable parallel resistance in the megohm region is minimal. In some cases—micropower circuits and transducer inputs, for example—you do not have the option. In other cases a change in circuit philosophy could be beneficial; a long time constant analogue integrator or sample and hold might be replaced by its digital equivalent, with an improvement in accuracy and repeatability.

If there is a particular circuit node, which must be maintained at a high impedance, the wiring to this can be kept off the board by taking it to a PTFE stand-off insulator. PTFE has excellent surface resistance properties even in the presence of contamination. Alternatively, reducing the length of high impedance PC tracks and increasing their distance from other tracks may offer enough improvement, at the expense of packing density. Simply rerouting an offending track might help: if a power rail is run past a high impedance node, which is biased near 0 V, you have an unwanted potential divider that will pull the bias voltage up by an unpredictable amount. Put the power rail track elsewhere.

2.4.1 CONFORMAL COATING

If none of the above methods are sufficient, or if they are inapplicable, or if the working environment of the board is severe (relative humidity approaching 100%, conductive or organic contamination present, corrosive atmosphere), then you will have to go for conformal coating. This is *not* a decision to be taken lightly; try everything else first. Coating adds pain, sweat, and expense to the production process; and the following discussion will outline why.

Coating Versus Encapsulation

Note that conformal coating is not the same as encapsulation, or potting, which is even less desirable from a production point of view. Encapsulation fills the entire unit with solid compound so that the end result looks rather like a brick and is used to prevent third parties from discovering how the circuit works, or to meet safety approvals, or for environmental or mechanical protection. A badly potted unit will probably work like a brick too: differential thermal expansion as the resin cures can crack poorly anchored pc tracks, and the faulty result is unrepairable. Conformal coating covers the board with thin coats of a clear resin so that the board outlines and components are still visible. It provides environmental protection only. Occasionally it may help in meeting safety approval clearances but this is rare.

The main environmental hazard against which conformal coating protects is humidity. The popular coating types, acrylics, polyurethanes, epoxy resins, and silicones are all moistureproof. Most offer protection against the common chemical contaminants: fluxes, release agents, solvents, metal particles, finger grease, food and cosmetics, salt spray, dust, fuels, and acids. Acrylics are rather less resistant to chemical attack than the others. A conformal coating will not allow closer track spacing when this is determined by the surface insulation properties of the board, but it does eliminate degradation of these properties by environmental factors: *if it is used properly.*

Steps to Take Before Coating

The first point to remember is that a conformal coating seals in as well as out. The cleanliness of the board and its low moisture content are paramount. If residual contaminants are left under the film, corrosion and degradation will continue and will eventually render the coating useless. A minimum of three steps must be followed immediately prior to coating:

- vapor degrease in a solvent bath (note that with increasing concern about environmental pollution and ozone depletion, traditional cleaning fluids, particularly CFCs, are being phased out)

- rinse in deionized water or ethyl/isopropyl alcohol to dissolve inorganic salts
- oven bake for 2 h at 65–70°C (higher if the components will allow it) to remove any residual solvent and moisture.

After cleaning and baking, the assembly should only be handled with rubber or lint free gloves. If the cleaned boards are left for any appreciable time before coating, they will start to reabsorb atmospheric moisture, so they should be packed in sealed bags with .

Application

The coating can be applied by dipping or spraying. The application process must be carefully controlled to produce a uniform, complete coat. Viscosity and rate of application are both critical. At least two and preferably three coats should be applied, air- or oven-drying between each, to guard against pinholes in each coat. Pot life of the coating material—the length of time it is useable before curing sets in—is a critical parameter since it determines the economics of the application process. Single-component solvent-based systems are preferable to two-part resins in this respect and also because they do not need metering and mixing, a frequent source of operator error. On the other hand, solvent-based systems require greater precautions against operator health hazards and flammability.

Nearly all boards will require breaks in the coating, for connectors or for access to trimming components and controls. Aside from the question of the environmental vulnerability of these unprotected areas, such openings require masking before the board is coated, and removal of the mask afterward. Manual application of masking tape and semiautomatic application of thixotropic latex-based masking compound are two ways to achieve this.

Test and Rework

Finally, once the board is coated, there is the difficult problem of test, rework, and repair. By its nature the coating denies access to test probes, so all production testing must be done before the coating stage. Acrylics and polyurethanes can be soldered through or dissolved away to achieve a limited degree of rework but other types cannot. After rework, the damaged area must be cleaned, dried, and recoated to achieve a proper seal. Ease of rework and ease of application are often the most important considerations in choosing a particular type of coating material.

You might now appreciate why conformal coating is never welcome in the production department. Because it is labor-intensive, it can easily double the overall production cost of a given assembly. Specify it only after long and careful consideration of the alternatives.

2.5 SOURCING BOARDS AND ARTWORK

Before we leave the subject of pcbs, a short discussion of board procurement is in order. There are two stages involved: generation of the artwork and associated documents and production of the boards. The two are traditionally separated because different firms specialize in each, so that having generated the artwork from one source you would take it to another to have the boards made. Some larger pcb firms have both stages in-house or have associated operations for each.

2.5.1 **ARTWORK**

The artwork includes the track and solder resist patterns, the hole drilling diagram, the component legend, and a dimensional drawing for the board. The patterns are generated photographically, generally direct from CAD output. To create the artwork you can either do it yourself in-house using your CAD system or you can take the work to a bureau that specializes in PC artwork and has its own CAD system. There are advantages and trade-offs both ways.

Using a Bureau

So you may be faced with the choice of going to an outside bureau. Reasons for doing this are if

- your own company does not have any artwork facilities at all;
- you do have the facilities but they are insufficient for or inappropriate to your design;
- you have all the facilities you need but they are not available in the timescale required.

There are less tangible advantages of a bureau. Its staff must by nature all be highly skilled and practiced layout draftsman, and they should be using reasonably up-to-date and reliable CAD systems, so the probability of getting a well-designed board out in a short time may be higher than if you did it in-house. They will also be able to give you guidance in the more esoteric aspects of pcb design that you may not be familiar with. This can be particularly important for new surface mount designs, or those with severe performance requirements such as high-speed critical track routes. A bureau that is interested in keeping its customers will be flexible with its timescales. Even the cost should be no more than you would have to pay in-house, since the bureau is using its resources more efficiently. Finally, a bureau that works closely with a board manufacturer will be able to generate artwork that is well matched to his requirements, thereby freeing you of an extra and possibly unfamiliar specification burden.

Disadvantages of a Bureau

There are three possible disadvantages. One is confidentiality: if your designs go outside, there is always the possibility that they will be compromised, whatever assurances the bureau may give. Another is that you are then locked in to that bureau for future design changes, unless you happen to have the same CAD system (remembering that even different software versions of the same system can be incompatible) in-house. The risk is that the bureau might not be as responsive to requests for changes in the future as it would be to the first job. You have to balance this against what you know of the bureau's past history.

Finally, you may feel that you have less control over the eventual layout. This is not such a threat as it may seem. You always have the option to go and look over the layout designer's shoulder while they are doing the work; select a bureau nearby to avoid too much travel. Or, use e-mail to review the design as it continues, at defined stages in the process. Even if you do not have the same CAD system in-house, the artwork can be viewed in standard Gerber format using the freely available GCPrevue software.

But more importantly, giving your design to an outsider imposes the very beneficial discipline that you need to specify your requirements carefully and in detail. Thus you are forced to do more rigorous design work in the early stages, which in the long run can only help the project.

2.5.2 BOARDS

PCB suppliers are not thin on the ground. It should be possible to select one who can offer an acceptable mix of quality, turnround, and price. There is, though, a lot of work involved in doing this, and once you have established a working relationship with a particular supplier it is best to stay with him, unless he or she lets you down badly. Any board manufacturer should be happy to show you round his or her facilities and should be able to be specific about the manufacturing limits to which he or she can work. You may also derive some assurance from going to a quality-assessed supplier, or it may be company policy to do so. While market competition prevails, using a quality-assessed supplier should be no more expensive than not.

The major difference lies between sourcing prototype and production quantity boards. In the prototype stage, you will want a small quantity of boards, usually no more than half a dozen, and you will want them fast. A couple of weeks later you will have found the errors and you will want another half a dozen even faster because the deadline is looming. The fastest turnround offered on normal doublesided PTH boards is four or 5 days, and the cost for such a "prototype service" is between two and five times that for production quantities with several weeks' lead time.

Sometimes you can find a manufacturer who can offer both reasonable production and prototype services. Usually though, those who specialize in prototype service do not offer a good price on production runs, and vice versa. If you manage to extract a fast prototype service on the promise of production orders, you then have to explain to the purchasing department why they cannot shop around for the lowest price later. This can be embarrassing; on the other hand they may be operating under other constraints, such as incoming quality. There are several trade-offs that the purchasing department has to make, which may not be obvious to you, and it pays to involve them throughout the design process.

The board supplier will want a package of information, which is more than just the board artwork. A typical list would include the following:

- Artwork for all copper layers
- Artwork for solder mask layers
- Ident layers, peelable layers, carbon layers—where applicable
- Drilling data
- Drawings for a single pcb or pcb panel if applicable
- Details of material specifications to be used—usually you will refer to the board manufacturer's own standard materials
- Metallic finish required
- Solder mask and ident—type and color
- Layer build details
- Testing requirements

It would be typical to have a "boilerplate" specification, which applies to all boards used in-house and which covers all these points in a standard way, so that only the differences need to be considered in depth for a particular board design.

Passive Components

CHAPTER OUTLINE

The Circuit Designer's Companion. http://dx.doi.org/10.1016/B978-0-08-101764-7.00003-7

3.1 **RESISTORS**

Resistors are ubiquitous. Because of this, their performance is taken for granted; provided they are operated within their power, voltage, and environmental ratings. This is reasonable, since after millions of accumulated resistor-years experience, there is little left for their manufacturers to discover. But there are still applications where specifying and applying resistors needs to be handled with some care.

Let us start with an appreciation of the different varieties of resistor that are available. Table 3.1 is a guide to the common types that will be encountered in general circuit design. There are more esoteric types, which are not covered.

3.1.1 **RESISTOR TYPES**

Surface Mount Chip

The most common general-purpose resistor is the thick film surface mount chip type. Available in huge quantities and very low prices, it is the workhorse of the resistor

Table 3.1 Survey of Resistor Types

Type	Ohmic Range	Power Range	Tolerance	Tempco Range	Manufacturers	Applications	Cost
Carbon film	2.2 –10 M	0.25–2 W	5%	−150 > −1000 ppm/°C	Neohm, Rohm, Piher	General purpose/ commercial	<1 p
Carbon composition	2.2 –10 M	0.25–1.0 W	10%	+400 > −900 ppm/ °C	Neohm, VTM, Welwyn	Pulse, low inductance	1–3 p
Metal film (standard)	1–10 M	0.125–2.5 W	1%, 2%, 5%	±50 > 200 ppm/°C	BC, Neohm, Vishay, Piher, Rohm, Welwyn	General purpose/ industrial and military	1–3 p
Metal film (high ohm)	1 –100 M	0.5–1 W	5%	±200 > 300 ppm/ °C	BC, Neohm	High voltage and special	5–20 p
Metal glaze	1 –100 M	0.25 W	2%, 5%	±100 > 300 ppm/ °C	Neohm	Small size	5 p
Wirewound	0.1 –33 K (0.01...)	2–20 W (10 –100 W aluminum)	5%, 10%	±75 > 400 ppm/°C	BC, Welwyn, CGS, VTM, Erg	High power	15–50 p [50 p–£1 (aL)]
Metal film (precision)	5–1 M	0.125–0.4 W	0.05% > 1%	±15 > 50 ppm/°C	BC, Welwyn, Beyschlag	Precision	10–50 p
Wirewound (precision)	1–1 M	0.1–0.5 W	0.01% > 0.1%	±3 > 10 ppm/°C	Rhopoint, Vishay	Extra precision	£2–£20
Bulk metal (precision)	1–200 K	0.33–1 W	0.005% > 1%	±1 > 5 ppm/°C	Vishay, Welwyn	Extra precision	
Resistor networks and arrays	10–1 M	0.125–0.3 W per element	2%	±100 > 300 ppm/ °C ±50 ppm/°C tracking	Bourns, Dale, CTS, Beckman	Multiresistor	10–35 p
SM chip film resistors	0–10 M	0.1–0.5 W	1%, 2%, 5%	±100 > 200 ppm/ °C	BC, Welwyn, Rohm,	Surface mount, hybrids	0.2–2 p

(1) This survey does not consider special resistor types.
(2) Manufacturers quoted are those widely sourced in the United Kingdom at time of writing.
(3) Quoted ranges are for guidance only.
(4) Costs are typical for medium quantities.

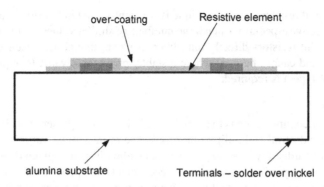

FIGURE 3.1 **Chip Resistor Construction.**

world. The construction is very simple (Fig. 3.1) and hardly varies from manufacturer to manufacturer. An alumina (aluminum oxide ceramic) substrate with nickel-plated terminations has a resistive ink film printed or otherwise deposited on its top surface. The terminations are coated with a solder dip to ensure ease of wetting when the part is soldered into place, and the top of the part is coated with an epoxy or glass layer to protect the resistive element.

Different manufacturers make various claims for the ruggedness and performance of their parts, but the basic features are similar. Power dissipation is largely controlled by the thermal properties of the printed circuit board (pcb) pads to which the chip is soldered; and if you are running close to the rated power of the part, it will be necessary to confirm that your pad design agrees with the manufacturer's recommendations.

Common, standardized chip sizes are shown in Table 3.2. You can also get metal film chip resistors for higher performance applications, but these are more expensive than the common thick film.

The resistive ink technique used for chips can also produce standard axial-lead resistors (metal glaze) of small size and can be used directly onto a substrate to

Table 3.2 Chip Resistor Sizes

Size	Dimensions mm L × W × H
0201	0.6 × 0.3 × 0.25
0402	1.0 × 0.5 × 0.25
0603	1.6 × 0.8 × 0.45
0805	2.0 × 1.25 × 0.5
1206	3.2 × 1.6 × 0.6
1210	3.2 × 2.6 × 0.6
2010	5.1 × 2.5 × 0.6
2512	6.5 × 3.2 × 0.6

generate printed resistors. This technique is frequently used in hybrid circuits and is very cost-effective especially when large numbers of similar values are required. It is possible to print resistors directly onto fiberglass pcb, though the result is of very poor quality and cannot be used where a stable, predictable value (compared with conventional types) is required.

Metal Film

The next most common type is the metal film, in its various guises. This is the standard part for industrial and military purposes. The most popular varieties of leaded metal film are hardly any more expensive than carbon film and, given their superior characteristics, particularly temperature coefficient (tempco), noise, and power handling ability, many equipment manufacturers do not find it worthwhile to bother with carbon film.

Variants of the standard metal film cater for high- or low-resistance needs. The "metal" in a metal film is a nickel—chromium alloy of varying composition for different resistance ranges. A film of this alloy is plated onto an alumina substrate. For leaded parts, the end caps and leads are force fitted to the tubular assembly, and the resistance element is trimmed to value by cutting a helical groove of controlled dimensions in it, which allows the same film composition to be used over quite a wide range of nominal values. The whole part is then coated in epoxy and marked. The disadvantage of the helical trimming process is that it inherently increases the resistor's stray inductance and also limits its pulse handling capability.

Chip metal film resistors use the same film-on-alumina technique, but the trimming, if necessary, is done by cutting small, controlled segments out of the film element to increase its resistance slightly.

Carbon

The most common leaded resistor for commercial applications is the carbon film. It is certainly cheap—less than a penny in quantity. It also has the least impressive performance in terms of tolerance and tempco, but it is normally adequate for general-purpose use. The other type that uses pure carbon as the resistive element is carbon composition, which was the earliest type of resistor but nowadays finds a use in certain applications that involve an assured pulse withstand capability.

Wirewound

For medium and high-power (>2 W) applications the wirewound resistor is almost universally used. It is fairly cheap and readily available. Its disadvantages are its bulk, though this allows a lower surface temperature for a given power dissipation; and that because of its construction, it is noticeably inductive, which limits its use in high-frequency or pulse applications. Wirewound types are available either with a vitreous enamel or cement coating or in an aluminum housing, which can be mounted to a heat sink. Aluminum housings can offer power dissipations exceeding 100 W per unit.

Precision Resistors

Once circuit requirements start to call for accuracy and drift specifications exceeding the usual metal film abilities, the cost increases substantially. It is still possible to get metal film resistors of "precision" performance up to an order of magnitude better than the standard, though at prices an order of magnitude or more higher. Drift requirements of less than 10 parts per million per °C (ppm/°C) introduce many more significant factors into the performance equation, such as thermal emf, mechanical and thermal stress, and terminating resistance. These can be dealt with, and the resistive and substrate materials can be optimized, but the unit costs are now measured in pounds, and delivery times stretch to months.

Resistor Networks

Thick film resistor networks are manufactured like chips. A resistive ink is screen printed onto a ceramic substrate to form many resistors at once, which is then encapsulated to form a multiresistor single package. The resulting resistors have the same performance as a single thick film chip, though with reduced breakdown voltage and power handling ability. However, it is also possible to manufacture metal film networks with very good accuracy and stability, and more importantly the property of closely tracking values over temperature.

3.1.2 TOLERANCING

Perhaps the most fundamental application question is that of accuracy and tolerances. Process variations apply to all standard resistor manufacturers; for example, a single 68 K resistor will not have an absolute value of 68,000 Ω. If it is a 5% part, its value when supplied should lie between 64.6 and 71.4 K (and occasionally outside, if the manufacturer's quality assurance is not up to scratch). What are the consequences of this, for example, for the humble potential divider shown in Fig. 3.2, when the input voltage (V_i) is defined as 10 V?

The unloaded value of V_O is not 5 V. In the real circuit there are two worst-case values to take into account: when R_1 is at the upper limit of its tolerance and R_2 is at the lower limit, and vice versa. These two cases yield

$$V = \frac{10 \times 64.6}{64.6 + 71.4} = 4.785 \text{ V } (-5\%) \tag{3.1}$$

$$V = \frac{10 \times 71.4}{64.6 + 71.4} = 5.25 \text{ V } (+5\%) \tag{3.2}$$

The general case is

$$\frac{V'}{V} = 1 - \frac{2K \times R_1}{[R_2 \times (1 - K) + R_1 \times (1 + K)]} \tag{3.3}$$

where V is the output voltage with both resistors nominal and V' is the output voltage when R_1 is at its high tolerance K and R_2 is at its low tolerance K (for 5% resistors, $K = 0.05$).

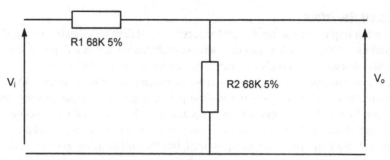

FIGURE 3.2 Simple Potential Divider.

If the resistors are equal, the voltage variation is the same as the resistor tolerance.

You have to check that neither of these cases lead to out-of-limits circuit operation; and to be thorough, this must be done for all critical resistor combinations in the whole circuit. The trick, of course, is to know which combinations are critical. In a complex resistor network, it is not always easy to see which permutations produce worst-case results. In such cases a circuit simulator can quickly prove its worth.

Basic Statistical Behavior

If we consider any parameter of a device or circuit as being defined by a single number, then this is defined as the nominal value. It is easy, particularly when an optimization has been used, to consider this an exact and precise definition of the model behavior. This is, however, an error in many cases, as in fact the nominal value is often simply the *mean* value obtained from multiple tests with a complete batch of components to establish the values.

For example, consider a batch of 1000 resistors that have been manufactured to have a designed nominal value of 100 Ω. Each device is measured on a resistance bridge to establish its actual value; and if we plot the results, we would in fact see a "spread" of results due to the intrinsic variability of the manufacturing process.

Clearly it can be seen that the results of the measurement show a spread around the desired value of 100 Ω, but that in fact most of the results are not exactly the correct value and some results are quite some distance from the nominal value. If we take the same results and calculate a histogram of the values, then we can measure the mean value and also the standard deviation.

As we have seen in the discussion of noise earlier in this chapter, the calculation of the mean value for a continuous signal is the integration over a time period; however, in this case we are looking at a number of discrete values, so the integration becomes a summation as defined in Eq. (3.4).

$$\mu = \frac{1}{n}\sum_{i=1}^{n} x[i] \tag{3.4}$$

where n is the number of samples, $x[i]$ is the individual sample, and μ is the mean value.

We also need to calculate the variability of the measurement, i.e., how much do the measurements deviate from the nominal value? To do this we need to calculate the variance from the nominal value using the expression defined in Eq. (3.5).

$$V = \frac{1}{n} \sum_{i=1}^{n} (x[i] - \mu)^2 \tag{3.5}$$

The variance is not a particularly useful number as it is scaled (squared in fact) from the original units, and so to get a measure of the variability we take the square root of the variance to obtain a measure of the typical deviation of a sample from the mean value, and this is called the standard deviation (σ) as defined in Eq. (3.6).

$$\sigma = \sqrt{V} = \sqrt{\frac{1}{n} \sum_{i=1}^{n} (x[i] - \mu)^2} \tag{3.6}$$

If we take the results shown in Fig. 3.3 and calculate the mean and standard deviation, we can superimpose them on the statistical "count" plot in Fig. 3.4. We can see that the mean value is measured as 99.949 Ω, and the standard deviation is measured as 3.3854 Ω.

A valid question that the circuit designer could ask at this point is what these numbers actually represent? Clearly the mean value is useful as it demonstrates that on average the device is very close to our specified value of 100 Ω; however, in many cases the results deviate from this. The standard deviation is a useful measure in that it can be used to estimate the proportion of devices within a certain deviation from the nominal value. We can only make this assumption, however, if we

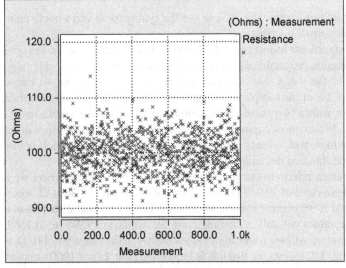

FIGURE 3.3 Random Spread of Resistance Measurements.

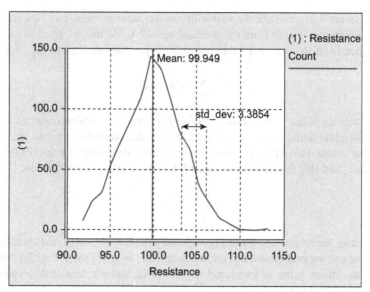

FIGURE 3.4 **Mean Value and Standard Deviation.**

have a well-defined statistical variation behavior. In practice we can say that a truly random variation will follow a statistical function called a "normal" distribution. This has, if there are enough samples, a bell curve shape with the probability density function (PDF) as defined by Eq. (3.7).

$$PDF = \frac{1}{\sigma\sqrt{2\pi}}\exp\left(-\frac{[x-\mu]^2}{2\sigma^2}\right) \tag{3.7}$$

If we plot this graphically, we can see the symmetrical and smooth nature of the PDF of the normal distribution.

A key significant aspect of the PDF shown in Fig. 3.5 is that we can quantify using this function proportionally how much of the values will be within one standard deviation of the mean, two standard deviations, and so on. As we can see from Fig. 3.6, 68.2% of the samples are within one standard deviation of the mean value, 95.45% are within two standard deviations, and 99.73% are within three standard deviations. In fact, most engineering designers will work to a tolerance of three standard deviations, which leads to the term "six-sigma design," which is to plus and minus three standard deviations (sigma).

If we return briefly to our previous example of the resistor, where we have obtained a mean value of 99.949 Ω and a standard deviation of 3.3854 Ω, we can therefore say that if we assume that the variation follows a normal distribution that over 99% of the values will fall in the range $\mu \pm 3\sigma$, which is 99.949 $\pm$ 10.1562 Ω. This indicates that we will see a spread of values mostly in the range 90–110 Ω, and if we return to Fig. 3.3, we can see that this is indeed the case. In our 1000 samples, we can see that there is only one measurement that falls outside this range.

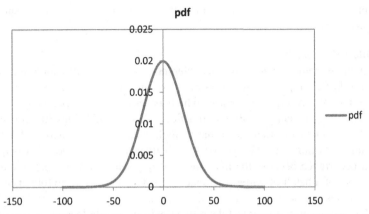

FIGURE 3.5 Probability Density Function (PDF) of Ideal Normal Distribution With Mean Value of 0 and Standard Deviation of 20.

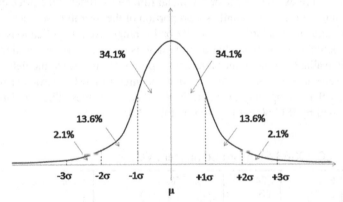

FIGURE 3.6 Proportions of Samples With a Normal Probability Density Function Relative to the Standard Deviation.

Modeling Distributions

We can implement a probability distribution function on a parameter by applying a "normal" distribution directly to the parameter and passing this to the model. For example, if the resistance parameter is normally passed to the model using the mean value (in our example $r = 100\,\Omega$), we can instead pass the output from a normal distribution function (with the same mean value and also with a "tolerance" value, which is the same as the 3σ value we calculated in the previous section). In this example we could therefore round the measured values to a mean value of $100\,\Omega$ and a tolerance (3σ) value of $10\,\Omega$. This is implemented in the model as a normal function, as shown in Eq. (3.8).

$$r = \text{normal}(100, 90, 110) \tag{3.8}$$

where in this case the normal function parameters are the mean value, mean $- 3\sigma$, and mean $+ 3\sigma$.

Tolerance Variations

It is sometimes tempting to hope that tolerance variations will cancel out over a medium-to-large sample, so that if a circuit parameter depends on several resistor values a sort of "average" tolerance, which is less than the specified tolerance, can be used. This is dangerously bad practice. Process similarities (the flip side of process variations) can often mean that a single batch of 5% resistors all have the same value to within, say, 1%, yet be 4% off their nominal value. This is quite a frequent occurrence because the manufacturer may select parts for close tolerance purposes out of a batch of wider tolerance. What is left in the wide tolerance batch then has "holes" in the middle of the tolerance range. The manufacturer is perfectly entitled to ship them, but your test department will find a whole batch of assemblies built with perfectly good components and all showing the same fault.

If the standard 5% thick film tolerance is not good enough, there is the option of the (slightly) more expensive 2% or 1% metal film types. Indeed, the price differential between these is now so small as a proportion of the overall cost of an assembly that it is common to standardize on the 2% or 1% ranges for all applications, regardless of the actual requirement. One point to note is that extreme low or high values may not be available in the tighter tolerances. Below 1% tolerance the field is taken over by precision resistors, which are often specified to have a particular required value, rather than being only available in the standard values. The standard value range specified by IEC 60063 is given in Table 3.3.

Table 3.3 IEC 60063 Standard Component Values

E6 ±20%	E12 ±10%	E24 ±5%	E48 ±2%	Additional E96 ±1%
1.0	1.0	1.0, 1.1	1.00, 1.05, 1.1, 1.15	1.02, 1.07, 1.13, 1.18
	1.2	1.2, 1.3	1.21, 1.27, 1.33, 1.40, 1.47	1.24, 1.30, 1.37, 1.43
1.5	1.5	1.5, 1.6	1.54, 1.62, 1.69, 1.78	1.50, 1.58, 1.65, 1.74
	1.8	1.8, 2.0	1.87, 1.96, 2.05, 2.15	1.82, 1.91, 2.00, 2.10
2.2	2.2	2.2, 2.4	2.26, 2.37, 2.49, 2.61	2.21, 2.32, 2.43, 2.55, 2.67
	2.7	2.7, 3.0	2.74, 2.87, 3.01, 3.16	2.80, 2.94, 3.09, 3.24
3.3	3.3	3.3, 3.6	3.32, 3.48, 3.65, 3.83	3.40, 3.57, 3.74
	3.9	3.9, 4.3	4.02, 4.22, 4.42, 4.64	3.92, 4.12, 4.32, 4.53
4.7	4.7	4.7, 5.1	4.87, 5.11, 5.36	4.75, 4.99, 5.23, 5.49
	5.6	5.6, 6.2	5.62, 5.90, 6.19, 6.49	5.76, 6.04, 6.34, 6.65
6.8	6.8	6.8, 7.5	6.81, 7.15, 7.50, 7.87	6.98, 7.32, 7.68, 8.06
	8.2	8.2, 9.1	8.25, 8.66, 9.09, 9.53	8.45, 8.87, 9.31, 9.76

3.1.3 TEMPERATURE COEFFICIENT

Where precision resistors are needed, usually in measurement applications, another parameter becomes important: their resistive tempco, expressed in parts per million per °C (ppm/°C).

Standard metal film and chip resistors have tempcos of the order of ±50 to ±200 ppm/°C. The value of a 200 ppm/°C part could change by up to 1% over a temperature range of 50°C. This does not mean that every resistor quoted at 200 ppm/°C will change by this much, only that this is the maximum that you can expect. The actual tempco depends on the manufacturing process and also on the value. For instance, carbon film tempco varies from −150 to −1000 ppm/°C depending on value. Precision wirewound, metal film, and bulk metal resistors can achieve orders of magnitude better than this—1 ppm/°C is achievable—but are correspondingly expensive.

A typical use for precision resistors is in dividing down a stable voltage reference—in Fig. 3.7; for example, there is no point selecting a voltage reference with a tempco of 30 ppm/°C and then dividing it down with 200 ppm/°C resistors. The type used for R_1 in this circuit is not critical as input voltage regulation will usually be far more significant than value changes. On the other hand, if V_{out} is to be used as a reference derived from V_{ref}, then R_2 and R_3 must be of comparable stability to the reference voltage. Suppose that V_{out} is required to be 1.00 V ± 1.5% and to show 30 ppm/°C temperature stability. The reference is an LM385B−1.2, which has a specified voltage of 1.235 V ± 1% and an average tempco of 20 ppm/°C.

To get 1.00 V with nominal values and assuming R_3 = 10 K with no load current, then R_2 will need to be 2.35 K (not a standard value, though the closest in the E96 series is 2.37 K). Taking worst-case tolerances for the LM385 and both resistors (V_{ref} high, R_2 low, R_3 high; V_{ref} low, R_2 high, R_3 low) and assuming both resistors have the same tolerance, then some calculation shows that the specified resistor tolerance should be better than 1.4%. Similarly the tempcos should be better than 26 ppm/°C. These requirements would point to the use of 1% 25 ppm metal film types.

Temperature changes can come either from ambient variations (including other nearby components) or from self-heating due to dissipated power. Any application

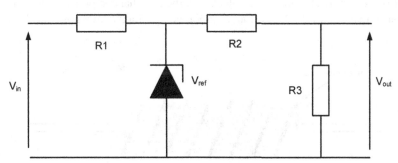

FIGURE 3.7 Stable Voltage Reference Divider Circuit.

that requires good resistance value stability should aim for minimum, or at least constant, power dissipation in the component. Manufacturers' data normally show a graph of temperature rise versus power dissipation for a given resistor type, and this should be checked if value stability has to be maintained through changes in dissipation.

3.1.4 POWER

Power dissipation is, of course, one of the most important specifications a designer has to check for each component. Power dissipation results in temperature rise, which is determined by how fast the heat is conducted away from the body. The maximum body temperature usually occurs in the middle of the resistor, and this is known as the hot spot temperature. Prolonged high temperatures (remembering that the quoted hot-spot temperature must be added to the maximum ambient) cause two things: a reduction in reliability, not only of the resistor but also of components near it, and a resistance shift.

A good rule of thumb for a reliable circuit is never to allow a dissipation of more than half the rated power in each component; many companies have their own in-house rules based on experience or customer's specifications.

Power should be calculated under worst-case operating conditions: for example, a resistor may be placed across a nominally 12 V supply rail, which under extreme conditions could reach 17 V. The difference in power dissipation is nearly double.

3.1.5 INDUCTANCE

In some applications, other performance factors must be taken into account. The construction of carbon and metal film-leaded resistors is basically a helix cut into a resistive film on a tubular ceramic substrate (Fig. 3.8). The dimensions of the helix, together with the bulk resistance of the film, determine the actual resistance element value. Such a form effectively provides a low-Q inductor. When the frequency of circuit operation extends into the RF region, the reactance can become a significant part of the total impedance, and noninductive resistors must be used.

The easiest way to obtain this is to choose a carbon composition type, where the resistive element is a homogeneous block of carbon, so that component inductance is primarily that due to the leads. Early carbon compositions have now mostly been

FIGURE 3.8 Helical Construction of Film Resistors.

superseded by ceramic—carbon types that, rather than a hot-molded solid carbon block, use a carbon conductor mixed with a ceramic filler; different values are obtained by altering the ratio of filler to conductor. For more demanding requirements you need to use special noninductive metal film or foil resistors. Chip resistors exhibit inherently low inductance because of their small size and are usually adequate for RF applications if their handling requirements and power ratings are suitable.

3.1.6 PULSE HANDLING

Another application where conventional helical-cut film resistors, and indeed ordinary small chip resistors, are unsuitable is in pulse applications, where a high voltage must be withstood for short durations, so that the average power dissipation is small though the peak power is many times larger. A common application of this sort is in thyristor, triac, or power transistor snubbers for high-voltage switching. The standard snubber circuit is shown in Fig. 3.9.

The RC combination restricts the rate of rise of voltage across the device during inductive switch-off, but in doing so the resistor is faced with a momentary fast voltage spike, which approaches the power supply voltage. This can cause arcing between adjacent turns of the helix of a film resistor and swift breakdown of the whole unit. Use of a wirewound high-power component in this application is ill-advised because the high self-inductance of the resistor creates a tuned circuit, which, although it has a low Q, can actually *increase* the transient voltage seen by the device. Again, a carbon composition type suffers less from both the breakdown mode and the inductance. Other chip and metal glaze resistors are available with specifically characterized pulse response. Applications such as telecoms protection, where a series resistor is used to protect an input circuit from applied high voltage surges, which may occur only very occasionally, also call for such pulse characterization.

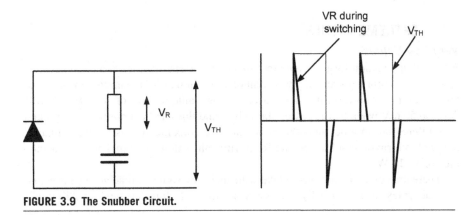

FIGURE 3.9 The Snubber Circuit.

Limiting Element Voltage

The limiting element voltage (LEV) is the maximum continuous voltage that can be applied to a resistor. For lower values the power rating is exceeded before the LEV is reached, but with higher values the LEV imposes limits on the applied power. For instance, a 470 K 0.33 W 1206-size resistor requires 394 V to dissipate its rated power; but a commercially available such part has an LEV of 200 V, so this can in fact only dissipate (continuously) 85 mW. The LEV limit becomes more significant with pulsed applications of lower-valued resistors where the average power dissipated may be very low, but the peak voltage applied can exceed the LEV.

Pulse applications may stress the power rating of the resistor and the voltage; the average power dissipated is equal to the peak power times the duty cycle of the pulses, but for pulse duration longer than a millisecond or duty cycles in excess of 10 or 20, the average power permissible has to be derated from the theoretical. Different types of resistor construction suffer in different ways. For instance, wire-wound or film types must dissipate all the applied power in the conductor itself, rather than letting the heat out through the body of the resistor, since it takes a finite time for the heat to pass from the conductor into the body. As the mass of the wire or film is low, the energy handling capability is also low and the derating is significant. Some manufacturers publish curves to allow this derating to be calculated.

For repetitive pulses you need to relate the average power in the pulse to the rated power of the resistor. This can be done for rectangular or exponentially decaying impulses as follows:

$$\text{For a rectangular pulse}: \ P_{\text{ave}} = \left(\frac{V^2}{R}\right) \times \left(\frac{\tau}{T}\right) \leq P_{\text{rated}} \qquad (3.9)$$

$$\text{For an exponential pulse}: \ P_{\text{ave}} = \left(\frac{V^2}{R}\right) \times \left(\frac{\tau_0}{2T}\right) \leq P_{\text{rated}} \qquad (3.10)$$

where V is the peak pulse voltage, R is the nominal resistance, T is the pulse period (so $1/T$ is the pulse repetition rate), τ is the impulse duration of a rectangular pulse, and τ_0 is the time constant to 0.37 V of an exponential pulse.

3.1.7 EXTREME VALUES

Very Low Values

A typical application for low value resistors is in current sensing. You often need to control or monitor a power supply current of a few amps with the minimum possible voltage drop, for power efficiency reasons; a low value resistor of, say 10 mΩ will develop 50 mV when 5 A is passed through it, and this can be amplified by a precision differential amplifier and used for monitoring. Bulk metal resistors are available in flat chip form or as solder-in wire form with values down to 3 mΩ and power ratings of 1−10 W.

There are some precautions to take with such low values. To achieve an acceptable accuracy it is normally necessary to make a four-terminal or "Kelvin"

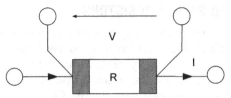

FIGURE 3.10 The Kelvin Connection.

connection, so that the current-carrying tracks and the voltage sense tracks go separately to the pads (Fig. 3.10). An easy way to do this for through-hole components is
to use opposite sides or different layers of the pcb for the two purposes, or at least to
connect to opposite sides of the pad. Even when this is done, there is still some pad
area and solder in series with the actual resistor element, which could compromise
accuracy and/or tempco. To get around this, specify a component actually designed
to have four terminals.

The low-impedance, low-level voltage sense is susceptible to magnetic field
interference, and to control this you should minimize the loop area between
the sense resistor and the sense circuit input. When you are monitoring AC signals with a few mΩ of resistance, the self-inductance of the resistor itself may
become significant. For instance, at, say, 400 Hz (a common aerospace AC power
frequency) the impedance of a stray inductance of 100 nH is 0.25 mΩ, which
could give a 5% error on a voltage measured across a 5 mΩ sense resistor. This
makes wirewound types unattractive, and bulk metal or chip resistors are to be
preferred.

Finally, a metallic element with a high dissipation and a low sense voltage may
give rise to thermoelectric errors. The junction between the element and its termination is a thermocouple, generating a voltage across it proportional to temperature. So
a sense resistor in fact includes two thermocouples back-to-back, one at each terminal. As long as the temperatures at each termination are the same, their errors cancel
out. This means that you should aim for thermal symmetry in the layout, by offering
similar heat sinking to each terminal (via the pcb tracks, usually) and by keeping
other heat sources distant.

Very High Values
Different considerations apply to multimegohm resistors. Here, the main problem is
leakage due to contamination across the terminals. The highest values (up to 10^{14} Ω)
are encapsulated in glass envelopes, and it is necessary to avoid handling the glass
and to touch only the leads, to prevent finger grease from affecting the resistance
across the terminals. Electrically, it is helpful to have a "guard" electrode around
the resistor to act as an electrostatic shield and to reduce or null out the effects of
leakage current into the terminals. High value resistors have long time constants,
and even a small amount of self-capacitance can have a significant effect: for
instance, a 100 GΩ resistor with a self-capacitance of only 1 pF has a time constant
of 0.1 s.

3.1.8 **FUSIBLE AND SAFETY RESISTORS**

There are some applications where a resistor is used in series with a power circuit to provide both a circuit function and a safety function. For instance, it may provide an inrush current limiter in a power supply input, so that a short duration surge current at switch on is limited in value; and it may also provide a sacrificial component that fuses in the presence of a prolonged fault current, for instance, if the input diode bridge or electrolytic capacitor fails. Under these circumstances a resistor should be specified principally for a predictable fusing characteristic, i.e., the time it takes to go open circuit for a given energy duration (see also Section 7.2.3). It is often necessary to be sure that the component will be flameproof when it fuses so that there is no fire hazard. Metal oxide and metal film components are available with these parameters fully characterized.

3.1.9 **RESISTOR NETWORKS**

A section on resistors would be incomplete without a mention of the resistor network. There are two main advantages to resistor networks: production efficiency and value matching/temperature tracking.

Production Efficiency

Design for production is a subject in itself, more fully covered in Chapter 9. For leaded components, handling and insertion costs can be a significant fraction of the total cost of a pcb assembly. A single resistor whose purchase cost is less than 1 p may cost as much as 5–10 p once it has been inserted, depending on how production costs are calculated. If several resistors can be combined in one package, then the overall cost of the package plus one insertion can easily be less than the cost of several resistors plus several insertions. To properly evaluate this trade-off you need to have an accurate knowledge of your particular production costs. Chip components in surface mount assemblies have a quite different arithmetic, since handling and placement on the board are usually automatic and the cost per part may well be negligible.

Usually resistor networks combine several resistors of one value in a package, dual-in-line or single-in-line, either all separate or with one terminal commoned. An obvious application for the latter type is for digital bus or I/O pull-ups. Linear circuits can benefit as well, especially if they can be designed to use several resistors of one value rather than a mixture of similar values.

One point to be wary of is the occasional temptation to use resistor networks too widely, so that they "gather up" what would otherwise be widely separated individual components. This is attractive for production but often disastrous for the board layout, since undesirably long tracks might be needed to get the signal to and from the network. There will often be a trade-off to make between this aspect and the demand to minimize insertion costs.

Value Tracking: Thick Film Versus Thin Film

Resistor networks are available using two technologies, the universal thick film type and the less widely available thin film. Thick film networks have no better tolerance and drift specifications than conventional resistors, and so they cannot be used in demanding applications. However, because all resistors in a package are constructed by simultaneously screen printing a resistive ink onto a substrate, manufacturers can guarantee a better tempco tracking between resistors than an absolute tempco for each resistor. A typical performance is 250 ppm/°C individually, but a tracking of 50 ppm/°C, i.e., all resistors within the package will exhibit the same tempco to within 50 ppm/°C. Thin film types can show an order of magnitude better performance.

This feature can be made use of in precision amplifier circuits, and in some instrumentation amplifiers it is essential to meet performance requirements. Consider the differential op-amp configuration of Fig. 3.11.

For optimum common-mode rejection the ratios R_1/R_2 and R_3/R_4 must be equal; for unity gain all resistor values should be equal. Equality must be maintained over the operating temperature range. While it would be possible to use separate precision resistors of sufficient stability, the absolute value of the resistors is not critical, only their ratios. Since resistor networks can have a much better tempco tracking performance than absolute value, they are highly suited to this type of application. Indeed, some manufacturers offer special networks with different values in the package with guaranteed ratios, especially for such circuits.

Multiples of the package value can be easily obtained by parallel or series connection, and the overall tracking of resistor values is not impaired. If, for example, both half and quarter of the same reference voltage are required, the best circuit for stability and accuracy would have a potential divider in which all resistors were part of the same package (Fig. 3.12).

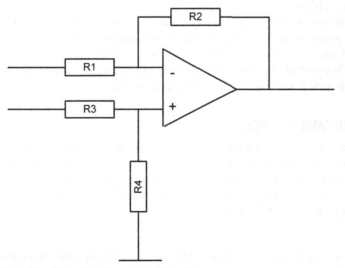

FIGURE 3.11 Basic Differential Op-amp Configuration.

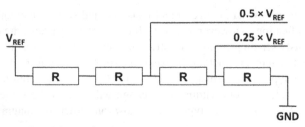

FIGURE 3.12 Potential Divider Using Equal Value Resistors.

3.2 POTENTIOMETERS

Potentiometers are one of the last bastions of electromechanical components in the face of the avalanche of digital silicon. They are bulky, unpredictable, and unreliable and can be a problem in fast circuitry; parasitic effects abound, and you can really only use them where the signal frequency times the circuit resistance is less than 10^6 Hz-Ω. They add time and money at the test and calibration stage of the production cycle. Many of their functions can be taken over by microprocessor-controlled digital equivalents. For instance, a classical use for the standard trimpot is to null out that bugbear of analog amplifiers, the offset voltage. In the days when op-amps had offsets measured in tens of millivolts, this was a very necessary function. But op-amps are now available at reasonable prices with offsets below half a millivolt, and if this is not enough, it is often feasible to use a chopper-stabilized device whose effective offset is limited only by thermal effects to a few microvolts. Alternatively, the intelligence of a microprocessor can be used to dynamically calibrate out the offset of a cheap op-amp by frequently auto zeroing the entire A–D subsystem. Equally, the gain of a network can be trimmed using a resistive D–A converter rather than a trimpot.

Nevertheless, there will remain applications where a pot is overall the best solution to a given circuit problem: digital implementations have been eschewed for other reasons, or the inherent linearity and lack of distortion of a passive resistor element are essential, or assured nonvolatility of a setting is required. For these purposes, you need to consider the different types available.

3.2.1 TRIMMER TYPES

Potentiometers can be divided into two classes. Those types that are mounted on the circuit board and are only intended for adjustment on test and calibration, or possibly by maintenance technicians, are known as trimmers and are distinguished by their small size: quarter-inch diameter units are commonplace and 3 mm square surface mount types are now on the market.

Carbon

The cheapest and lowest performance types have a molded carbon film track and are of open construction (sometimes known as "skeleton" construction). They are prone to mechanical and environmental degradation and are therefore not suitable for professional applications, but their low cost (approaching 5 p in quantity) makes them popular in noncritical areas.

Slightly more expensive (about 20% more) are the enclosed carbon versions, which protect the track from direct contamination but are otherwise similar to the skeleton type. Value tolerance for all carbon film trimmers is normally ±20%.

Cermet

The most popular type for commercial and professional purposes is the cermet. Several mounting versions and sizes, all multiple-sourced, are available, with costs ranging from 20 to 80 p for single-turn variants. The term "cermet" refers to the resistive element, which is a metal film deposited on a ceramic substrate.

The cermet offers a wide range of resistance, from 10 Ω to 2 MΩ, with tolerances usually of ±10%, though cheaper parts offer ±20% and tighter tolerances can be ordered. Because it can be obtained in small sizes with low self-capacitance, it is useful at higher frequencies than other types.

Wirewound

The main advantages of wirewound trimmers are their low tempco, higher-power dissipation, lower noise, and tighter resistance tolerance. When used as a variable resistor, their lower contact resistance improves the current-carrying capability through the wiper (cf pot applications). Their resistance stability with time and temperature is slightly better than cermet, but the high-resistance extreme is comparatively low (50 kΩ), and the type is not suitable for high frequency. Their resolution is poor, and they are also somewhat more expensive and less widely sourced than cermet.

Multiturn

Both cermet and wirewound trimmers are available in multiturn configuration. This means that more than 360 degrees mechanical adjustment is needed to cause the wiper to traverse the total resistance element. (Note that a single-turn trimmer normally offers less than 270 degrees adjustment angle.) You will commonly find 4, 10, 12, 15, 20, and 25 turn units. These are used when better adjustability is required than can be offered by a single-turn, but at somewhat greater cost and size, typically twice as much.

3.2.2 PANEL TYPES

The other potentiometer classification applies to those that are to be adjusted by the user and therefore have a spindle, which protrudes through the equipment panel. These can be mounted on the panel using an integral bush or directly to the pcb. The latter tends to put extra strain on the pot's terminals, and requires careful consideration of mechanical tolerances, but is popular because it dispenses with a wiring loom and therefore speeds up production time. The spindle can be insulated or of metal, with or without a locating flat for the knob; insulated types offer a potential safety and electromagnetic compatibility advantage but are also less mechanically rigid.

Carbon, Cermet, and Wirewound

The electrical characteristics of these types are similar to those of single-turn trimmers of the same construction. Because the size of panel pots is necessarily larger to allow for mechanical strength and easy mounting, they can have higher-power ratings than trimmers. 0.4 W is typical for carbon types, whereas cermets and wirewounds offer 1−5 W.

Conductive Plastic

When you need a very high-quality track construction, then the conductive plastic type is suitable. These offer very long life and low torque and are also suitable for position transducers. High-accuracy components are very expensive, but general-purpose ones can be competitive with good-quality cermet types.

3.2.3 POT APPLICATIONS

Firstly, remember that the wiper contact is the weakest point. Enormous advances have been made in potentiometer reliability over the years, and the cermet construction has proved itself in many applications. But the wiper is still essentially an electromechanical component and as such is the source of virtually all pot problems.

The golden rule is draw as little DC through the wiper as you possibly can. If you have to draw significant current, use a wirewound device. If the pot is used as a variable potential divider to set a circuit voltage, ensure that it is operating into a high impedance. If it is used in a signal path to vary signal amplitude, incorporate a DC blocking capacitor to prevent any DC flow through the wiper. The wiper/element contact has an unpredictable resistance of its own (Fig. 3.13), which is affected by oxidation and electrochemical corrosion, and the effect of this resistance on the circuit (usually manifested as noise) must be minimized.

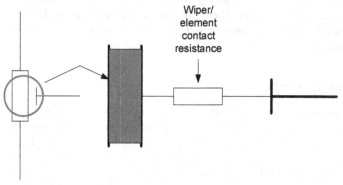

FIGURE 3.13 Potentiometer Contact Resistance.

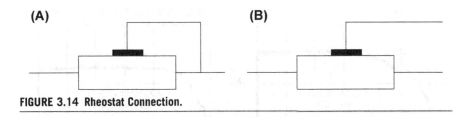

FIGURE 3.14 Rheostat Connection.

The less current that is drawn through the wiper, the less it will contribute to any noise voltage. Unfortunately, for some types a minimum current (such as 25 µA) *should* be drawn through the wiper, to "wet" the contact. This should be kept low.

Use as a Rheostat

Secondly, if the pot is being used as a variable series resistor (historically called a "rheostat"), connect the wiper to one end of the track. This is a very simple precaution—one short length of pc track. The reason is that under conditions of age, dirt, or extreme vibration the wiper can become temporarily (or even permanently) disconnected from the element track. If it is connected as in Fig. 3.14A, the maximum circuit resistance is limited to the end-to-end resistance of the pot; and if you connect it as in (B), then the device can become open circuit. In some circuit configurations this is merely a nuisance, but in others it could be catastrophic.

In this mode, the current passes almost entirely through the wiper and this can be a limiting factor in the circuit. Make sure the wiper current is limited to that given in the pot's specification. If this is not available, you can safely assume that it is the current that would produce maximum power dissipation, if applied through the wiper only, with as a rule of thumb an absolute maximum of 100 mA for small trimmers. In addition, remember that pots exhibit an "end resistance," which prevents wiper access to the ends of the resistance element. This restricts the minimum and maximum range that can be obtained: it is impossible to get a potentiometric division ratio (at "minimum volume") of zero.

Adjustability

Do not expect infinite adjustability from your pot. A cermet element is theoretically capable of infinite adjustment, but test and calibration technicians will find a center-zero reading elusive, and the test and calibration labor costs will multiply, if you try to obtain more resolution from the pot than is realistically achievable. A multiturn trimmer offers a better solution in this respect (Fig. 3.15). Resolution is also compromised by shock- and vibration-induced jumps. If you use a wirewound element, do not even think about high resolution, think of it more like a 100-way resistance switch.

As is emphasized in the section on design for production, one of the major aims for any designer must be to make their design cheap to produce. Any trimming adjustment is a production cost, and the time taken to do it must be minimized. The fewer trim adjustments there are, the better the design. But, when you are selecting a trimmer and determining its placement on the board, keep in mind the people

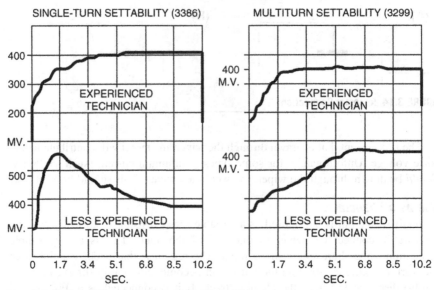

FIGURE 3.15 Settability of Single- and Multiturn Trimmers.

From Bourns.

who will have to use it and ensure that the adjustment screw is accessible. A useful rule of thumb is to place side-adjustment pots on the edge of the board and top-adjustment ones in the middle.

For best performance, use a series fixed resistance with a trimmer to provide only the range of adjustment required by the application—do not use the trimmer to give you all the necessary resistance just to avoid one extra fixed resistor.

Law Accuracy

The two common potentiometer laws are linear and logarithmic. These refer to the law that relates angular displacement to proportion of total track resistance at the slider. For linear pots, the law accuracy is specified as linearity and is closely related to cost; a low-cost panel component will probably not specify linearity at all, and it is likely to be no better than 10%. Logarithmic pots are generally only intended for use as audio volume controls and the adherence to a log law will be even less accurate. If you need high law accuracy, for instance, because you are using the pot as a position transducer, you will need to specify and pay for it. Linearities better than 1% are possible, but a typical high-quality component will offer no better than 5%. Note that the specification tolerance is a different parameter, as it refers only to the toler-ance on the end-to-end resistance of the track.

Manufacturing Processes

Another factor that works against potentiometers is their dislike of board soldering and cleaning processes. This of course applies to any electromechanical component:

relays and switches are equally susceptible. When a board is being soldered, it and the components it carries are exposed to extreme temperature shocks; once it has been soldered, it carries flux residues that must be removed by washing in water or solvent. Electromechanical components can be sealed or open. If they are sealed, there is the danger of a damaged seal allowing ingress of washing fluid, which then remains and causes an early failure of the component. If open, the danger is that the washing fluid will directly affect the component's operation. Either case is unsatisfactory, and many manufacturers prefer to add electromechanical components by hand, for reliability reasons, after the rest of the board has been populated, soldered, and cleaned—which is clearly an added production cost.

3.3 CAPACITORS

Like resistors, capacitors can tend to be taken for granted. There is as much a profusion of capacitor types and subvariants as there is of resistors, and it is often hard to select the optimum part for the application. Table 3.4 shows the characteristics and applications of the more common types.

Capacitors can be subdivided into a number of major types and subheadings within those types. The divisions are best made by dielectric:

Film: Polyester, Polycarbonate, Polypropylene, Polystyrene
Paper Ceramic: Single layer, barrier layer, high-K, low-K
Multilayer: COG, X5R, X7R, Z5U
Electrolytic: Nonsolid and solid aluminum, solid tantalum

This list covers all of the common types likely to be used in general-purpose circuit design. There are certain special or obsolete types—porcelain, trimmer, air dielectric, silver, mica—which are not included because their applications are too specialized. The divisions listed above are further subdivided depending on their construction—chip, radial lead, axial lead, disc, or whatever—but this does not affect their fundamental circuit characteristics, although it becomes important when pcb layout and production processes are considered.

3.3.1 METALIZED FILM AND PAPER

For a survey of the applications, let us start with the film types. These all have the same general construction, a sandwich of dielectric and conductive films wound into a roll and encapsulated along with their connecting wires, as shown in Fig. 3.16. There are two common methods of providing the electrode; one has a separate metal foil wound with the film dielectric and the other has a conductive film metalized onto the dielectric directly. The film and foil construction requires a thicker dielectric film to reduce the risk of pinholes and therefore is more suitable to lower capacitance values and larger case sizes. Metalized foil has self-healing properties—arcing through a pinhole will vaporize the metallization away from the pinhole area—and

Table 3.4 Survey of Capacitor Types

Type	Capacitor Range	WV Range	Tolerance	Temperature coefficient Range	Manufacturers	Applications	Unit Cost
Metalized film:							
Polyester	1 nF–15 µF	50 –1500 V	5%, 10%, 20%	Nonlinear	EPCOS, Wima, BC Components	General purpose, Coupling and decoupling	5–50 p
Polycarbonate	100 pF –15 µF	63 –1000 V	5%, 10%, 20%	±5% over −55 > 100°C	Arcotronics, Evox-RIFA	Low tc, timing, and filtering	10 p –£3
Polypropylene	100 pF –10 µF	63 –2000 V	1%, 5%, 10%	<−1% over −55 > 100°C	ICW, LCR, Roederstein	High power, high frequency	10 p –£1.50
Polystyrene	10 pF–47 nF	30 –630 V	1%–10%	±2% over −55 > 100°C; −125 ppm/°C	LCR	Close tolerance, low loss	7–50 p
Metalized paper	1 nF –0.47 µF	250 V AC	±20%	—	Wima, Evox-RIFA	Mains RFI suppression	20 p –£1.00
Ceramic:	10–220 nF	12–50 V	−20% + 80%	(Barrier layer type)	Beck, Taiyo Yuden, Panasonic, Murata		3 p –£1.50
Single layer	1 pF–47 nF	50 V –6 KV	2% > 20/80%	Dependent on dielectric	Dubilier, BC components	General purpose and HV	
Multilayer:							
COG/NPO	1 pF–27 nF	50 –200 V	2%, 5%, 10%	0 ± 30 ppm/°C	Syfer, Sprague, Kemet	Low tc, frequency	30 p –£2, Molded
X7R	1–680 nF	50 –200 V	5%, 10%, 20%	Nonlinear; ±15% over −55 > 125°C	BC Components, Murata	Sensitive and timing, General purpose, Coupling and decoupling	10 p –£1.50, Dipped

Type	Capacitance	Voltage	Tolerance	Temperature/nonlinearity	Manufacturers	Application	Cost
Y5V, Z5U	1 nF—10 µF	10, 16 V	20%	Nonlinear: +22%, −56%	AVX, Vitramon, EPCOS		3–60 p Chip
Electrolytics:							
Aluminum oxide	1–4700 µF (0.1 –68,000 µF)	50, 100 V 6.3 –100 V (Up to 450 V)	−20% + 80% (−10% + 50%)	Over +10 > 85°C –	BC Components, Dubilier, Elna, BHC Rubycon, NCC, Panasonic, EPCOS	General purpose Reservoir and decoupling	4 p–£3 (10 p –£10)
Solid aluminum	0.1–68 µF	6.3–40 V	±20%	–	BC Components	High performance	20 p –£2
Tantalum bead	0.1–150 µF	6.3–35 V	±20%	–	Sprague, Dubilier	General purpose	6 p–£1
Chip					AVX, Kemet	Small size	

(1) This survey does not consider special types.
(2) Manufacturers quoted are those widely sourced in the United Kingdom at time of writing.
(3) Quoted ranges are for guidance only.
(4) Costs are typical for medium quantities.

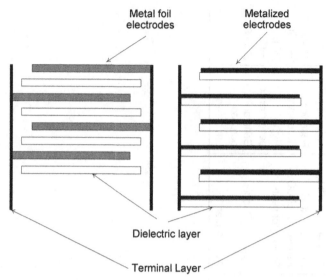

FIGURE 3.16 Film Capacitor Construction.

can therefore utilize thinner dielectric films, which leads to higher capacitance values and smaller size. The thinnest dielectric in current use is of the order of 1.5 μm.

Capacitor Specifications:

Temperature coefficient is the change in capacitance C with temperature and may be quoted in parts per million (ppm)/°C or as a percentage change of C over the operating temperature range

tan δ is a measure of the lossiness of the component and is the ratio of the resistive and reactive parts of the impedance, R/X, where $X = \omega C$

Insulation resistance or **time constant** is a measure of the DC leakage and may be quoted as the product $R_{dc} \cdot C$ in seconds or as R_{dc} in MΩ. **Leakage current** describes the same characteristic, more usually quoted for electrolytics

Dielectric absorption is a measure of the "voltage memory" characteristic, which occurs because dielectric material does not polarize instantly but needs time to recover the full charge; it is defined as a percentage change in stored voltage a given time after the sample is taken.

Polyester

Of the film dielectrics listed the most common is polyester. This has the highest dielectric constant and so is capable of the highest capacitance per unit volume. It approaches multilayer ceramic capacitors for volumetric efficiency and in fact can be used in most of the same applications: decoupling, coupling, and bypass, where the stability and loss factor of the capacitor are not too important.

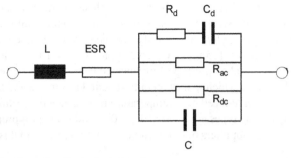

FIGURE 3.17 The Capacitor Equivalent Circuit.

C is the "ideal" capacitance of the device

R_{ac} is the equivalent resistance due to AC dielectric losses and may vary nonlinearly with frequency and temperature; it is usually combined with equivalent series resistance for electrolytic capacitors

R_{dc} is the leakage or isolation resistance of the dielectric and may vary with temperature

ESR is the equivalent series resistance due to the electrode, lead, and terminating resistances

L is the equivalent series inductance due to the electrodes and leads

R_d and C_d are equivalent components to represent dielectric absorption properties

Polyester has a nonlinear and comparatively high tempco. Its dissipation factor tan δ (see Fig. 3.17) is also high, of the order of 8×10^3 at 1 kHz and 20°C, and varies markedly with temperature and operating frequency. These factors make it less useful for critical circuits where a stable, low-loss component is needed.

Polycarbonate

For these cases polycarbonate is best suited. This has a near flat temperature— capacitance characteristic at room temperature, with a decrease of about 1% at the operating extremes. It also has a lower tan δ, typically less than 2×10^{-3} at 20°C, 1 kHz. Polycarbonate would normally be specified for frequency-sensitive circuits such as filters and timing functions. It is also a good general-purpose dielectric for higher-power use, but in these applications polypropylene comes into its own.

Polypropylene and Polystyrene

Polypropylene has a lower dielectric constant than the others and does not metalize so easily, and so gives a larger component for a given CV product. Also it exhibits a fairly constant negative tempco of −200 ppm/°C, which restricts its use in frequency-critical circuits, although a defined tempco can be useful for temperature compensation in some instances. Its main advantage is its very low dissipation factor of around 3×10^{-4} at 20°C and 1 kHz, almost constant with temperature. This allows it to handle much higher powers at higher frequencies than the other types, so it is suitable for switch-mode power supplies, TV line deflection circuits, and other high-power pulse applications.

Polypropylene capacitors can also be made to close tolerances, and this makes them competitors to polystyrene in many tuned circuit and timing applications. Additionally, both polypropylene and polystyrene have similar tempcos and loss factors (polystyrene tempco -125 ppm/°C, tan δ typically 5×10^{-4}). They both also show a better dielectric absorption performance ($0.02\%-0.03\%$ — see Fig. 3.17) than the other film types, which makes them best suited to sample and hold circuits. Both types suffer from a reduced high temperature rating, generally limited to 85°C, though some polystyrene types are restricted to 70°C and some polypropylene types are extended to 100°C. Suppliers of polystyrene types are scarce and it is rarely used.

Metalized Paper

A further capacitor type that is often bracketed with film dielectrics is the metalized paper component. Paper was historically a very widely used dielectric, particularly in power applications, before the technical developments in plastic films superseded it. The great advantage of plastic films is that they absorb moisture very much less readily than paper, which has to be impregnated to prevent moisture ingress from destroying its dielectric properties. Paper is now mostly reserved for use in special applications, particularly for across-the-line mains interference suppressors. When a capacitor is used directly with a continuous connection across the mains, a fault in the dielectric or a transient overvoltage stress can lead to localized self-heating and eventually the component will catch fire, without ever blowing the protective mains fuse. This has been found to be the cause of many electrical equipment fires and much investigation has been carried out (prompted by insurance claims) into the question of capacitor flammability. Paper has excellent regenerative characteristics under fault conditions; very much less carbon is deposited by a transient dielectric breakdown than is the case for any of the plastic film dielectrics, so that self-heating is minimal and the component does not ignite. Metalized paper is therefore the preferred construction for this application, although polyester, polypropylene, and ceramic suppressor types are available.

3.3.2 MULTILAYER CERAMICS

Multilayer ceramics have a superficially similar construction to film capacitors, but instead of being wound, layers of dielectric and electrode material are built up individually and then fired to produce a solid block with terminations at each end (Fig. 3.18). For this reason they are often called "monolithic" capacitors. They can be supplied either in chip form or encapsulated with leads; the encapsulant can be a molded body or a dip coating.

Of the many manufacturers of monolithic multilayer ceramics, virtually all offer products in three grades of dielectric: COG, X7R, and Y5V. Two other less common varieties are Z5U and X5R. COG is also known as NP0, referring to its tempco. These classifications are standardized internationally by the IEC (Europe) and EIA (America), so allowing direct comparison of different manufacturers' offerings. The three grades are quite different.

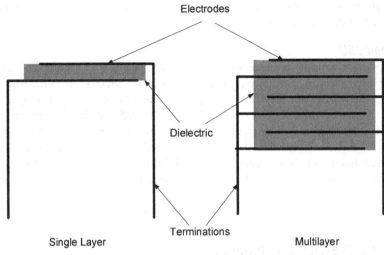

FIGURE 3.18 Ceramic Capacitor Construction.

COG

COG is the highest quality of the three but has a lower permittivity, which means that its capacitance range is more restricted. It exhibits a near-zero tempco, negligible capacitance and dissipation factor change with voltage or frequency, and its tan δ is around 0.001. These features make it the leading contender for high-stability applications, though polycarbonate can be used in some cases.

X5R and X7R

X7R is a reasonably stable high-permittivity (Hi-K) dielectric, which allows capacitance values up to 1 μF to be achieved within a reasonable package size. It can be used over the same temperature range as COG, but it exhibits a nonlinear and quite marked change of both capacitance and tan δ over this range. X5R has a lower high temperature limit than X7R (see Table 3.5). Tan δ at 20°C and 1 kHz is 0.025. Capacitance and tan δ also change with applied voltage and frequency by up to

Table 3.5 Temperature Characteristics of Capacitance for Class 2 Ceramics (to EIA 198—1; —2; —3)

Lower Temperature Limit	Upper Temperature Limit	Maximum Deviation of Capacitance, % Ref 25°C
$X = -55°C$ $Y = -30°C$ $Z = +10°C$	$4 = +65°C$ $5 = +85°C$ $6 = +105°C$ $7 = +125°C$ $8 = +150°C$	$P = \pm10\%$ $R = \pm15\%$ $S = \pm22\%$ $T = +22/-33\%$ $U = +22/-56\%$ $V = +22/-82\%$

10%, which rules out many applications, really leaving only the general-purpose coupling and decoupling area.

Y5V and Z5U

In comparison with the previous two, the Y5V/Z5U dielectrics show a very much worse performance. Capacitance changes by over 50% with changes in temperature and applied voltage; the Z5U rated temperature range is only $+10°C$ to $+85°C$, though it can be used at lower temperatures. The Y5V dielectric extends this down to $-30°C$. Tan δ is similar to X7R. The initial tolerance can be as wide as -20%, $+80\%$. Working voltages are restricted to 100 V. Virtually the only redeeming feature of these ceramics is their high permittivity, which allows high capacitance values, up to 2.2 μF, to be achieved. Their performance limitations mean that the only real application is for IC decoupling; however, this market alone guarantees sales of millions of units, so they remain widely available.

3.3.3 SINGLE-LAYER CERAMICS

By contrast with multilayers, single-layer ceramic capacitors are mostly of European or Japanese origin. There is a wide range of thicknesses and types of dielectric material, and so the varieties of capacitor under this heading are correspondingly wide.

Barrier Layer

Barrier layer ceramics use a semiconducting dielectric with a surface layer of oxide, which results in two very thin dielectric layers, effectively connected in series. The thickness of the formed layers determines the capacitance and working voltage, so that for a given disc size C and V are inversely proportional. Breakdown voltage is low, tan δ is high, and capacitance change with temperature, voltage, and frequency are all high. Their only practical advantage over Y5V multilayers (see above) is cost.

Low-K and High-K Dielectrics

Other single-layer ceramics use low-K (permittivity) or high-K dielectric materials and are generally distinguished as Type 1 or Type 2 (or Class 1/Class 2).

Type 1 dielectrics have a tempco ranging from $+100$ to -1500 ppm/°C or greater, depending on the ceramic composition. The tempco is reasonably linear, and the capacitance value is stable against voltage and frequency. Low tan δ at high frequencies (typically 1.5×10^{-3} at 1 MHz) allows their use extensively in RF applications. Because of their low permittivity the capacitance range extends from less than 1 pF to around 500 pF.

Type 2 dielectrics use ferroelectric materials, usually barium titanate, to allow much higher permittivities to be obtained, at the cost of variability in capacitance and tan δ with voltage, frequency, temperature, and age. The X7R and Z5U dielectrics used in multilayer components are also available among the very many type 2 materials used for single layers. Each manufacturer offers his or her own particular brew of ceramic and if your application requires critical knowledge of C and/or tan δ, then inspect the published curves closely. Capacitance ranges from 100 pF to 47 nF are common. Working voltages can be extended into the kV region by simply increasing the dielectric thickness, with a corresponding increase in electrode plate area.

3.3.4 ELECTROLYTICS

Electrolytic capacitors represent the last major subdivision of types. Within this subdivision the most popular type is the nonsolid aluminum electrolytic. These are available from numerous suppliers and are used for many applications. There are two major characteristics common to all electrolytics: they achieve very high capacitance for a given volume, and they are polarized. General-purpose aluminum electrolytics span the capacitance range from 1 to 4700 μF. Large power supply versions can reach tens of thousands of μF; subminiature versions can be had down to 0.1 μF, where they compete with film and ceramic types on both size and price.

Construction

The nonsolid aluminum electrolytic is constructed from a long strip of treated aluminum foil wound into a cylindrical body and encased (Fig. 3.19). The dielectric is aluminum oxide (Al_2O_3) *formed* by electrochemically oxidizing the aluminum. It is contacted on one side by the base metal (the anode) and on the other by an ionic conducting electrolyte contained within a spacer of porous paper. The electrolyte is itself contacted by another electrode of aluminum, which forms the cathode contact.

The electrodes are normally etched to increase their effective surface area and so increase the capacitance per unit volume, hence modern electrolytics are sometimes known as "etched aluminum." Because the electrolyte is an ionic conductor, the polarity of the applied voltage must not be reversed, or hydrogen will be dissociated at the anode and create a destructive overpressure. The thickness of the Al_2O_3 dielectric is determined by the applied voltage during the electrochemical forming process and to avoid further thickening of the dielectric during use (again with destructive results) the operating voltage must be kept below this voltage by a suitable safety

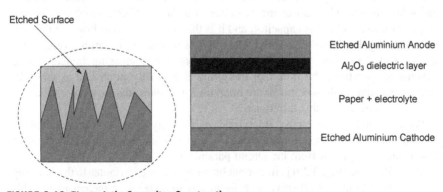

Etched Surface

Etched Aluminium Anode

Al_2O_3 dielectric layer

Paper + electrolyte

Etched Aluminium Cathode

FIGURE 3.19 Electrolytic Capacitor Construction.

factor, which determines the rated voltage of the component. Some manufacturers may specify a "surge" voltage rating, which is effectively the forming voltage without a safety factor.

Solid aluminum electrolytics are a variant in which the cathode is formed from a manganese dioxide semiconductor layer, contacted by an etched aluminum electrode. Although in principle some reverse polarization is possible with this system, it is not recommended because some ionic reactions due to trapped moisture can still occur.

Leakage

The important electrolytic characteristics tend to be dependent on application, and two broad areas can be discussed: for general-purpose coupling and decoupling and for power supply reservoir purposes. In the first field, the most important factor is usually leakage current, which becomes especially critical in timing circuits where it will determine the maximum achievable time constant. Leakage currents for general-purpose components are usually between 0.01CV and 0.03CV μA where C and V are the rated capacitance and working voltage. Many manufacturers offer low leakage versions, which are usually specified at 0.002CV μA. Alternatively, leakage current is a fairly well-defined function of applied voltage and usually drops to around a 10th of its rated value at about 40% of rated voltage, so that a low-leakage characteristic can be obtained by underrunning the component. Leakage is also temperature dependent and can be 10 times its rated value at 25°C when run at maximum operating temperature. It is also a function of history (see later under "lifetime"): when voltage is first applied to a new component its leakage is higher.

Ripple Current and Equivalent Series Resistance

For power supply reservoir applications, leakage current is unimportant and instead two other factors must be considered, ripple current (I_R) and equivalent series resistance (ESR). The ripple current is the AC current flowing through the capacitor as the reservoir charges and discharges, usually at 100/120 Hz for AC mains supplies or at the switching frequency for switch-mode supplies. It develops a power dissipation across the resistive part of the capacitor impedance (ESR), which results in a temperature rise within the capacitor, and it is this dissipation that limits the capacitor's I_R rating. Published data for all electrolytics include an I_R rating, which must be observed. The rating increases to some extent with increasing frequency and reducing temperature. But note that the rating is normally published as an RMS value, and actual ripple waveforms are often far from sinusoidal, so a correction factor must be derived for this difference.

Such thermal considerations imply that, particularly for reservoir applications, you may need to select a capacitor with a higher voltage or capacitance rating than would be expected from the circuit parameters.

The ESR value (Fig. 3.20) is important both because it contributes to the I_R rating and because it limits the effective high-frequency impedance of the capacitor. This

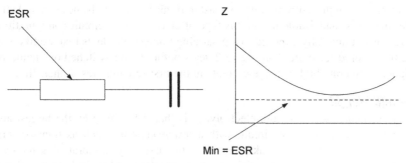

FIGURE 3.20 Capacitor Equivalent Series Resistance.

point has become of increasing importance with the advent of high-frequency switching power supplies where the output ripple voltage is determined by the output capacitor's ESR rather than its absolute capacitance value. Some manufacturers now offer special low-ESR versions specifically for these applications. ESR of nonsolid electrolytics increases dramatically as the operating temperature is reduced below 0°C, which can be a problem in circuits where actual dissipation is low. The better-quality ranges of electrolytics include this factor in their specification of "impedance ratio," the ratio of ESR at some subzero temperature to that at 20°C, which is usually around three or four but may be much worse. Solid electrolytics do not exhibit this behavior to the same extent.

Temperature and Lifetime

The capacitor characteristics as discussed for ceramic and film types are generally worse for electrolytics. Capacitance/temperature curves are rarely published but nonsolid types can vary nonlinearly by around ±20% over the operating temperature range, capacitance reducing with lower temperatures; solid types are better by a factor of two. Tan δ is around 0.1–0.3 at 100 Hz and 20°C but worsens dramatically with lower temperature and increasing frequency, higher voltage ratings having lower tan δ. Temperature ranges are typically −40 to 85°C, with some types being rated for extended temperature of −55 to 105 or 125°C. Lifetime is an issue with electrolytics, in two respects. Nonsolid electrolytics suffer from eventual drying out of the electrolyte, which is a function of operating temperature and the integrity of the component seal. In general, the life of these types can be doubled for each 10°C drop in operating temperature. Solid electrolytic types do not show this failure mechanism.

The second problem is that of shelf life. Nonsolid aluminum electrolytics are one of the few types of electronic component that degrade when not in use. The dielectric Al_2O_3 film can deteriorate, leading to increased leakage, if the component is maintained for long periods without a polarizing voltage. The effect is dependent on temperature, and shelf life is usually measured in years at 25°C. Capacitors that have suffered this type of degradation can be "reformed" by applying the forming voltage across them through a current limiting resistor, if this is found to be necessary. At the same time, it is inadvisable to run this type of electrolytic in circuit configurations where it is not normally exposed to a polarizing voltage. Products that are likely to have been stored for more than one or 2 years before being switched on should be designed to tolerate high leakage currents in the first few minutes of their life.

Size and Weight

One disadvantage of aluminum electrolytics is that they are often the largest and heaviest components in the circuit, with a few exceptions such as transformers. This means that they are a weak point when the assembly is vibrated. Some care should be taken in choosing the right part not only for its electrical characteristics but also for the mechanical strength of its terminals, if these are the only means of attachment or in providing alternative means of mounting.

3.3.5 SOLID TANTALUM

Solid tantalum electrolytics are generally used when the various performance, construction, and reliability limitations of aluminum electrolytics cannot be tolerated. The construction is similar to the solid aluminum type, with a manganese dioxide electrolyte and sintered tantalum powder for the anode. They can be supplied with a temperature range of $-55-85°C$ or up to $+125°C$, and have a very much greater reliability than aluminum, and so are favored for military use.

Leakage current is around $0.01CV$ μA, which is comparable to the better aluminum types, and tan δ is between 0.04 and 0.1, about twice as good as aluminum. Capacitance change with temperature can vary from $±15\%$ to as good as $±3\%$ across the working temperature range. Some proportion of the working voltage can be tolerated in the reverse direction, which relaxes the application constraints. The resin-dipped bead tantalum construction offers usually the best trade-off between price, performance, and size in a given application, and tantalum beads are available from a wide range of manufacturers.

Tantalum Chip Capacitors

A major advantage of tantalum capacitors is that they can be packaged in much smaller sizes than aluminum electrolytics. This means that they are better suited to surface mount production and indeed the majority of small SM electrolytics are tantalum types. Capacitance values from 0.1 to 470 μF are available.

One unusual problem does affect these components: tantalum is a rare material and there have been supply problems, compounded by their popularity, resulting in quoted lead times being pushed out to half a year or more. For this reason some designers look on tantalum capacitors unfavorably and try to minimize their use, or at least make sure they have multiple sources available for the chosen types. A newer material, niobium oxide, is offering itself as an alternative to tantalum and does not suffer from these supply problems, though its electrical characteristics are slightly less attractive.

3.3.6 CAPACITOR APPLICATIONS

As with resistors, the actual capacitance that a component can exhibit is only mildly related to its marked value. The art of circuit design lies in knowing which components must be carefully specified and which can have wide tolerances.

Value Shifts

The actual capacitance will vary with initial tolerance, temperature, applied voltage, frequency, and time. If we define the coefficients for temperature (C_t), voltage (C_v), frequency (C_f), and aging (C_a), then we can calculate the actual value of the capacitance using the following equation:

$$C_{\text{actual}} = C \times [\pm\text{tolerance}] \times [\pm\Delta T \times C_t] \times [\pm\Delta V \times C_v] \times [\pm\Delta f \times C_f]$$
$$\times [\pm t \times C_a] \tag{3.11}$$

Take a nominally 0.1 μF Z5U multilayer ceramic capacitor, rated at 50 V. It has an initial tolerance of −20%, +80%; a tempco of +22%, −56% maximum over the temperature range +10−+85°C; a capacitance voltage coefficient that reduces by 35% of its value at 60% of rated voltage; a frequency characteristic that reduces capacitance by 3% of its value at 10 kHz and 6% at 100 kHz; and an aging characteristic that reduces its value by 6% after 1000 h. It is run in a circuit with an operating frequency between 10 and 100 kHz, over its full temperature range, and with an applied voltage that varies from 5 to 30 V. The worst-case limits of its actual value will be as follows:

(a) 0.1 μF × 1.8 [max tolerance] × 1.22 [max pos temp coeff] × 1.0 [max voltage coeff] × 1.0 [freq coeff, DC] = **0.219 μF**

(b) 0.1 μF × 0.8 [min tolerance] × 0.44 [max neg temp coeff] × 0.65 [min voltage coeff] × 0.94 [min freq coeff] × 0.94 [1000 h aging] = **0.0202 μF**

In other words, an 11:1 variation. Where is the 0.1 μF capacitor now? At the very least, you can see that it is most unlikely to be 0.1 μF!

Now repeat the calculation for a 10% polycarbonate component of the same value and 63 V rating, which will be larger; and a 20% tantalum bead electrolytic rated at 35 V, which will be roughly the same size but polarized. All three types are roughly the same price.

Polycarbonate:

(a) 0.1 μF × 1.1 [max tolerance] × 1 [max pos temp coeff] × 1 [max voltage coeff] × 1 [max freq coeff] = **0.11 μF**

(b) 0.1 μF × 0.9 [min tolerance] × 0.994 [max neg temp coeff] × 1 [min voltage coeff] × 1 [min freq coeff] × 1 [aging] = **0.089 μF**

Tantalum bead:

(a) 0.1 μF × 1.2 [max tolerance] × 1.05 [max pos temp coeff] × 1 [max voltage coeff] × 1 [max freq coeff] = **0.125 μF**

(b) 0.1 μF × 0.8 [min tolerance] × 0.99 [max neg temp coeff] × 1 [min voltage coeff] × 0.5 [min freq coeff] × 0.95 [aging] = **0.038 μF**

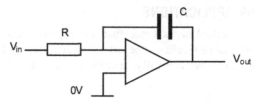

FIGURE 3.21 Single Op-amp Integrator.

The polycarbonate performance is dominated by its initial tolerance; the tantalum bead shows a worse performance at the higher frequency. Clearly some 0.1 μF types are better than others!

The circuit conditions quoted above are fairly extreme, particularly the wide voltage range and the frequency excursions. But there are some applications where the subtleties of capacitor behavior have a serious effect on circuit operation. Consider a simple op-amp integrator (Fig. 3.21).

The output voltage follows the equation defined below:

$$V_{out} = -V_{in} \times \frac{t}{CR} \tag{3.12}$$

Usually you will assume that C and R are constant and so if V_{in} is also constant the output ramp is linear with time. But if the integrator output swings over a wide range, say the 5–30 V considered in the last example, C may not be constant but will change as the voltage across it changes. This effect is most marked in accurate timing circuits or in circuits where a linear ramp is used to measure another voltage, such as in some A–D converters. A Z5U ceramic will introduce an enormous nonlinearity; even X7R will have a poor showing.

The effect is best combated by choosing the proper dielectric type and/or under-running it, i.e., using a 100 V rated capacitor with no more than a 1 V ramp. Plastic film would be a better choice, polycarbonate being the preferred type for tempco and frequency stability and would introduce negligible nonlinearity.

NPO/COG ceramic, polycarbonate, polystyrene, and polypropylene in roughly that order are most suitable for any circuit, which relies on the stability of a capacitor, especially timing, tuning and oscillator circuits, as their capacitance values are least subject to temperature and aging. Of course, these types are restricted to the lower capacitance ranges (pF or nF—only polycarbonate stretches up to the μF range, and here size is usually a problem) and so are more suitable for higher frequencies. If long, stable time periods are needed it is normally better to divide down a high frequency using a digital divider chain than it is to use large-value capacitors or to expect stability from an electrolytic.

3.3.7 SERIES CAPACITORS AND DC LEAKAGE

Capacitor voltage ratings can hide pitfalls. As discussed earlier, it is always better to underrun the working voltage of a capacitor for reasons of reliability. If a particular

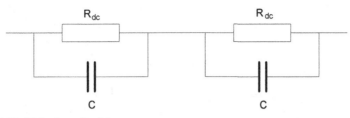

FIGURE 3.22 DC Leakage Resistance.

working voltage is just too high for the wanted capacitor type, it may seem reasonable to simply put two or more capacitors in series and add up the overall voltage rating accordingly, always taking into account the reduced total capacitance.

This is certainly possible but more is required than just multiple capacitors. The capacitor equivalent circuit (Fig. 3.17) also includes the DC leakage resistance R_{dc} of each capacitor, as shown in Fig. 3.22. The DC working voltage impressed across the terminals is divided between the capacitors not by the ratio of capacitance but by the ratio of the two values of R_{dc}. These are undefined (except for a minimum) and can vary greatly even between two nominally identical components. Because R_{dc} is usually high, of the order of tens to thousands of megohms, other leakage resistance factors—particularly pc board leakage (see Section 2.4)—will also have an effect. The result is that the actual voltage across each capacitor is unpredictable and could be greater than the rated voltage. The problem is at its worst with electrolytics whose leakage current is large and varies with temperature and time.

The situation is to a certain extent self-correcting because an overvoltage will in most cases result in increased leakage, which will in turn reduce the overvoltage. The major consequence is that the actual capacitor working voltage will be unpredictable and therefore reliability of the combination will suffer. Once one component goes short-circuit the other (or others) will be immediately overstressed and rapid failure of all will follow.

Adding Bleed Resistors

The solution is simple and consists of placing resistors across each capacitor to swamp out the DC leakage resistance (Fig. 3.23). The resistors are sized to be comfortably below the minimum specified leakage resistance so that variations in working voltage are kept below the rated maximum for each capacitor. Naturally this increases the leakage current of the combination but this is often an acceptable price, particularly if the application is for a high-voltage reservoir where some extra drain current is available.

Indeed, it is often necessary for safety reasons to have a defined "bleed" resistance across a high-voltage reservoir capacitor. If the load resistance is very high, it can take seconds or even minutes for the capacitor voltage to discharge to a safe level after power is removed, with a consequent risk of shock to repair or test technicians. A bleed resistor (or resistors, if the voltage rating of a single unit is inadequate) is a simple way of defining a maximum discharge time to reach a safe voltage.

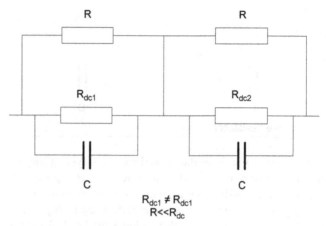

FIGURE 3.23 Swamping Leakage Resistance.

3.3.8 DIELECTRIC ABSORPTION

Another effect mentioned earlier is the phenomenon of dielectric absorption. If a capacitor is charged to a given voltage, discharged by shorting it, and then open circuited again, its voltage will begin to creep up from zero toward the original voltage. The capacitor exhibits a "voltage memory" because the dielectric molecular dipoles need time to align themselves in an electric field.

This effect is of most concern to designers of sample and hold circuits. If a capacitor has been holding a voltage V_A and then samples another voltage V_B at the other end of its range, when returned to hold mode (effectively open-circuit), its voltage will drift exponentially toward the old voltage V_A (Fig. 3.24).

The effect can be modeled by an extra circuit in parallel with the main capacitor, $R_d C_d$ in the equivalent circuit (Fig. 3.17). $R_d C_d$ has a long time constant and transfers charge slowly to C when the capacitor is open circuit. The dielectric absorption

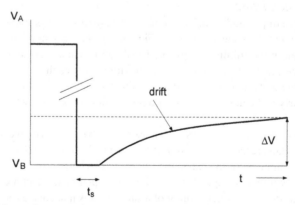

FIGURE 3.24 Dielectric Absorption Drift Characteristic.

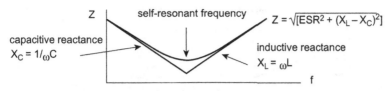

FIGURE 3.25 Capacitor Self-resonance.

figure, quoted for $t \gg t_s$, is $\Delta V/V_A - V_B$. The error due to dielectric absorption in a typical circuit will be reduced if the hold time of the old voltage (V_A) is short or if the measurement is made just after the sample is taken rather than many multiples of t_s later. At the same time, selection of capacitor type plays a part, and of the readily available dielectrics polystyrene and polypropylene are the best, exhibiting dielectric absorption factors of 0.01%−0.02%.

3.3.9 SELF-RESONANCE

When capacitors are used at high frequencies another factor comes into play, and this is their self-resonant frequency (SRF). The capacitor equivalent circuit includes the ideal capacitance C, the ESR and the equivalent series inductance (ESL). These three components form a low-Q series tuned circuit whose impedance versus frequency curve has the characteristic shape of Fig. 3.25, showing a minimum impedance at self-resonance.

The term "high frequency" here is relative. All capacitors exhibit the same basic curve, but the minimum for say a 47 μF tantalum electrolytic could be at 500 kHz and the null could be very flat, or for a 100 pF COG chip ceramic the null could be at 100 MHz and very well defined.

The ESL is determined by the lead length and the body size. (Lead length includes the lengths of connecting pc track to the adjoining circuit nodes.) Therefore small, leadless chip capacitors show the lowest inductance and highest SRF whereas large, leaded capacitors have high inductance and low SRF. Many manufacturers publish impedance/frequency curves if they expect their components to be used at high frequencies. A typical comparison, comparing small polyester film, tantalum electrolytic and multilayer ceramic of the same value, is shown in Fig. 3.26.

Consequences of Self-Resonance

A capacitor used above its SRF is effectively a low-Q inductor and so RF circuits using inappropriate components can show somewhat unpredictable behavior. The problem also appears more frequently as the speed of digital circuits increases, and the clock frequencies approach or exceed the SRF of the capacitors that are used for supply rail decoupling (cf Section 6.1.4). Clearly a tantalum electrolytic with an SRF of 1 MHz is not much use for decoupling clock transients of 10−20 MHz, but you can parallel large-value electrolytics with small-value, e.g., 10 nF, ceramic or film capacitors whose SRF is in the 10−100 MHz region. The

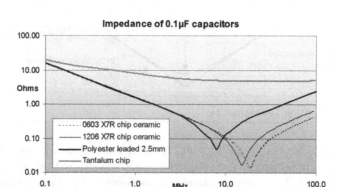

FIGURE 3.26 Impedance—Frequency Characteristics for Plastic, Ceramic, and Tantalum 0.1 μF Capacitors.

combination is then effective over a much wider range of frequencies, and the technique is common practice in wideband amplifier circuits, although you need to beware of intercomponent resonances, where for instance the high value component's self-inductance resonates with the low value capacitance.

3.4 INDUCTORS

Hardly surprisingly, the inductor's equivalent circuit (Fig. 3.27) is very similar to the capacitor's. All practical components contain the three passive circuit elements, R, C,

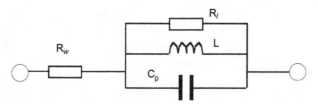

FIGURE 3.27 The Inductor Equivalent Circuit.

L is the "ideal" inductance of the device

R_w is the series resistance due to the winding wire and terminations, increasing with temperature

R_l is an equivalent parallel resistance due to the magnetic core losses and is variable with frequency, temperature, and current

C_p is the self-capacitance of the winding, determined by the method of construction of the component

R_w and R_l may be lumped together into an equivalent series resistance R_{eq}, in which case the overall lossiness of the component is quoted as its Q, which is given by $\omega L/R_{eq}$. The reciprocal of Q is tan δ, which is analogous to the capacitor tan δ.

and L, in their makeup. Both capacitors and inductors are fundamentally energy storage devices, but the practical inductor departs further from the ideal than does the practical capacitor, and so is less universally used. In addition, off-the-shelf inductors are less common than capacitors, and the prospects for multiple sourcing are slight.

For this reason a tabular survey such as those already given for resistors and capacitors is not really feasible. Inductors are generally made by winding wire around a magnetically permeable material, and it is the performance of the material, along with the DC resistance of the wire, which determines the performance of the finished component.

3.4.1 PERMEABILITY

Permeability is the measure of a material's ability to concentrate lines of magnetic flux. Air (and other nonmagnetic materials) has a relative permeability, μ_r, of 1. Inductors wound on a nonmagnetic core are inherently low-loss, the only losses being due to the wire resistance. Unfortunately they are also inherently low inductance, which generally limits their use to the hf and vhf region. Achieving an inductance greater than 100 μH without a magnetic core requires a large number of turns, so that either the component becomes unmanageably large or the wire becomes unmanageably thin. At low frequencies the winding resistance approaches the inductive reactance, making the coil practically useless. Large air-cored coils do have applications where a stable, low-loss inductance is needed in the presence of high DC currents.

Most inductors for low- and midfrequency use call for higher values and must use permeable cores to achieve this. Relative permeabilities of several thousand are possible, allowing small inductors of several henries to be wound fairly easily. However, there are disadvantages, as you might expect the following:

- high-μ materials introduce losses of their own, reducing achievable Q factors;
- the materials reduce in permeability as the magnetic field increases. This is known as "saturation," and means that the inductance drops off at high power levels or bias currents;
- a slight change in the material's molecular structure on magnetization results in "hysteresis," which shows itself as a remanent magnetic flux when the magnetic field reduces to zero, depending on the level of previous magnetization;
- the absolute magnetic and physical properties of the core are hard to control closely, so that the tolerance on the inductance value of the finished component is wide, though this can be controlled by an air gap in the magnetic circuit;
- the permeability and loss vary with temperature; above the "Curie point" the magnetic properties vanish almost completely.

Fig. 3.28 shows a typical curve of magnetic flux versus applied magnetic field (the "B−H" curve) for a ferrite material, illustrating the effects of saturation and hysteresis. You will use such published curves in inductor design, primarily to determine the power handling capability of the component.

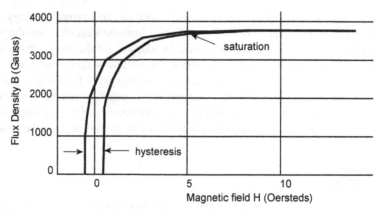

FIGURE 3.28 Typical BH Curve for a Ferrite.

Ferrites

The properties of permeable cores are reminiscent of the dielectric characteristics—variability with voltage, temperature, and frequency—of the hi-K ceramics that we looked at earlier, and indeed there is much in common. The most common core material, ferrite, is in fact a type of ceramic and is produced in at least as many varieties as are the capacitor ceramic materials. Ferrites are a metal oxide ceramic made of a mixture of Fe_2O_3 and either manganese–zinc or nickel–zinc oxides pressed or extruded into a range of core shapes. Every manufacturer offers a wide variety of shapes and will usually also offer a custom service for large volumes, but for most uses a selection from a relatively small range of standard types is adequate and offers the benefit of sourcing from different suppliers. Two of the most popular core types are the RM series to IEC 60431 and the E, EP, and EC series for medium-power high-frequency transformers.

Manganese–zinc ferrites not only have a high permeability but also their losses increase rapidly with frequency, making them more suitable to low-frequency applications. Nickel–zinc ferrites are of lower permeability, but their lower high-frequency losses make them useable up to about 200 MHz. Their resistivity is higher by several orders of magnitude, and their Curie point temperature is higher. The ratio of manganese to zinc or nickel to zinc in either case determines the grade of material.

The use of the terms "hard" and "soft" when applied to magnetic materials does not refer to their physical hardness but to their coercivity (although there is a correlation and is in fact where the term comes from originally). Soft magnetic materials are generally defined to have coercivity less than 1 kA/m and hard magnetic materials greater than 10 kA/m. Ferrites are chemically inert ceramic materials, which have the general structure XFe_2O_4 where X is one of the transition metals given in Table 3.6.

Combinations of powders of these metals are then mixed according to specific proportions, milled to ensure the required grain size and then pressed into shape.

Table 3.6 Transition Metals for Magnetic Materials

Metal	Symbol
Manganese	Mn
Zinc	Zn
Nickel	Ni
Cobalt	Co
Copper	Cu
Iron	Fe
Magnesium	Mg

Sintering, heating to between 1150 and 1300°C, hardens the material physically and ensures the desired magnetic properties. The sintered core may still require further finishing ensuring accurate dimensions or smooth surfaces for mating core halves and if so the material is finely ground. This grinding is also used to accurately implement center core air gaps.

The sintered and finished material contains thousands of small crystals or grains of the ferrite material, usually 10s of μm across. Fig. 3.29 shows a scanning electron microscope image of a soft ferrite material magnified 2500 times illustrating the shape and size of the grains in the material. Each grain has one or more magnetic regions called domains that are already randomly magnetically oriented due to the material's intrinsic magnetization after heating. These domains come into progressive alignment with the application of an external magnetic field as shown in Fig. 3.30.

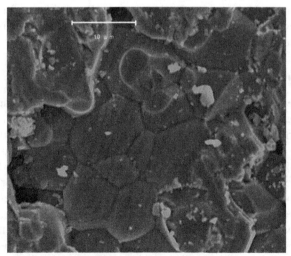

FIGURE 3.29 Scanning Electron Microscope Image of Philips 3F3 soft Ferrite Material With a Magnification of 2500 Showing Ferrite Grains.

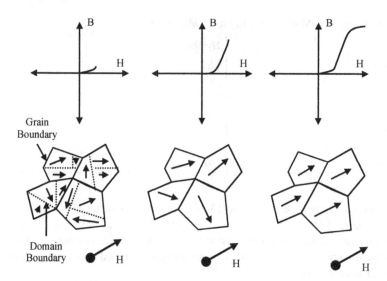

FIGURE 3.30 Progressive Alignment of Magnetic Domains and Grains, and the Resulting Magnetization Curves for Applied Field Strength H.

Iron Powder

The other core material in wide use is iron powder. Such cores are pressed from a very fine carbonyl iron powder mixed with a bonding material. Eddy current losses in the core are minimized by creating an insulating layer on the surface of each particle before pressing, but this introduces minute gaps in the magnetic circuit, which restricts the permeability of the material to a maximum of around 30. The same effect makes iron powder cores very hard to saturate. The main uses for iron powder are for high-frequency tuned circuit cores, and suppressor chokes where low saturability is more important than high inductance.

3.4.2 MAGNETIC MATERIAL DEFINITIONS AND METRICS

To assist in the definition of appropriate magnetic materials and components, it is important to define the requirements. It is quite normal to classify the differences between materials simply using the overall major loop characteristics, but it is also important to consider specific metrics, such as remanence or coercive force. The behavior of smaller loops, called minor loops, is significant for wideband transformers. The accurate characterization of the energy loss is critical for energy conservation and power calculations especially with arbitrary applied fields. With the use of specific metrics clear distinctions can be drawn between materials and their equivalent models under any applied field conditions. In this section, the important aspects of the behavior of magnetic material hysteresis are defined. Metrics are also specified for the measurement of, and comparison between, different materials and material models.

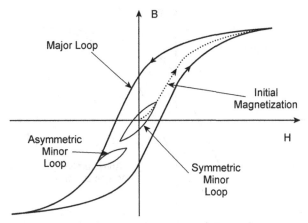

FIGURE 3.31 Types of Magnetization Curve and Hysteresis Loop.

Hysteresis Loop Definitions

To understand the different requirements of magnetic models, it is useful to distinguish between different types of hysteresis behavior in magnetic materials. The basic types of magnetization curves are shown in Fig. 3.31.

The major loop is the magnetization curve when the applied field (H) is high enough to force the material into saturation and the BH loop closes at the tips. The initial magnetization curve is the transition of the flux density (B) from the initial condition of zero flux to the tip of the major loop. Any loop inside the major loop is defined as a minor loop, and these may be asymmetric or symmetric about the origin as shown in Fig. 3.31. It is obvious that for accurate simulation of an arbitrary applied field, the initial conditions of flux or magnetization in the model must be correct. The anhysteretic curve is that which the magnetization would relax to for any specific value of applied field (H). This is effectively the minimum state of energy and is shown in Fig. 3.32.

When considering the minor loop behavior of magnetic materials, it is necessary for a model to be capable of exhibiting noncongruency. This means that the minor loop shape changes depending on the initial flux level.

Hysteresis Loop Metrics

It is possible to compare the performance of magnetic materials using numerical methods such as the least squares approach; however, it is more useful to compare standard measured characteristics and use them as metrics, which define the performance of the material and allow better classification of the materials. Fig. 3.33 shows a basic set of the metrics for a major hysteresis loop. The definitions of the metrics are given in Table 3.7.

While these metrics are straightforward to measure directly from a BH curve, there are other metrics that are also of use including the area of the BH loop. This defines the energy lost per cycle and in conjunction with the frequency can be

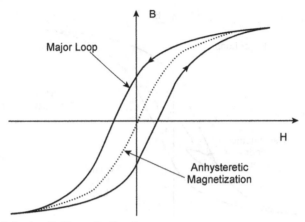

FIGURE 3.32 Anhysteretic Magnetization Curve.

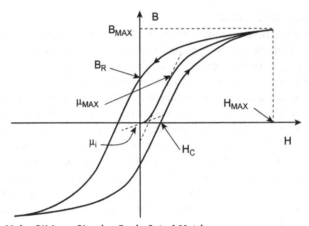

FIGURE 3.33 Major BH Loop Showing Basic Set of Metrics.

Table 3.7 Definition of Magnetic Material Metrics

Metric Symbol	Metric Definition
B_R	Remanence—B axis crossing point
H_C	Coercive force—H axis crossing point
B_{MAX}	Maximum flux density
H_{MAX}	Maximum applied field
μ_{MAX}	Maximum permeability
μ_i	Initial permeability

used to estimate the power lost by the material. Examples of the estimated power loss for a range of materials can be found in typical manufacturer's data. The maximum permeability μ_{MAX} can refer to the maximum initial permeability (as shown), the maximum incremental permeability (when a small AC signal is applied on top of a DC offset signal and the permeability measured), or the maximum effective permeability (measured at the extremities of the BH loops).

3.4.3 INDUCTOR APPLICATIONS

There are three major applications for inductors: as frequency-determining components in tuned (resonant) circuits, as energy storage components, usually in power supplies, and as filter components in suppression circuits. Each application emphasizes different inductor characteristics and calls for a different approach to inductor design.

Tuned Circuits

Signal tuned circuits demand predictable inductance values and high Q (low losses). They do not normally see high bias currents, so core saturation and hysteresis loss is not a problem. For low frequencies, ferrite pot cores are the most popular, but for RF use (above 1 MHz or so) other types of ferrite core, or iron dust cores, are better. For the best stability and initial tolerance, a lower permeability material is preferable.

As well as the intrinsic stability of the material, for these applications it is important to consider mechanical stability. Any movement or distortion of the core, or movement of the winding relative to the core, will affect the magnetic path and hence the inductance. In addition, any mechanical, magnetic, or thermal shock to the core causes an immediate change in permeability followed by a long, slow relaxation toward the original value. This is known as "disaccommodation." These effects mean that the core characteristics have to be very carefully considered when a stable inductance is required in a high-shock or high-vibration environment. It is common for the winding, bobbin, and core to be encapsulated in varnish to enhance mechanical stability. Hard encapsulating compound should not be used as the high shrinkage could mechanically damage the brittle core.

Power Circuits

Energy storage chokes and power transformers, as used for example in switching power supplies, have a quite different set of important parameters. In these, inductance stability is not required but high volumetric efficiency is. Energy stored in the choke is given by $L \cdot I^2$ and so a material that shows a high saturation flux density, allowing a higher magnetizing current, is to be preferred. At higher operating frequencies hysteresis becomes the dominant loss mechanism and limits the power-handling capacity of the core. A small gap in the magnetic circuit, usually obtained by grinding away a part of the core, allows higher saturation at the expense of lower effective permeability. These considerations point to the use of gapped manganese—zinc ferrites or iron dust cores in which the air gap is inherent in the material.

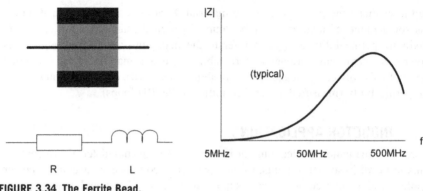

FIGURE 3.34 The Ferrite Bead.

Suppression

In contrast to the previous applications in which low core losses were required for high Q or highpower—handling, suppression chokes work best if they have high losses. A suppression circuit has to reflect or absorb high-frequency interference energy and prevent it from being propagated beyond the suppressor. The more energy is absorbed within the choke the better will be its circuit performance. Clearly, high-loss ferrites are the best type for these applications; all ferrites when used well above their intended frequency range exhibit high losses, but materials specifically designed and characterized for this purpose are available. The ferrite bead (Fig. 3.34) is an extreme example, in which a straight piece of wire is transformed into a high-frequency choke merely by stringing a bead onto it. The losses induced in the ferrite at high frequencies give the assembly a complex impedance (resistance + reactance) of several tens of ohms. The same principle is applied to monolithic ferrite chip components, in which the conductor is passed between layers of ferrite to create a surface mount part.

Leakage Inductance

When a magnetic component is constructed from a number of windings and a core, the main design aim is usually to maximize the magnetic flux through the core, as this allows maximum energy transmission from winding to winding and the maximum inductance from the component. Unfortunately, not all the flux passes through the core and the flux that passes through the windings themselves or the winding space is called the leakage flux. This flux gives rise to an inductance, which is called the leakage inductance. Fig. 3.35 shows a simple two winding transformer with the ideal and leakage flux paths defined.

The leakage inductance can be a vitally important parameter in switching power systems as it can give rise to resonance and ringing effects—causing losses. Equally leakage inductance can cause problems with line transformers as this parameter is crucial in defining the bandwidth attainable for a transformer.

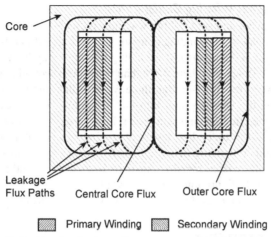

Core

Leakage
Flux Paths   Central Core Flux   Outer Core Flux

Primary Winding   Secondary Winding

FIGURE 3.35 Flux Paths Showing Leakage Flux.

Measurement Methods

The standard method for measuring leakage inductance of magnetic components is based on the fact that if a winding is shorted, the field applied by another winding will then be automatically canceled out by the opposite field generated by the shorted winding (this is the assumption, but in practice the mutual flux is not canceled out completely). The IEEE standard practice for the measurement of leakage inductance using this method is given in "Recommended Standard Practice for testing electronic transformers and inductors," IEEE Standard 389-1979.

In practice the simplest method is to short the secondary winding(s) and measure the inductance across the primary winding using either an inductance bridge or an impedance analyzer. It is important to ensure that measurements are taken across the frequency range of interest and at a low signal level to reduce the effect of core saturation reducing the effective inductance.

Numerical Methods

The use of finite element analysis to calculate the leakage inductance in transformers is in some ways even more accurate than measuring using the shorted winding approach described previously. In a finite element analysis, it is possible to define the applied fields so that the mutual flux is *exactly* zero thus ensuring maximum accuracy of the calculated leakage inductance. The accuracy of the calculated value is primarily affected by the numerical accuracy in the basic finite element analysis and any approximations made in the construction of the model.

The procedure is to apply the equal and opposite fields to the windings to make the mutual flux zero. This is obviously straightforward in the finite element analysis software as the applied winding currents can be specified exactly, and from the resulting energy stored in the finite element model, the leakage inductance can be directly calculated.

Issues With Unusual Winding Configurations

Multiwire, interleaved, and split bobbin windings present a more difficult problem. In these cases, instead of the winding consisting of an integral number of turns in a single section of the winding area, the winding topology may be much more complex. Multiwire windings are where more than one wire is wound together to create a winding. For two wires this is bifilar, three wires trifilar, and so on. Split bobbins are used to divide the winding area into lateral regions so that windings can be placed side by side. Interleaving can occur in a variety of ways, but the basic approach is to split the individual layers of a winding and intersperse the layers with those of another winding. This approach is sometimes called sectionalized winding. The main reason for this method is to reduce leakage inductance. Fig. 3.36 shows the configurations for a two winding transformer with (A) a simple side by side arrangement, (B) bifilar windings, (C) split bobbin, and (D) interleaving.

3.4.4 SELF-CAPACITANCE

Generally, do not wind directly onto the core. Apart from mechanical instability, the very high dielectric constant of the ferrite will increase the self-capacitance C_p of the winding several times if the two are in close contact. The bobbin serves to keep the winding well spaced from the core and minimizes self-capacitance. The way the winding is built up also influences self-capacitance; a single layer has the lowest capacitance, but if multiple layers are necessary then two possibilities exist to reduce it:

- "scramble" or "wave" wind rather than build up in discrete layers. This can reduce C_p by around 20% over a layer winding but uses more winding space;
- use a multisection former for a single winding. Again this uses more space, but for example a two-section former can reduce C_p by a factor of 3.

Generally the capacitances are not considered frequency or temperature dependent (i.e., the dielectric material is largely of a constant relative permittivity) within

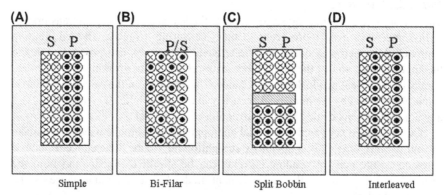

FIGURE 3.36 (A) Simple, (B) Bifilar, (C) Split Bobbin, and (D) Interleaved Windings.

the range of interest of SMPS and communications components. This assumption cannot be made for higher frequency RF components however. It is also assumed for the purposes of calculation that the voltage across a winding is distributed evenly, proportional to the number of turns.

Measurement Methods

The measurement of the winding capacitances of a transformer is not an easy task to perform but can be carried out using the procedure defined in IEEE Standard 389-1979 for a two winding transformer. More complex winding structures can also be measured using variations of this approach. The technique uses the resonant frequency of the transformer in various configurations to extrapolate the winding capacitances based on the equivalent circuit shown in Fig. 3.37 and using separately obtained values for the inductive elements.

3.4.5 WINDING LOSSES

Background

Losses in magnetic component's windings occur for a variety of reasons. The wires have a finite resistivity which causes a dc resistive loss. This can be quite easily calculated from the resistivity of the wire and its dimensions. As the frequency of an applied signal increases eddy currents are generated inside the winding leading to further losses. These eddy currents occur in two forms, skin effect and proximity effect. The skin effect is a function of the frequency of the applied signal and is calculated easily. The effect occurs as the signal tends to flow nearer the surface of the conductor as the frequency increases. The second is the proximity effect of the conductor strands themselves. This is a result of eddy currents induced from the overall magnetic field and is dependent on the wire layout, and whether stranding, twisting, or bunching has taken place. This can be more difficult to calculate

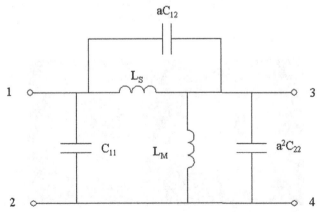

FIGURE 3.37 Equivalent Circuit Assumed for Measurement of Winding Capacitance.

analytically as there may be a significant variation in the layout of the individual conductors.

Theoretical Methods

The low-frequency resistance of a conductor can easily be calculated using the length of the conductor, the cross-sectional area, and the resistivity (for copper this is $1.709 \times 10^{-8}\ \Omega m$). It is important to note that the resistivity is temperature dependent. For copper the thermal coefficient is $Tc = 0.00393\ \Omega m/°C$. This is an important consideration for power devices where a 40°C temperature rise would not be uncommon. This approach can be used for a winding with N turns and a Mean Turn Length l_w to estimate the dc resistance as given below, where d is the diameter of the conductor and ρ_c is the conductivity.

$$R_{dc} = \frac{4\rho_c N l_w}{\pi d^2} = N l_w R_c\ \Omega \tag{3.13}$$

The losses due to eddy currents in windings can be divided into skin and proximity effects. As the frequency of an applied signal increases, the current flows in a progressively thinner layer at the surface of the conductor. Fig. 3.38 shows the change in the penetration depth in a conducting copper wire at different frequencies. The thickness of this layer is called the penetration depth, denoted by Δ, and can be calculated using the equation below, where ρ_c is the resistivity of the conductor, μ_c is the relative permeability of the conductor, and f is the frequency of the signal.

$$\Delta = \sqrt{\frac{\rho_c}{\pi \mu_0 \mu_c f}} \tag{3.14}$$

The skin depth is related to the conductor diameter d to obtain a relative measure of the skin depth that results in a "skin effect factor" F. If d/Δ is <2 then the skin effect is negligible and can be ignored. As Δ decreases and d/Δ increases above five then the skin effect factor, F, can be approximated using

$$1 + F \approx \frac{1}{4}\left(\frac{d}{\Delta} + 1\right) \tag{3.15}$$

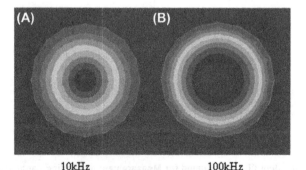

10kHz   100kHz

FIGURE 3.38 Variation in Conductor Skin Depth With Frequency.

Measurement Methods

To measure the winding losses, several steps need to be taken. The losses can be generally lumped into two parts, *Rdc* that is the dc winding resistance and *Rac* that is the frequency dependent loss, comprising mainly the eddy current losses in the winding.

The *Rdc* measurement can be carried out easily using a measurement bridge or an impedance analyzer at very low frequency. The *Rac* measurement can best be carried out using an impedance analyzer, sweeping the frequency of the applied signal over the required range. The impedance and phase can be used to calculate the AC resistance, or the analyzer may output the real and imaginary parts directly. Care must be taken to ensure that low applied signal levels are used so that the core of the device under test is not saturated.

Proximity Effect

With more than one winding in the component, the magnetic field in other windings causes eddy currents in the winding of interest and this is called the proximity effect. This is illustrated with the finite element analysis of two windings with currents flowing in the opposite directions shown in Fig. 3.39. This shows the distribution of the magnetic field and the current inside the conductors. (The light colors indicate areas of high current density.)

This can be approximated for idealized winding structures leading to the expression for the power loss due to the proximity effect for round conductors given by:

$$P_{pe} = \frac{\pi \omega^2 B_{max}^2 l d^4}{128 \rho_c} \qquad (3.16)$$

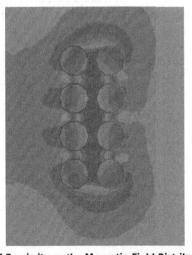

FIGURE 3.39 The Effect of Proximity on the Magnetic Field Distribution in Windings.

Further dependence on frequency in the form of a skin effect factor for this power loss is introduced similarly to the basic skin effect factor F, but this time alters the proximity effect. This factor, G_r, tends to unity for low frequencies, but as the ratio of d to Δ increases beyond 4, the factor takes the form shown in:

$$G_r \rightarrow \frac{32}{(d/\Delta)^4}\left(\frac{d}{\Delta}-1\right)$$

(3.17)

3.4.6 THE DANGER OF INDUCTIVE TRANSIENTS

One of the fundamental circuit laws pertaining to inductors is the relationship:

$$V = -L \times \frac{di}{dt}$$

(3.18)

This says, in effect, that the voltage across an inductor is proportional to the rate of change of current through it, and it is the basis for an enormous number of unreliable circuit designs.

Consider a simple circuit: an inductor in series with a resistor and a switch, connected to a DC voltage source, as in Fig. 3.40.

When the switch is closed, the current through the inductor will build up according to the above equation until it is limited by R to the steady-state value of V/R. So far so good. But what happens when the switch opens?

The current through the inductor is cut off instantaneously with nowhere to go. But this means that di/dt is infinite. So, according to the equation, should be the voltage across it. And this is indeed what happens; at the instant of switch-off, a large voltage transient is induced across the inductor. In practice, its amplitude is not infinite but is determined by the Q and the self-capacitance of the inductor, which forms the only path (apart from leakage) for diversion of the stored current. Or, if the self-capacitance is small and the transient is large enough, its amplitude is

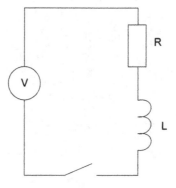

FIGURE 3.40 Series Inductor—Switch Circuit.

limited by breakdown across the switch. This is the source of the unreliability of any circuit design, which uses this configuration, of which there are many. (It is also the principle of operation of the car ignition circuit, of which there are millions.) As an example, consider the relay coil.

Relay Coils

A typical relay drive circuit has the coil driven by a transistor switch (Fig. 3.41). The waveform seen at A on switch-off is a damped sinusoid that has a high initial value but dies away rapidly. The equivalent circuit of the relay coil is an *RLC* tuned circuit where the *R*, *L*, and *C* values are the winding resistance, inductance, and self-capacitance, respectively. The period of the damped oscillation is determined by the *L*−*C* resonant frequency, and the amplitude and decay time constant of the waveform are determined by the Q of the circuit, i.e., by *R*, *L*, and *C*. The resistance of a relay coil is always specified by its manufacturer; the inductance is sometimes specified, the self-capacitance never. So the only way to determine the peak value of the transient is to measure it in circuit.

Transient amplitudes of several times the supply voltage are common. It is quite possible to get transients of hundreds of volts from a 12-V supply, as automotive electronics designers know only too well. If the transistor switch has a collector–emitter breakdown voltage less than the transient peak, the transistor will suffer avalanche breakdown and the transient will be limited. Transistors can withstand repeated low-energy avalanche breakdowns before failure and so, if the circuit is not fully investigated, it will appear to work quite satisfactorily on the bench; it will only be after some time in the field that a high incidence of transistor failures will be noticed. By that time, the product's reputation for unreliability has been established.

This problem is not limited to transistor drive circuits. There are many applications where a small switching contact is driving a larger switching coil, for example,

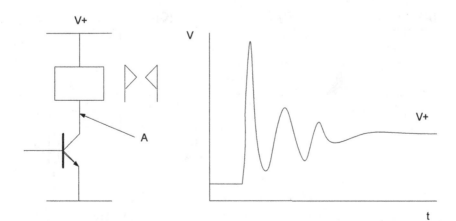

FIGURE 3.41 Relay Driver and Turn-off Waveform.

a reed relay driving a larger relay or a large relay driving a contactor. The same inductive spike phenomenon will lead to spark erosion of the low-power switching contacts, and early failure, if it is not prevented.

Transient Protection

Unfortunately, all protection methods involve extra components. Each method seeks to divert away the energy storage current from the inductor without creating a large transient voltage. The diode method of Fig. 3.42A is the simplest and often the best. It clamps the positive-going spike to the supply rail and for all practical purposes limits the positive voltage seen by the switch to that of the supply. The diode need only be sized to withstand the surge of flyback current limited by the coil resistance, and its voltage rating need be no more than the supply. This circuit does, though, lengthen the turn-off time of the coil, since the stored inductive current continues to flow through the diode for a short time after the switch is opened, and this may limit its applicability.

Protection Against Negative Transients

The single diode clamp does not protect against a negative-going transient that drives the switch voltage below the 0 V rail. This can be prevented by including another diode in series with the switch or by the Zener circuit shown in Fig. 3.42B. The Zener clamps both negative and positive transients, but its positive clamping action is somewhat less effective than the single diode. This is because its specified breakdown voltage must ensure that worst-case tolerances on supply voltage, and Zener voltage do not result in continuous conduction, and so the actual clamping voltage is inevitably higher than optimum. This is not usually a problem unless the breakdown voltage of the switch is already close to the supply voltage, which is bad practice anyway. The Zener has the additional advantage of protecting the switch against transients on the supply rail. It can also be sized to achieve a compromise with respect to the turn-off time of the coil.

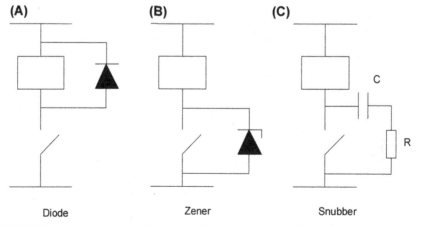

(A) **(B)** **(C)**

Diode   Zener   Snubber

FIGURE 3.42 Inductive Transient Protection.

AC Circuits

Both the diode and Zener methods are only applicable when the supply is a polarized DC voltage. An AC coil requires a different protection method, and for this the snubber circuit Fig. 3.42C is used. This essentially places an *RC* network in the path of the inductive current—the network can equally well be across the switch or the coil, provided the supply impedance is low—so that the current is absorbed by the action of charging *C*. The *C* is effectively in parallel with the self-capacitance of the coil, which it swamps. The resistor limits the switch current when *C* is discharged at turn-on.

In this circuit, *C* should be sized carefully to ensure it is no greater than required to reduce the transient to a manageable level, since it slows down the response of the switch and also allows some current into the load when the switch is open. Similarly *R* should be kept as high as possible consistent with the snubbing action since power is lost both in it and in the switch, as *C* is discharged. The snubber is a popular network both for AC inductive clamping and in many other circuits as a *dV/dt* limiter. Calculation of snubber values is outlined in Section 4.2.6.

3.5 CRYSTALS AND RESONATORS

The quartz crystal has been widely used as a frequency-determining component for many years. It is small, robust, accurate, and stable. In addition, like other components, it has its vices. This section will look briefly at crystal theory before reviewing some application pitfalls and also cover its cheaper but more popular cousin the ceramic resonator.

Quartz (silica, SiO_2) exhibits a piezoelectric effect whereby mechanical stress generates a directionally related electric field, and conversely an applied electric field causes a directionally related force across the crystal. An alternating voltage applied to the crystal will cause it to vibrate, and if its frequency is close to the mechanical resonance, the generated electric field will be amplified and can be used to stabilize the applied frequency.

Angle of Cut

The crystals used in electronic circuits are in the form of plates or elements cut from a synthetic crystal. The resonant properties vary depending on the angle of cut referred to the base crystal's major axis. X- and Y-cut units, where the direction of cut is perpendicular to the major axis, show subsidiary responses, which can reduce the Q factor of the element and impose a fairly low upper limit on the achievable frequency range. In addition, the tempco of these cuts is large.

Happily, a particular angle of cut of 35 degrees 21' from the major axis known as the AT-cut shows very small coupling between the principal and other modes of vibration, therefore lacks subsidiary resonances and is capable of very high-frequency operation. Its resonant frequency is governed directly by a fraction of the element thickness, and its tempco follows a cubic-plus-linear law whose actual slope varies according to the deviation of angle of cut (see Fig. 3.46). The AT-cut crystal is the most widely available crystal unit for general-purpose use. Other cuts are available for specialized applications.

3.5.1 RESONANCE

The crystal equivalent circuit is a series *LCR* tuned circuit together with a parallel capacitance (Fig. 3.43); *C*, *L*, and *R* are functions of the mechanical resonance properties of the crystal element, and C_0 is the static capacitance due to the electrodes and terminations. *C* is very low (of the order of femto-farads) and *L* is very high (of the order of henries), whereas *R* is generally tens or hundreds of ohms for high-frequency units, and the Q of the resulting combination is very high (30,000–100,000).

Because of this, the phase angle changes very rapidly with changes in frequency near resonance. So as an oscillator feedback component, the crystal will correct amplifier phase deviations with only a slight frequency shift.

C_0 is several hundred times larger than *C* and is also increased by external circuit capacitance. The crystal shows two resonant modes, series and parallel, as indicated in Fig. 3.44. Their resonant frequencies are very close together and are given by

$$f_s = \frac{1}{2\pi \times \sqrt{LC}} \tag{3.20}$$

$$f_p = \frac{1}{2\pi \times \sqrt{LC_x}} \tag{3.21}$$

where C_x is the series combination of *C* and C_p, and $C_p = C_0 +$ external capacitance.

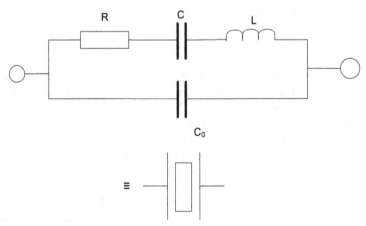

FIGURE 3.43 Crystal Equivalent Circuit.

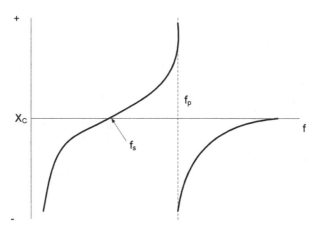

FIGURE 3.44 Series and Parallel Resonance.

The crystal can be operated in either mode. In series mode, the element is operated at a low impedance, which is equivalent to R; it can be "pulled" upward in frequency slightly from f_s by inserting external series capacitance. In parallel mode the element operates at high impedance and can be pulled downward in frequency from f_p by adding external parallel capacitance. The frequency shift from f_s in either case is the same for a given external capacitance. Clearly a given resonant frequency is only obtainable if the external capacitance is known, and in fact all crystals are supplied for a quoted "load" capacitance. The unit will only operate at the marked frequency (within the given tolerance) if the actual circuit capacitance is as specified. Conversely, the frequency can be trimmed if absolute accuracy is needed by using a variable load capacitance.

3.5.2 OSCILLATOR CIRCUITS

There are two common circuits for digital clock oscillators (Fig. 3.45), one operating in each mode. The parallel circuit is only suitable for high-impedance (CMOS) devices, whereas the series circuit can be used for high- or low-impedance devices. The parallel circuit can be run at very low-power levels (down to 1 µA) but is slow to start. It is commonly used by on-chip microprocessor clock oscillators and other CMOS oscillator/divider ICs, such as real-time clocks.

R_f biases the inverter to linear operation and should be low enough for input bias current to have negligible effect but high enough not to load the crystal. Generally 10–15 MΩ is reasonable. The crystal appears primarily inductive and provides 180 degrees phase shift in the feedback loop. C1 and C2 in series together with circuit strays (amplifier input and pc track capacitance, amounting to at most 10 pF with good layout) form the crystal load capacitance. The ratio C2:C1 should generally be of the order of 3:1, C2 being variable if frequency trimming is desired.

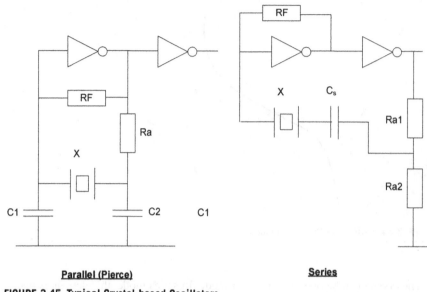

Parallel (Pierce)   **Series**

FIGURE 3.45 Typical Crystal-based Oscillators.

Drive Level Resistance

R_a is an important component and should not be omitted without proper consideration. It sets the drive level to the crystal. Too high a drive will lead to frequency instability and possible damage to the element. Too low a level will make the oscillator slow to start, perhaps impossible to start with low-activity units and susceptible to interference. Typical AT-cut crystals have a maximum drive level of 0.5–1 mW. Some circuits (for example, low-power on-chip CMOS oscillators) have a high enough output impedance to make R_a unnecessary, but this is not normally the case with discrete-gate oscillators. For watch crystal units (32.768 kHz) R_a should be tens or hundreds of kΩ.

Series Circuit

Some applications may be embarrassed by the slow starting time (possibly up to 1 s) of the parallel oscillator circuit. Crystals have a very high Q, and if the drive level is low, for frequency stability or to conserve current, the time taken to reach working level is appreciable. This may be unacceptable in microprocessor clock circuits where the clock is expected to be present immediately on power-up. For these purposes the series oscillator, in which the crystal is operated at a low impedance with minimal phase shift across it, is preferable. Its main disadvantage is its higher supply current. The same strictures on drive level apply. Note that the effective series resistance of the element, which is equivalent to its motional resistance R, can vary widely from unit to unit. A spread in this parameter of two- or three-to-one is not uncommon, so it is wise to design the circuit for assured start-up with a three times higher R than quoted.

Layout

Circuit board layout is important, particularly for the parallel mode. Extra capacitance across the crystal should be minimized, as this will increase loop gain and short-term stability. So should coupling between the oscillator circuit and other circuits, especially logic switching circuits, as this decreases the likelihood of spurious oscillation. Ground traces around the crystal to buffer other tracks are advisable; on no account route logic signals near or through the oscillator circuit as they will couple into the high-impedance nodes and cause frequency instability or jitter.

3.5.3 TEMPERATURE

Lastly, beware of tempcos. The temperature law of the AT-cut is cubic (Fig. 3.46) and if the cut angle is chosen carefully, can be fairly flat at room temperature, but it worsens rapidly as the temperature limits are neared. A crystal will oscillate outside its rated temperature (usually), but the frequency stability will be impaired.

Tuning fork crystals (the ubiquitous 32.768 kHz type, universally used for real-time clocks) show a parabolic curve, of around -0.04 ppm/$°C^2$. The turnover temperature is around 25°C, which means that for digital watch applications, where the wrist temperature remains around this value, it is ideal and very stable. Transfer this type of crystal to an industrial real-time clock (for example) and its timekeeping at the extremes of the range is hopeless: at $+85°C$ and at $-35°C$ it is 144 ppm low, which represents a loss of 12 s per day. Be warned: use an AT-cut!

3.5.4 CERAMIC RESONATORS

A cheaper alternative to the quartz crystal is the ceramic resonator. This device uses the mechanical resonance of a piezoelectric ceramic, typically lead zirconate titanate, which vibrates in various mechanical modes depending on the chosen resonant frequency. The frequency ranges are approximately as follows:

30 kHz–1 MHz: longitudinal mode
100 kHz–2 MHz: area mode
1–10 MHz: shear thickness mode
2–100 MHz: expansion thickness mode
10 MHz–1 GHz: surface acoustic wave mode

The equivalent circuit of the resonator is identical to that of the quartz crystal (Fig. 3.43), but the component values give it orders of magnitude lower Q. In terms of oscillation frequency accuracy, the resonator sits between the quartz crystal and the LC resonant circuit. The tempco of a resonator is of the order of 10^{-5}/°C compared to the better than 1 ppm/°C achievable with quartz, and the 10^{-3}–10^{-4}/°C of *LC* circuits. Its initial frequency tolerance is of the order of $\pm0.5\%$, whereas quartz routinely achieves $\pm0.003\%$; to achieve these figures using *LC* circuits would need a trimming adjustment. On the other hand, the resonator is cheaper and smaller than quartz crystals and can use the same or similar oscillator circuits.

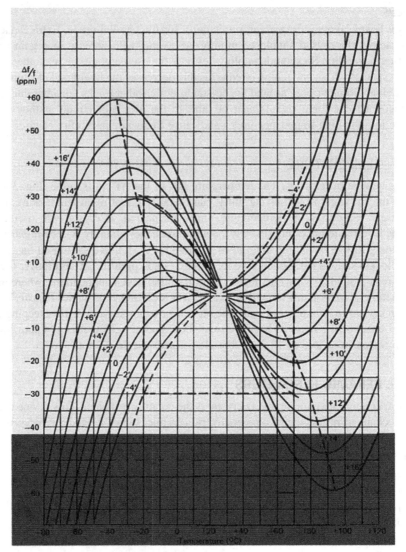

FIGURE 3.46 AT-cut Frequency/Temperature Curves.

From ECM Electronics.

Load capacitors as in Fig. 3.45 are necessary to prevent spurious oscillation modes and the manufacturer's recommendation for the oscillator circuit should be followed. A further advantage of the resonator is that because of its lower Q, the oscillation will start up more quickly than for an equivalent crystal circuit, which makes it attractive for applications that spend a lot of their time in "sleep" mode with the oscillator powered off.

All of these characteristics make the ceramic resonator the component of choice for frequency control of low- and midperformance digital products, where a stable clock frequency is needed but where absolute accuracy or close control of tempco is not a requirement. It is mass produced and available in a wide range of standard frequencies, matched to particular consumer applications such as DTMF (telephone dialing tone) generators, remote control units, and TV and audio systems.

Active Components

4

CHAPTER OUTLINE

The Circuit Designer's Companion. http://dx.doi.org/10.1016/B978-0-08-101764-7.00004-9

159

In this chapter we shall concentrate on discrete semiconductor components that are in wide use throughout electronics design. Although there is a continuing trend toward "gathering up" discrete semiconductors into application-specific ICs, and a parallel trend toward replacing as many analog functions as possible by digital signal processing, discrete analog circuits are still needed when these solutions are impossible or uneconomical. It is as well to be familiar with the characteristics of practical components as even when integrated they show the same fundamental properties.

This chapter covers the common two- and three-terminal devices: diodes, thyristors, triacs, transistors, FETs, and insulated gate bipolar transistors (IGBTs).

4.1 DIODES

The diode is a two-terminal device whose function is to pass current in one direction but not in the other. A conventional diode is formed from the junction of p-type and n-type silicon. The ideal device has a "brick-wall" V—I characteristic: the practical silicon diode has an exponential characteristic, which approximates to the brick wall, if viewed on a large enough scale (Fig. 4.1). Also notice that at a significant reverse voltage the device will "breakdown" across the junction and start to conduct.

4.1.1 FORWARD BIAS

The first thing to notice is that the forward voltage v_f is not constant nor is it zero. It has two determinants, forward current i_f and temperature T. They are related by the classic equation for the diode forward current:

$$i_f = i_s \left[e^{(v_f q / kT)} - 1 \right] \tag{4.1}$$

Eq. (4.1) is known as the "diode equation" or the "Ebers—Moll equation", arguably the most fundamental mathematical expression in the whole of semiconductor electronics. The parameters q and k are the electron charge 1.6×10^{-19} coulombs and Boltzmann's constant 1.38×10^{-23} joules per degree Kelvin respectively. T is

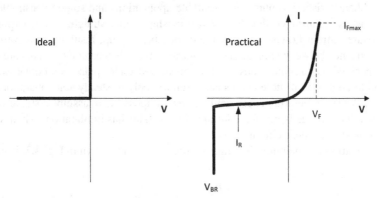

FIGURE 4.1 Ideal and Practical Diode Characteristics.

the absolute temperature. The expression kT/q evaluates to 0.025 V at 20°C and is the source of many of the properties of the silicon p-n junction. i_s is the saturation current and depends on the device, and on temperature.

A frequent rule of thumb allows 0.6 V for the v_f drop of any silicon diode. This is because over the most common range of i_f, v_f remains fairly close to this value. But if i_f is in the microamp or nanoamp region, v_f is close to zero and the slope resistance v_f/i_f is high. The diode behaves more like a nonlinear resistor than a rectifying device. At room temperatures the slope resistance can be taken to be $0.025/i_f$ ohms. If i_f is nearing its maximum limit, which for low-level signal diodes is in the hundreds of milliamps region, v_f can approach 1 V and the device begins to dissipate significant power.

Forward Current

The maximum forward current, $i_{f(\text{max})}$, is limited by power dissipation, $i_f \times v_f$. This leads to a rise in junction temperature, which must not exceed some maximum value, usually between 125 and 200°C. Diodes are rated for continuous use, but it is possible to exceed the rating for pulse applications, when the average power dissipation depends on the duty cycle ($P_{\text{avg}} = D \times P_{pk}$). Rectifier diodes are also characterized for "surge" current, which can exceed the average current by 30—70 times. The specification is linked to a typical surge duration. Frequently, for US-manufactured rectifiers, this is quoted at 8.33 ms, which is one half-cycle at 60 Hz, the US mains frequency. If no current—time curves are shown, the value can be extrapolated to a limited degree for other time values by taking a constant I^2t product. This specification is important when considering power supply switch-on surges, when a reservoir capacitor is being charged from zero. Remember that v_f carries on rising with i_f as $i_{f(\text{max})}$ is exceeded.

Temperature Dependence of Forward Voltage

The other determinant of forward voltage is temperature. i_s in the diode equation has an exponential temperature dependence, which dominates the voltage temperature coefficient of the device, which for silicon is around -2 mV/°C for a constant current (Fig. 4.2).

This characteristic has numerous desirable applications and some undesirable effects. It means, for instance, that the v_f: i_f relationship is not a straightforward exponential, because current flowing through the device heats it up, resulting in a complex interdependence between temperature and current. For this reason v_f/i_f curves are normally specified as "instantaneous" and are measured under pulse conditions, which can lead to confusion if these curves are taken to apply to steady-state operation.

The voltage temperature coefficient makes it impossible to assume a stable value for v_f even if the diode is run at a constant current. This has implications whenever a diode is used in a linear circuit.

The equation for the output voltage rectifier circuit, as shown in Fig. 4.3, is given by Eq. (4.2):

$$V_0 = \frac{R_2}{R_1 + R_2} \times (V_{\text{in}} - V_F) \tag{4.2}$$

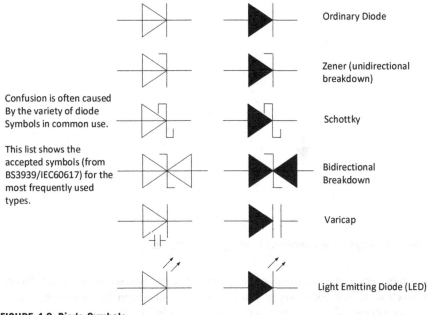

Confusion is often caused
By the variety of diode
Symbols in common use.

This list shows the
accepted symbols (from
BS3939/IEC60617) for the
most frequently used
types.

Ordinary Diode

Zener (unidirectional
breakdown)

Schottky

Bidirectional
Breakdown

Varicap

Light Emitting Diode (LED)

FIGURE 4.2 Diode Symbols.

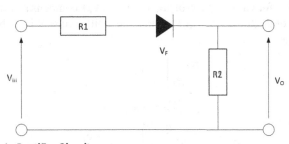

FIGURE 4.3 Diode Rectifier Circuit.

As a simple example, you may want to use a diode to give a rectifying (unipolar) characteristic to a potential divider (Fig. 4.3).

The potential divider relationship is immediately complicated by the addition of V_F. Over the commercial temperature range of $0-70°C$ it will vary by about 150 mV (for more explanation about the effects of component tolerances and variations generally, see Chapter 3 of this book).

If R_1 and R_2 are 10K and V_{in} is $+5$ V, V_F is taken to be 0.45–0.6 V, then V_0 will vary from 2.275 to 2.2 V over the temperature range—it is *not* half of V_{in}!

On the positive side, the silicon diode junction does form a cheap and fairly reproducible, if somewhat inaccurate, temperature sensor. Also, two junctions in close proximity can be expected to track changes in V_F repeatedly, which allows for fairly simple

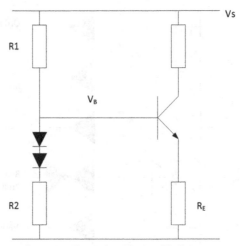

FIGURE 4.4 Temperature Compensation Using Biasing Diodes.

temperature compensation when necessary. This characteristic is common to all silicon p-n junctions so that, for example, you can use a pair of silicon diodes to compensate the dc conditions of a single-transistor gain stage, as in Fig. 4.4.

This circuit operates on the principle that as long as the biasing resistors R_1 and R_2 are equal, the forward voltage of the diodes ($2V_F$) compensates for the base–emitter voltage of the transistor V_{BE} such that the emitter current is set by the emitter resistor R_E alone.

$$V_B \approx \left[\frac{R_2}{R_2 + R_1}\right] \times [V_S - 2V_F] + 2V_F \qquad (4.3)$$

$$I_E = (V_B - V_{BE}) \times R_E \qquad (4.4)$$

$$I_E = \left[\frac{R_2}{R_2 + R_1}\right] \times V_S \times R_E \qquad (4.5)$$

if $V_{BE} \approx V_F$ and $R_1 = R_2$

Note that this circuit requires that the two diodes and the bias resistors are both equal, since the combined forward voltage is divided by the resistor ratio. The compensation is not accurate, because the diode and transistor junctions are not at identical temperatures, and they do not generally carry the same current. If $R_1 \gg R_2$ then it is possible to get away with one diode and accept rough-and-ready temperature compensation, which may be adequate for your application.

Alternatively, use dual transistors to ensure identical junction temperatures, with a more complex circuit arrangement to achieve very accurate compensation. This latter is the basis for many op-amp temperature compensation schemes, since extra transistors are essentially free and being on the same chip, are as close as it is possible to get to the required temperature.

4.1.2 REVERSE BIAS

So far we have only considered the forward characteristic, that is, for positive applied voltage. An ideal diode would block all current flow in the reverse direction. A practical diode does not. There are two main reverse characteristics, reverse leakage current I_R and reverse breakdown voltage V_{BR}. The diode Eq. (4.1) holds good in the reverse direction until V_{BR} is approached; in the low-voltage region I_R is almost equal to I_S.

Breakdown

V_{BR} is that voltage at which the reverse-biased junction can no longer withstand the applied electric field. At this point, avalanche breakdown occurs and a current limited mainly by the external source impedance will flow. If the device maximum power dissipation is exceeded the junction will be destroyed. Diodes operated conventionally, as opposed to zener diodes to which we will return shortly, are always run at reverse voltages lower than V_{BR}.

A common overvoltage excursion is the inductive turn-off transient (see Section 3.4.4) where a diode is used, intentionally or not, to block the transient. It can be difficult to predict the maximum voltage of the transient and, since the energy dissipated by a breakdown may be much less than needed to destroy the diode, such breakdowns may go unnoticed during the evaluation of the design. Diodes are available that are characterized for the amount of avalanche breakdown energy they can withstand and should be used if a circuit is expected to deliver predictable transients above the normal breakdown voltage.

4.1.3 LEAKAGE

Since reverse leakage current is of fundamental importance to circuit operation, all diode data sheets quote a specification for maximum leakage. Unfortunately, this hides as much as it reveals.

Leakage current I_R is relatively constant with voltage until V_{BR} is approached, at which point it starts to increase rapidly. It is not, however, constant with temperature, but roughly doubles for every $10°C$ rise of junction temperature. This characteristic, like the forward voltage temperature coefficient, is common to all reverse-biased p-n junctions, and we will meet it again later. Most diode leakage currents are specified both at $25°C$ and at a higher temperature, and the $25°C$ figure is highly misleading if you apply it over a typical temperature range. A leakage of 100 nA at $25°C$ translates to 2.2 µA at $70°C$, for instance. This is a common factor in the poor high-temperature performance of high-impedance circuits, or those which employ very low current levels.

Leakage Variability

To make matters more complicated, leakage is susceptible to process variations. It can vary by up to an order of magnitude from batch to batch under otherwise identical conditions. Therefore, manufacturers will put an artificially high maximum

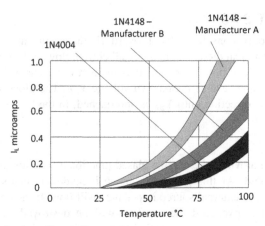

FIGURE 4.5 Diode Leakage Versus Temperature.

value of leakage in their specifications compared with the actual performance of the majority of delivered units, to have room for maneuver when a given batch shows a high value.

The consequence of this is that if your design is sensitive to leakage current then the prototype may work well while a production model does not. The probability is high that a device selected at random for the prototype will have a low leakage, whereas some production devices will come from high-leakage batches. If the design proceeded on the basis only of satisfactory measurements on the prototype then the seeds have been sown for production difficulties. To avoid them, always work on worst-case calculations even though these are not borne out by bench tests.

By way of illustration, a set of measured leakage characteristics is shown in Fig. 4.5. Several samples each of three types of diode were submitted to a temperature sweep from 0 to 100°C while their leakage currents were monitored. Two different manufacturers' versions of 1N4148, and one version of 1N4004, were tested. The curves show clearly the logarithmic relationship; all the samples had less than 10 nA leakage at 25°C. It is also clear that within a batch the variations are quite small, but between two manufacturers of a nominally identical device they are much larger. It is also interesting to see that the 1N4004 rectifier diodes, run at much less than their breakdown voltage, have a lower leakage than any of the small-signal diodes.

4.1.4 HIGH-FREQUENCY PERFORMANCE

Up to now the diode characteristics discussed have been those that apply at dc. As the frequency of use increases, ac characteristics become more important. The parameters of greatest interest are the equivalent capacitance and the turn-on/turn-off behavior.

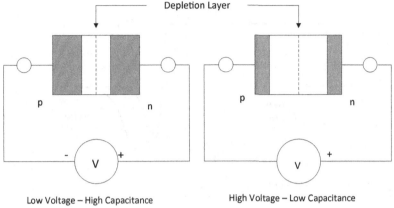

Depletion Layer

p   n   p   n

V   V

Low Voltage – High Capacitance   High Voltage – Low Capacitance

FIGURE 4.6 Diode Junction Capacitance Versus Applied Voltage.

The equivalent diode circuit includes a parallel capacitance. This is due to the depletion layer across the junction acting much as a dielectric separating two plates. As the applied reverse voltage changes, so does the width of the depletion layer, and so does the effective capacitance (Fig. 4.6). A low reverse voltage gives a thin depletion layer and a high capacitance; a high reverse voltage reduces the capacitance.

This effect is exploited in the so-called "varicap" diode for voltage control of tuned circuits, used in virtually every modern radio receiver, and also in the varactor diode, where the nonlinearity of the C/V law is used to generate RF harmonics. In most other applications it is either irrelevant or a nuisance. Actual capacitance depends on diode construction and varies from a few pF to hundreds of pF. If the circuit calls for high-speed or high-frequency operation then some allowance must be made for it in the design. Signal-switching applications, where the signal is small compared with the reverse biasing voltage, can assume a constant capacitance which can be reduced if necessary by increasing the bias voltage, other factors being equal. Large-signal or rectifying applications need to take into account the nonlinearity of the capacitance/voltage relationship, which most often manifests as unexpected waveform distortions.

The foregoing applies to diode capacitance under reverse bias. When the diode is forward biased the capacitance increases but the impedance is now low enough for this not to be a dominant issue.

4.1.5 SWITCHING TIMES

On turn-on, when reverse bias is changed to forward bias, the applied forward current has first to discharge the junction capacitance before the junction will conduct. Thus there is a delay in establishing the steady-state V_F, known as the forward recovery time, but no other adverse effects. Providing the diode is driven at turn-on from a reasonably low impedance, the turn-on time is much less than the turn-off time.

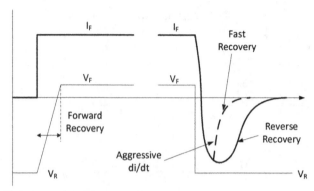

FIGURE 4.7 Diode Forward and Reverse Recovery.

On turn-off, the applied reverse voltage must "sweep" all the conducting minority carriers out of the junction before conduction can cease, and this takes a finite time, during which current continues. At the end of this period the current drops to the expected reverse leakage value. Since the mechanism is quite different, there is no direct correlation between this time and the junction capacitance. "Reverse recovery" time is directly related to the forward current before reverse voltage is applied and to the rate of change of current at turn-off. Fig. 4.7 illustrates recovery times.

Reverse Recovery

Reverse recovery time becomes an increasing embarrassment as switching speed and power increase because it represents dissipated power at quite a high level ($V_R \times I$). The faster the switching frequency, the greater the proportion of power that is dissipated in the reverse direction; in high-power circuits this becomes a limiting factor on the diode rating, especially at the higher voltages, and also contributes to inefficiency in power conversion. Conventional rectifiers have recovery times in the $1-20\ \mu s$ region.

To overcome this problem the "fast recovery" diode was developed, which by suitable processing reduces the reverse recovery time to a minimum, though not to zero. Typical recovery times are $150-200$ ns and fast recovery diodes are used extensively in high-voltage, high-speed switching. When even these speeds are too slow "ultrafast" diodes are also available which can have recovery times down to 20 ns.

Interference Due to Fast Recovery

Fast recovery brings its own problems, though. The characteristic tail on turn-off "snaps" back very quickly to I_R, producing a very high transient rate of change of current (di/dt). Usually it is the highest di/dt within the circuit and so is responsible for most of the unwanted electromagnetic interference (EMI) output. To save on the need for extra components to limit this current, yet another class of diodes has been developed, the "soft recovery" diodes, in which a compromise has been reached between speed of recovery and a comparatively gentle turn-off characteristic.

Table 4.1 Schottky Versus Conventional Diodes

Conventional	Schottky
Forward voltage typically 0.6 V	Forward voltage typically 0.4 V
At medium currents	At medium currents
Minority carrier charge storage	No minority carriers, no charge
Effects limit speed	Storage, high speed
High reverse breakdown voltage	Low reverse breakdown voltage
Achievable, in excess of 1 kV	Ranging generally from 30 to 100 V
Low reverse leakage current	Higher reverse leakage current

However, bear in mind that all p-n junction diodes exhibit some form of reverse recovery and are therefore capable of generating interference at harmonics of the switching frequency (even mains!) and of dissipating some power during this period.

4.1.6 SCHOTTKY DIODES

The p-n semiconductor junction is not the only arrangement to show rectifying properties. A metal—semiconductor junction also rectifies. Devices that use this property are known as Schottky diodes. The important differences between conventional p-n silicon diodes and silicon Schottky diodes can be summarized as shown in Table 4.1.

Schottky diodes are used primarily for their low forward voltage or for their high speed. Available types are characterized for three main areas:

- high-speed switching and general purpose
- RF and microwave mixers
- high-efficiency rectifiers

General Purpose

Small-signal Schottky diodes can be used in many of the same applications as conventional diodes as well as those where their lower V_F or high switching speed is essential. The shape of the V/I characteristic is the same though the values of V_F and V_{BR} differ. The temperature coefficient of V_F varies rather more with I_F and is around -1 mV/°C at the milliamp level. Leakage is up to an order of magnitude higher than typical p-n junctions and shows the same exponential temperature dependence. Schottky diodes are more expensive than their conventional counterparts—5 p versus 1 p unit cost—and this has restricted their widespread application.

RF Mixers

For RF applications, the Schottky diode is the almost ideal component for a mixer circuit, in which a deliberate nonlinearity is introduced to extract the sum or difference of two frequencies applied to its inputs. The high speed, low noise, and large signal-handling ability of the Schottky make it particularly suitable for wideband mixers. The earliest applications for them were in this field, and there is a range of devices characterized for such use.

Rectifiers

The largest growth area for Schottky rectifiers has been in the output stages of switch-mode power supplies. There is an enormous market for medium-to-high current 3.3-V or 5-V output switchers for supplying computer circuits, and the trend is toward ever-greater efficiencies and higher switching speeds, for which the Schottky is eminently suited. At higher currents the forward voltage of a conventional rectifier can approach 1 V, so that 20% of the total power of a 5-V switcher is lost in the diode alone; the Schottky V_F of 0.5 V cuts this to 10%. At the same time, the lack of a reverse recovery mechanism makes it designer-friendly at high speeds, and the low V_{BR} limitation is no hindrance at such a low output voltage. The higher unit cost of the Schottky is offset by the reduction in component count that can be achieved.

4.1.7 ZENER DIODES

By suitable selection of dimensions and impurities within the silicon it is possible to control the voltage at which reverse breakdown occurs. The slope of the diode V/I curve becomes quite flat in this region and the device can be used as a voltage regulator or clamp, and devices characterized for this purpose are called zener diodes. The breakdown voltage can be controlled from 2.4 V up to hundreds of volts, 270 V being a practical maximum. In the forward direction the zener functions just like an ordinary silicon diode, with a somewhat higher V_F and an unspecified V_F/I_F characteristic.

Just as with other components, the zener is not perfect. Its slope resistance is not zero, its breakdown knee is not sharply defined, it has a leakage current below breakdown, and its breakdown voltage has tolerance and temperature coefficient. Fig. 4.8 demonstrates these features.

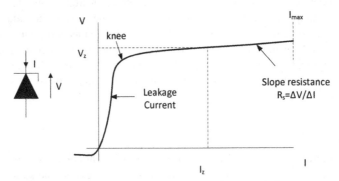

FIGURE 4.8 Zener Reverse Breakdown Voltage—Current Characteristic.

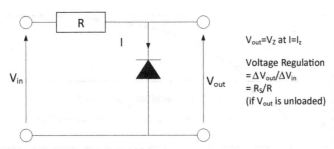

$V_{out}=V_Z$ at $I=I_z$

Voltage Regulation
$=\Delta V_{out}/\Delta V_{in}$
$= R_s/R$
(if V_{out} is unloaded)

FIGURE 4.9 Regulation of a Shunt Stabilizer.

Slope Resistance

Zeners are supplied for a quoted voltage, which is always defined at a given reverse current I_Z. At this current it will be within the specified tolerance, but at other currents it will differ, the difference being a function of the zener slope resistance R_s. The actual range of working voltage can be calculated by adding $(I - I_Z) \times R_s$ to the quoted voltage range, where I is the working current and I_Z is the current at which the zener voltage is quoted.

Over some range of I_Z, which you can determine from the published curves, R_s can be assumed to be fairly linear. As the current decreases the characteristic approaches the "knee" of the curve and R_s increases sharply. There is very little point in operating a zener intentionally on the knee. The actual knee current depends on the type and voltage but is rarely less than a few hundred µA. Consequently, zeners are not much use for micropower or high-impedance circuits. For shunt voltage regulator applications at low currents, circuits based on the wide range of band-gap reference devices (see Section 5.4.2) are preferable.

In a typical shunt stabilizer circuit (Fig. 4.9) the voltage regulation is directly related to R_s. Clearly, the lower R_s the better. Slope resistance falls to a minimum around 6.8 V and increases markedly at greater or lesser voltages. The lower voltage zeners have a much higher slope resistance than do those in the mid-range (Fig. 4.8). Below 5 V and above 100 V the simple zener shunt stabilizer exhibits poor voltage regulation as a consequence. If a high-voltage zener is needed, better performance can be had by putting two or more lower voltage devices in series to obtain the required voltage.

Leakage

Below the knee, when the reverse voltage is not sufficient to begin the breakdown, there is still some current flow. This is due to leakage in the same way, and with the same temperature dependence, as a conventional diode. Leakage is usually specified for a zener at some voltage below the breakdown voltage, of the order of 20%–30% less. It is an important specification when the zener is used as a clamp, so that the device's normal operating voltage is less than the breakdown value. A typical application is in protecting a circuit input from transient or continuous overvoltages (see Section 4.1.8).

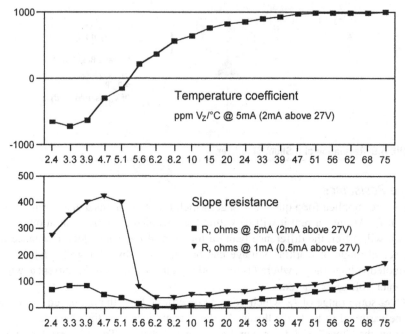

FIGURE 4.10 Zener Slope Resistances and Temperature Coefficients Versus Zener Voltage for the BZX79 Series.

Temperature Coefficient

Like all components, the zener's breakdown voltage exhibits a temperature coefficient; but the zener tempco is somewhat more subtle than usual. There are in fact two mechanisms for reverse breakdown in silicon. Electron tunneling is the dominant mechanism at low voltages and very thin junction barriers, while avalanche breakdown is dominant for higher voltages and thicker barriers. Depending on the required voltage, one mechanism or the other will predominate, and the crossover is at around 5 V. The practical significance is that the two mechanisms have opposite temperature coefficients. They are also the reason for the dramatic variations in slope resistance. The optimum zener voltage for minimum temperature coefficient is between 4.7 and 5.6 V, and where you have a choice of regulation voltage it is best to go for one of these values if the temperature coefficient is important.

The graphs shown in Fig. 4.10 illustrate the temperature coefficient and slope resistance variability for the Philips BZX79 range of zener diodes. Because these characteristics depend on the basic physics of the zener effect, other manufacturers' ranges will show similar performance.

Precision Zeners

A further quirk of the process means that a device with a breakdown voltage of about 5.6–5.9 V has a temperature coefficient of roughly +2 mV/°C, which balances the temperature coefficient of a conventional forward biased silicon junction. By putting the two in series a virtually zero temperature coefficient zener diode can be created, with an effective breakdown voltage of between 6.2 and 6.4 V. These are available off-the-shelf as "precision reference diodes" (the 1N821 series is the most common example) with a closely adjusted temperature coefficient and tolerance for use as voltage references. They are expensive, with the cost increasing in direct proportion to the tightness of specification. A similar effect can be achieved at around 8.4 V, by putting two junctions in series with a 7.5 V zener, which has a positive temperature coefficient of 4 mV/°C. These devices compete directly with band-gap reference ICs in most aspects of performance and price; usually the band-gap wins out due to its lower slope resistance, lower operating current, and more acceptable regulating voltage.

Zener Noise

Another feature of zener breakdown is that it is a noisy process, electrically speaking. In fact, a zener operated at a constant current, ac-coupled, and amplified is a good source of wideband white noise for calibration and measurement purposes. Zeners are not normally characterized for noise output so it is difficult to base a production design on them, but they can be used on a one-off basis. Noise is not usually a problem in voltage regulator applications, since it is many orders of magnitude below the dc zener voltage and can be virtually removed by the addition of a parallel decoupling capacitor. If the capacitor is omitted, either inadvertently or because a fast response is needed, then zener noise can be significant for precision references.

4.1.8 THE ZENER AS A CLAMP

In this application (Fig. 4.11) the zener is used to prevent input overvoltages from damaging subsequent circuitry. It must not affect the input for voltages within the

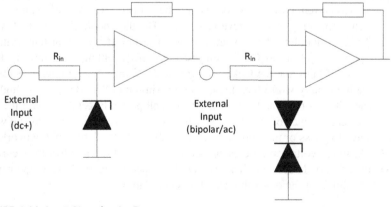

FIGURE 4.11 Input Clamping by Zener.

normal operating range but must clamp the op-amp input to a safe value when the external input is overloaded. The imperfections discussed in zener diodes mean that it is not very good at doing this. Leakage current and capacitance restrict its use to comparatively low-impedance and low-frequency circuits. The breakdown voltage knee, tolerance, and slope resistance mean that a practical limiting voltage must be substantially higher than the operating voltage and so the op-amp's full permissible input range is not utilized.

An Application Example

Take the bipolar input circuit of Fig. 4.11, assume that the op-amp has ±15-V supplies and that its input pins are not allowed to exceed the supply voltage under possible fault conditions of 100-V continuous input. The input itself can swing from +10 to −10 V under normal operation. A back-to-back zener pair is used for clamping. The input source impedance is 10 kΩ, and the accuracy required of the voltage follower is 0.1% over a temperature range of 0−50°C. The zeners must be cheap and easily available.

The absolute maximum clamping voltage must be $(15\ V - V_F)$ less the zener tolerance. We can take V_F to be 0.8 V ignoring the temperature coefficient and slope resistance (in defiance of the previous strictures about diode characteristics!) and if the tolerance is 5%, which is the most readily available, V_z should be a maximum of 13.5 V. The closest standard value is 13 V, but this does not allow for voltage rise with power dissipation and slope resistance, which will exceed 0.5 V. If we specify a BZX79 C12 (with a V_z of 12 V), then from the published curves the maximum of 13.5 V will be reached at around 25 mA. This is within the BZX79's dissipation rating of 400 mW at 50°C. This allows us to set the input resistor to be:

$$R_{in} = \frac{(100 - 13.5 - 0.8)}{25\ mA} = 3.4\ k\Omega \qquad (4.6)$$

Using preferred values, the nearest value will be 3K3.

Now assess the inaccuracy caused by diode leakage. The total input source resistance is 13.3 kΩ, and this is allowed to cause a 0.1% error on 10 V, i.e., 10 mV, due to leakage. This allows us a total leakage current of 10 mV/13.3 K = 0.75 μA. Neglecting op-amp bias current (which may or may not be realistic) we can assign all this to the 12-V zener operating at 10-V reverse voltage. The data book gives us 0.1 μA at 25°C and 8 V, so doubling the leakage current for every 10°C rise in temperature we can expect 0.56 μA at 50°C. We are just (*only* just!) within spec, and this has involved an act of faith that the leakage current at 10 V for a lowest tolerance zener will only be marginally worse than the quoted maximum at 8 V. Because the higher voltage zeners have a reasonably sharp knee we shall probably get away with it.

The diode capacitance of about 30 pF at 10 V (this is rarely specified, so you will probably have to guess or measure it) puts the 3-dB roll-off of the input network at 400 kHz. At lower voltages the capacitance increases and the bandwidth is correspondingly reduced. If the circuit is expected to operate up to these frequencies this will be a limiting factor on the use of a zener clamp.

However, despite its limitations the zener is a cheap, one- or two-component solution to input protection. Devices characterized especially as transient absorbers are also available, though not so cheap and can be designed-in in the same manner when high energy transients are expected. Another approach is to incorporate diodes between the input and the supply rails. This is discussed in Section 6.2.3.

4.2 THYRISTORS AND TRIACS

The thyristor, or silicon-controlled rectifier, is a four-layer diode whose conduction in the forward direction can be initiated by a trigger pulse or voltage at a third terminal called the gate. The triac is a similar device whose conduction can be initiated in both forward and reverse directions.

This class of devices (which includes further variants such as the diac, the unijunction transistor, and the gate turn-off thyristor) is extremely useful for power conversion and control and can be found in many other niche applications (Fig. 4.12). Both its utility and its limitations stem from the fact that once conduction has been initiated, it continues until the applied current has been removed. If a thyristor is used in a dc circuit, once triggered it will remain conducting. If it is used in an ac circuit, it stops conducting at every zero crossing and must be retriggered on every half-cycle.

4.2.1 THYRISTOR VERSUS TRIAC

Thyristors are used extensively both in light current applications and high-power switching and control, but triacs generally only find application in low-power (below 40 A) mains circuits. A triac costs much the same as a thyristor of a similar rating and is equivalent to two back-to-back thyristors, but this apparent advantage, in terms of component count, is also its greatest limitation. Since a triac conducts in both directions, it has only a brief interval during which sine-wave alternating current is passing through zero to recover to its blocked state. With inductive loads, the

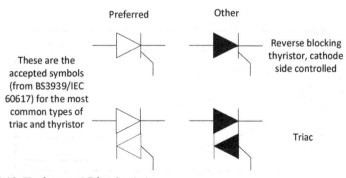

FIGURE 4.12 Thyristor and Triac Symbols.

phase shift between current and voltage means that when the current falls to zero and the triac stops conducting, there will be an applied voltage across it. If this appears too rapidly, the triac will carry on conducting and control will be lost. Reliable operation is therefore limited to mains line frequencies and lower, and even at these frequencies snubbing (see Section 4.2.6) is necessary for inductive loads.

As well as the characteristics that have already been discussed for conventional diodes —forward voltage and current, reverse breakdown voltage and leakage current, reverse recovery time—thyristors and triacs have another set of characteristics to consider, to do with triggering and conduction. These are trigger voltage and current, holding current and dV/dt.

The basic properties of silicon are the same for thyristors as for ordinary diodes. However, the thyristor construction is a p-n-p-n sandwich between its main terminals, so the forward voltage drop is higher than that of an ordinary diode, generally from 0.8 to 2 V depending on current. This restricts the thyristor's usefulness in low-voltage circuits. Reverse breakdown and leakage mechanisms are the same and so the reverse characteristics are similar.

4.2.2 TRIGGERING CHARACTERISTICS

Conduction is normally initiated by injecting energy into the gate terminal. The gate-cathode or gate-MT1 connection is a p-n junction and so is best driven from a low-impedance current source. Since triggering is energy dependent, a high current pulse can be applied for a short period or a lower current can be applied for longer.

If the total energy content of the trigger pulse is not sufficient then unreliable triggering will occur, particularly at the lower temperature extreme of operation when the energy required is greater. The typical behavior of minimum required gate trigger current with variations in temperature and pulse width is shown in Fig. 4.13.

Thyristors can only be triggered by one polarity of current applied to the gate terminal. Triacs, on the other hand, can be triggered on either polarity of main terminal voltage by either polarity of trigger current. This is known as "four-quadrant" triggering and is a useful property as it means that a unipolar pulse can trigger both

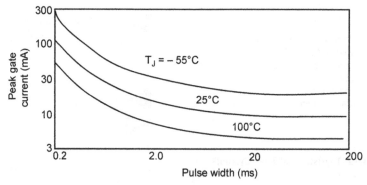

FIGURE 4.13 Gate Trigger Current Versus Pulse Width.

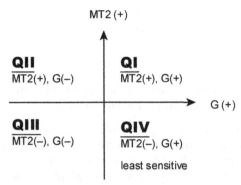

MT2 (+)

QII
$\overline{MT2}(+), G(-)$

QI
$\overline{MT2}(+), G(+)$

G (+)

QIII
$\overline{MT2}(-), G(-)$

QIV
$\overline{MT2}(-), G(+)$

least sensitive

FIGURE 4.14 Triac Triggering Quadrants.

positive and negative half-cycles of the ac waveform. The triac's construction makes for different gate sensitivities in different quadrants; usually quadrant IV (Fig. 4.14) is considerably less sensitive than the rest. A negative-going gate pulse, where possible, is preferable for equivalent triggering sensitivities under both main terminal polarities.

4.2.3 FALSE TRIGGERING

False triggering occurs when a spurious triggering pulse is coupled to the gate with sufficient amplitude to switch the device on, or if the main terminal (anode–cathode) blocking voltage is exceeded. It does not matter how conduction is initiated: once the device is conducting, it will continue to do so until the forward current is removed. The most usual coupling mechanism (Fig. 4.15) is through the main-terminal-to-gate capacitance when a high rate-of-change of blocked voltage, or a transient surge on the supply line, is present.

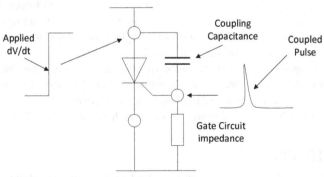

Applied
dV/dt

Coupling
Capacitance

Coupled
Pulse

Gate Circuit
impedance

FIGURE 4.15 False Triggering by Coupled Gate to Pulse.

The current coupled into the gate circuit can be calculated from the standard equation for the current in a capacitance:

$$i = C \times \frac{dv}{dt} \tag{4.7}$$

Although, in practice, you will often have to guess at C. There are two measures that you should take to guard against spurious coupling:

- Prevent high dV/dt, if necessary by snubber circuits (see later). Thyristor and triac data sheets include a maximum dV/dt specification, which should be observed.
- Reduce the gate input impedance by means of a low-value parallel resistor, or even a capacitor. This calls for increased gate drive. So-called "sensitive gate" devices, with low gate drive requirements, are more susceptible to spurious triggering.

Driving the wanted trigger pulse via a pulse transformer directly to the gate with no other gate components is bad practice because the leakage inductance of the transformer can present a high impedance to dV/dt coupled pulses; a low-value parallel resistor is still advisable. Noise coupled into the gate drive circuit can also cause false triggering. This can be a particular problem where the device is remote from its driver. Low gate impedance and/or a capacitor and good wiring layout (see Chapter 1) will reduce susceptibility. Overdrive the gate as far as is practicable (within dissipation limits) for the most reliable triggering.

4.2.4 CONDUCTION

Once the device is triggered it will stay in conduction until the current through it drops to the holding current, I_H. The value of I_H is given in data sheets but is dependent both on temperature and gate impedance. Even for small thyristors this current can be quite high, of the order of several milliamps to tens of milliamps, and it represents a limit on the minimum load that can be switched. Clearly, if in ac applications a lightly loaded device is triggered with a short pulse near the beginning of the half-cycle, the conduction current may not have built up to I_H before the trigger pulse ends (Fig. 4.16). The device will then not conduct for that half-cycle. A longer duration trigger pulse will overcome this, but you may also prefer not to attempt to use the first few tens of degrees of conduction angle, especially since the sine-wave power in this part of the waveform is minimal.

Reverse voltage on the gate increases the I_H value significantly, while forward bias will reduce it since the data sheet values are normally quoted for the gate open. Failure to appreciate this can cause latch and hold problems when thyristors are driven directly from a transistor, whose saturation voltage may reach hundreds of millivolts.

4.2.5 SWITCHING

At turn-on, the rate of change of forward current (di/dt) should be limited. When the device is triggered the conduction region spreads relatively slowly through the

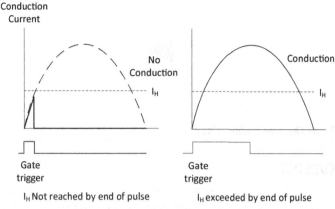

FIGURE 4.16 Effect of Thyristor Holding Current.

silicon. If the turn-on current rises too rapidly then a high current flow is concentrated into a small region near the cathode, causing localized overheating and eventual device destruction. Maximum *di/dt* is sometimes specified and can be met by incorporating a small amount of inductance. This is calculated from a variation of the standard inductance equation, as shown in Eq. (4.8):

$$L = -\frac{V}{di/dt} \qquad (4.8)$$

Looking into the load circuit; often the load itself is inductive enough for this. The permissible level of *di/dt* is strongly influenced by gate drive level and rise time, since a higher level of gate drive will spread the conduction region faster through the device. Higher gate drive also reduces the delay time from application of drive to turn-on, which is typically 1–2 µs.

Turn-Off

A reverse voltage cannot be used to turn off a triac, which conducts in both directions. However, thyristors *can* be turned off by reverse voltage, and their turn-off time has two components, reverse recovery time and forward blocking recovery time. The former has the same mechanism as a reverse-biased diode, i.e., removal of minority carriers from the reverse blocking junction with application of reverse voltage. The longer time constant is associated with forward blocking recovery and is the time required for the charge stored in the forward blocking junction to recombine. The total turn-off time is of the order of tens of microseconds and is increased by increasing junction temperature and on-state current. Negative gate bias will decrease it, as this will speed up the removal of charge from the forward blocking junction.

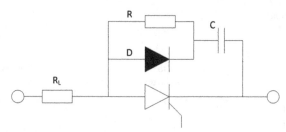

FIGURE 4.17 The Thyristor Snubber Circuit.

4.2.6 SNUBBING

Restrictions on *dV/dt* can be met by connecting a capacitor–resistor–diode network in parallel with the device. This technique is known as "snubbing," and it can apply to any switching circuit, not just to thyristor/triac circuits (earlier examples have been given for switching inductive circuits). The basic circuit is shown in Fig. 4.17.

The rate-of-rise of turn-off voltage is determined by the time constant $R_L \times C$. R_L is the circuit minimum load resistance, for instance the cold resistance of a heater or lamp, the winding resistance of a motor or the primary resistance of a transformer. The resistor R limits the surge current through the device at turn-on due to discharge of C, and the diode D removes the influence of R while the applied voltage is rising. If the calculated value of R is of the same order of magnitude as or less than R_L then the diode can be omitted. In triac circuits when a diode is required, the entire snubbing circuit can be put into a diode rectifier bridge across the triac.

Values for R and C

C is calculated from Eq. (4.9)

$$C = 0.63 \times \frac{V_{\text{peak}}}{\frac{dv}{dt}} \times R_L \tag{4.9}$$

where *dv/dt* is the device maximum specification, and V_{peak} is the maximum voltage to which it will be exposed, e.g., for 240-V phase-control applications V_{peak} will be 340 V (although it would be prudent to allow a higher value if frequent transient spikes are expected on the supply).

R is calculated from either Eq. (4.10) or (4.11)

$$R = \frac{V_{\text{peak}}}{[0.5 \times (I_{TSM} - I_L)]} \tag{4.10}$$

$$R = \sqrt{\frac{V_{\text{peak}}}{C\frac{di}{dt}}} \tag{4.11}$$

whichever is the larger, where I_{TSM} is the device half-cycle surge current rating, I_L is the maximum load current, and *di/dt* is the rate-of-rise of current rating if it is quoted. The factor 0.5 in the first equation is a safety factor. Remember to check the resistor's power rating, allowing for pulse derating. The diode should have the same voltage rating as the triac or thyristor, but the half-cycle surge current rating need only be two or three times I_L as it only conducts for a short period at each turn-off.

A triac is controlling a 1-kW cartridge heater at 240 V. Assume the heater cold resistance is roughly one-tenth its hot resistance, which translates to a R_L of 6 Ω and a peak turn-on current of 56 A. The triac could be a TIC226M, which has a noninductive dv/dt rating of 500 V/μs and an I_{TSM} of 80 A. This gives a capacitance C from Eq. (4.9) of

$$[0.63 \cdot 340/500 \cdot 6] = 0.07 \ \mu F,$$

so use the next highest value of 0.1 μF. R can be calculated from Eq. (4.10) to be

$$[340/(80 - 56) \cdot 0.5] = 28.3 \ \Omega$$

Using a 27-Ω resistor would give a di/dt of 4.7 A/μs. Because R is significantly higher than R_L, use of a parallel diode and putting the snubber in a bridge would be advisable.

Note that the selection of triac can have a large effect on the required values of the snubber components. For instance, a larger device, though unnecessary from the strict applications point of view, would have a larger I_{TSM} and could therefore get away with a lower R, which in turn might obviate the need for a diode. However, this might increase the stress on the capacitor, which would need to be a pulse-rated device in any case. Alternatively, other triacs of similar rating may have an order of magnitude less dv/dt specification, which would need a much larger (and more expensive) capacitor.

4.3 BIPOLAR TRANSISTORS

Much of what has been said about the characteristics of silicon diodes (see Section 4.1) applies also to silicon transistors. Because the underlying mechanisms are the same, the forward and reverse conduction and high-frequency characteristics of the p-n junctions in either bipolar or FETs will be the same as for the straightforward diode. The rest of this chapter will look at further characteristics which are peculiar to each type of device. Four families of transistor will be considered: bipolar, junction field-effect, MOS field-effect, and insulated gate bipolar (Fig. 4.18).

4.3.1 LEAKAGE

Leakage current is as much of a problem in transistors as it is in the diode. It is normally specified in transistor data sheets as I_{CBO}, collector cutoff current (the collector–base current with emitter open circuit). This ensures that the specification ignores collector current due to amplified base current. It is particularly important in

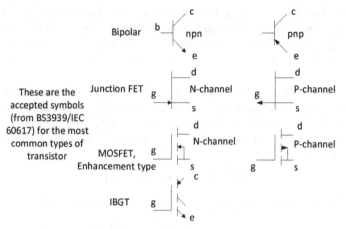

FIGURE 4.18 Transistor Symbols.

dc-coupled amplifier circuits, especially when collector leakage current in one transistor is injected into the base of the next and amplified.

A Simple Leakage Example

Consider the simple two-transistor noninverting buffer shown in Fig. 4.19. This basic configuration is used both in switching circuits and in linear amplifiers. As a digital level shifter it can be used to interface between logic circuits and high-voltage devices such as relays or stepper motors.

In the realization of the circuit shown at Fig. 4.19A TR1's entire collector current flows into TR2's base. This is fine when TR1 is fully conducting; the base current of TR2 is defined by Eq. (4.12):

$$I_B = \frac{(V^+ - V_{BE2} - V_{CAsat1})}{R_1} \tag{4.12}$$

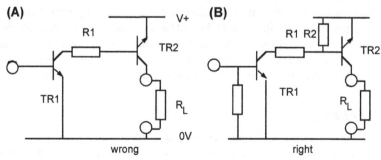

FIGURE 4.19 Biasing a Two Transistor Buffer.

(A) wrong (B) right.

This base current value can easily be set to turn TR2 fully on and apply V^+ to the load. When TR1 is off it is a different matter. Even with its base shorted to ground its collector leakage current is all injected into TR2's base and thence amplified into R_L. At high temperatures the collector leakage can reach several μA, or mA for a high-power device. If TR2 has a gain of a couple of 100, then upward of 1 mA could flow in R_L even when it is supposed to be off.

Worse, many applications will not tie the base of TR1 to ground when in the off-state. An offset of even a few tens of mV will allow a small base current to flow which will be amplified and quickly swamp the collector leakage. If, say, a base–emitter junction passes 1 mA at 600 mV, it will pass 2.5 μA at 100 mV (from the diode equation); if TR1 has a gain of 100 and TR2 a gain of 200, then the current through R_L will be a respectable 50 mA, quite significant for the off-state!

Adding a Base–Emitter Resistor

The simple solution, shown in Fig. 4.19B, is to add a base–emitter resistor to any transistor, which is threatened by leakage currents. The resistor is sized to divert only a modest proportion of the base current (typically one-tenth) when the transistor is being driven on. In the example above, assume that the base current of TR2 is set to 1 mA in the on-state; then taking $V_{BE} = 0.6$ V at this current, R_L is 0.6 V/0.1 mA = 6 kΩ. Now, if TR1 collector leakage reaches 10 μA, this will develop no more than 60 mV across TR2's base–emitter junction. From the diode equation again, only 45 nA of this will be diverted into the base, which will not give any significant off-state leakage into R_L. Similar design can be applied to the resistor across TR1's base, though this depends on knowledge of the output characteristic of whatever drive circuit is used.

4.3.2 SATURATION

Collector–emitter saturation voltage V_{CEsat} is the voltage that remains across the collector and emitter terminals when excess base current is applied to turn the device fully on. It is predominantly ohmic at higher collector currents, depending on the bulk silicon resistance between the terminals, but there is a residual voltage of between 50 and 200 mV even at low collector currents, which cannot be eliminated. V_{CEsat} is normally only specified at one or two values of collector and base current although data sheets will often give a graph of V_{CEsat} versus collector current. Increasing base current will only reduce V_{CEsat} marginally; setting it to more than a tenth of the operating collector current is pointless.

When a common emitter switching transistor drives the base of another transistor, its saturation voltage may be enough, especially if it is operated at high power, to keep the second transistor partially on. The cure for this, as in the previous section, is to divide the voltage at TR2's base with another base–emitter resistor (Fig. 4.20).

The collector saturation voltage becomes significant in very low voltage circuits, when it represents a major fraction of the overall available collector voltage, and in

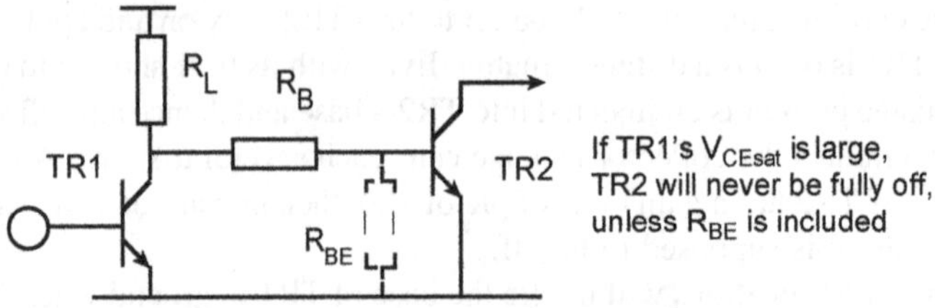

FIGURE 4.20 Minimizing the Effect of Collector Saturation Voltage.

high-current switching circuits, when it represents an appreciable power loss. It can also be a problem when the absolute value of the "on" level in switching circuits must be fixed. The temperature coefficient of V_{CEsat} is fairly low, generally less than 0.5 mV/°C, but it has a complex dependence on collector current and junction temperature.

4.3.3 THE DARLINGTON

High saturation voltage is one particular disadvantage of the Darlington transistor. The Darlington is essentially a transistor pair in a single package configured as in Fig. 4.21 so as to multiply the current gain of each, so that overall gains of over 1000 are readily achievable.

When it is driven into saturation the total voltage across the output transistor TR_B is the sum of TR_A's V_{CEsat} and TR_B's V_{BE}, because TR_A has to supply TR_B's base current. Normal operating V_{CEsat} of the Darlington is around 1 V.

Total base–emitter voltage, of course, is double that of a conventional transistor because the base–emitter junctions are in series. The internal base–emitter resistors are not present in all types of device; they represent a trade-off between high gain, high switching speed, and thermal stability. If there are no base–emitter resistors all the input current for each stage flows into the base, so the current gain is maximized. But this increases susceptibility to thermal variations in leakage and V_{BE}, and it lengthens the switching turn-off time. Devices characterized for power-switching use will normally include comparatively low-value base–emitter resistors.

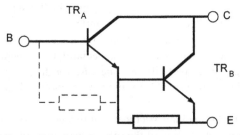

FIGURE 4.21 The Darlington Transistor Configuration.

4.3.4 SAFE OPERATING AREA

The safe operating area (SOA) of a bipolar transistor is determined by four limits:

- maximum collector current
- maximum collector—emitter voltage
- maximum power dissipation
- second breakdown

The first three are always defined in the data for any transistor but the fourth is normally only applied to power transistors. The power dissipation, defined in Eq. (4.13), is specified for a given ambient or package temperature, usually 25°C and must be reduced (derated) at higher temperatures using the quoted thermal resistance figures.

$$\text{Power} = I_c V_c + I_B V_B \qquad (4.13)$$

Use a heat sink (Section 9.5.2) to reduce the thermal resistance to ambient: do not expect a power transistor to achieve its maximum rated dissipation without one! Be careful even with small-signal devices if you are running them near their rated power at high ambient temperatures.

Second Breakdown

Second breakdown is a phenomenon peculiar to bipolar devices, which limits power dissipation further at high collector voltages. It is a thermal effect. If the transistor chip is thought of as a large number of elements in parallel, some of these will have a lower base forward voltage drop than others. Current will tend to concentrate in these, raising their temperature and lowering their voltage drop further. This will concentrate the current more, leading to local overheating and eventually a molten patch of silicon forming a short circuit between collector and emitter. Because it is a localized effect, it is independent of average junction temperature.

Safe Operating Area Curve

These four limits form the boundaries of the SOA for any transistor, and most manufacturers will provide a plot of the form shown in Fig. 4.22 for their power transistors. The second breakdown limit normally intersects the power dissipation limit at some point, although for small-signal devices the locus of second breakdown may lie outside the SOA altogether.

4.3.5 GAIN

Current gain—the ratio between collector (output) and base (input) current—is not an entirely straightforward transistor parameter. It is to be found in data sheets under the heading h_{FE}; note that this refers to dc current gain and is different from h_{FE}, which is small-signal ac current gain. It is normally specified between a minimum and a maximum, and for some devices graded versions are available, which have a tighter specification band. For example, the popular BC848 has a published h_{FE}

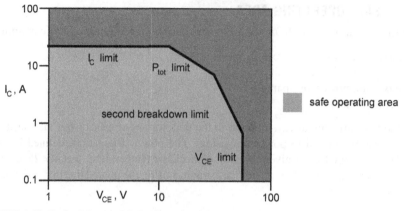

FIGURE 4.22 Typical Curve of Safe Operating Area.

band from 110 to 800 but is available in A, B, and C selections, which have bands from 110–220, 200–450, and 420–800, respectively. In many applications a very wide h_{FE} range is a real design headache, and in some instances—the BC848 is one—there is no cost penalty in selecting a graded device.

Current gain varies with collector current, voltage, and temperature. Most data sheets will present a curve of the form shown in Fig. 4.23 for gain versus current. Each transistor is optimized for a particular range of operating current and gain drops off substantially either side of this range, more so at the higher current end. When considering gain in the circuit design remember to allow for operating current if this is different from that at which the gain is specified. Similarly for collector–emitter voltage; lower voltages than quoted will cause a more dramatic fall-off in gain at the upper end of the operating current range. This is sometimes the cause of waveform distortion, as depicted in Fig. 4.24, due to apparent lack of drive in large-signal transistor amplifiers. There may in fact be enough base drive current

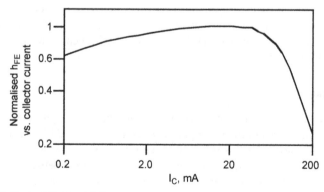

FIGURE 4.23 Typical Curve of Gain Versus Current.

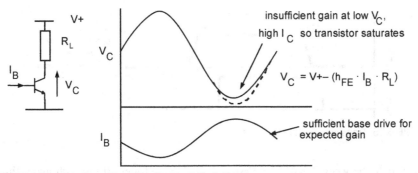

insufficient gain at low V_C,
high I_C so transistor saturates

$V_C = V+- (h_{FE} \cdot I_B \cdot R_L)$

sufficient base drive for
expected gain

FIGURE 4.24 Waveform Distortion Due to Gain Fall-Off.

for the expected gain, but the combination of high collector current and low collector voltage reduces the gain so far that it is insufficient to fully turn on the device.

Higher temperature increases the gain while lower temperature reduces it, by up to a factor of 2–3 for the widest temperature extremes. In fact, transistor gain is such a variable quantity—depending on the individual device, on operating conditions and on temperature—that no design should ever rely on it as a parameter for fixing circuit operation. Instead, seek to minimize the effect of its variations, and also remember that over at least some parts of the circuit's operating envelope, gain will be less than the minimum you find in the data sheet.

4.3.6 SWITCHING AND HIGH-FREQUENCY PERFORMANCE

The section on diodes (Switching Times) has already shown how the p-n junction turn-off differs from turn-on due to the need to sweep the minority carriers out of the junction. The same principle applies to transistor switching. Bipolar transistors have longer turn-off times than turn-on: the actual times can be modified by the transistor's construction. A typical small-signal switching transistor might have a turn-on time of less than 50 ns and a turn-off of 100–200 ns whereas a general-purpose amplifier device can be several times slower, and its data sheet will not include switching time figures. Do not use a general-purpose type (such as the BC84... series) and expect fast switching. The trade-off is between switching speed and gain.

Data sheets generally quote four switching time figures: delay, rise, storage, and fall times. Add the first two to get the turn-on time and the last two to get turn-off. The relationships are shown in Fig. 4.25.

Individual manufacturers may specify their switching times slightly differently. Also, check the switching test circuit: storage and delay times can be reduced by overdriving the base heavily, and the test circuit is usually quite different from the circuit that you will be using, so treat the quoted figures with caution. Turning off the base with a large negative voltage is common in test circuits, and is a good way to speed up the turn-off, but is often difficult to implement with the constraints of a real circuit. You should also bear in mind that reverse V_{BE} breakdown voltage is

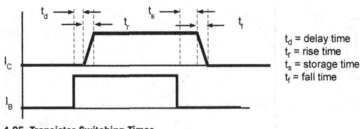

FIGURE 4.25 Transistor Switching Times.

low, usually between 7 and 10 V. Rise, fall, and delay times all fall with increasing collector current.

A major disadvantage of the Darlington compared with conventional bipolars is its low switching speed. This is because its configuration worsens the bipolar's already poor turn-off time. To improve storage time, and hence turn-off, it is necessary either to prevent saturation, or to provide a way to reverse the direction of base current on turn-off; the Darlington can do neither.

Speeding Up the Turn-Off

Transistor switching is conventionally taken to mean between fully off and fully on (saturated). If you prevent the device from saturating then the storage time, which is due to excess base current, is reduced to zero. This can be done in one of two ways: use an emitter-coupled pair as the basic switch, or divert the base current through a base–collector Schottky diode (Fig. 4.26).

The first is the principle behind the emitter-coupled logic IC family, and the second is the principle behind Schottky/low-power Schottky TTL logic (S/LSTTL—now

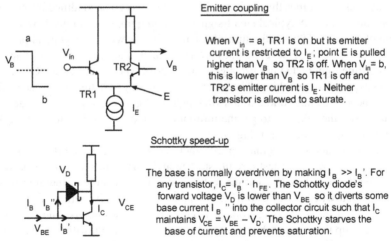

FIGURE 4.26 Speedup Methods for Transistor Switches.

largely of historical interest). The second technique can be applied less effectively with a conventional diode; this circuit arrangement is known as the "Baker clamp." Both techniques give much faster switching for a given collector current.

4.3.7 GRADING

In talking about gain, we have already touched on the subject of transistors selected and marked for a specific range of one particular parameter. In fact, this is the way that most transistors are made. A given transistor die may end up bearing any of a whole variety of different part numbers, depending on how the manufacturer tests and characterizes it. Grading for gain is one aspect; the same applies when testing collector breakdown voltage, so that for example a BC847 and a BC848 can be from the same batch but one breaks down at a higher voltage than the other. Devices can also be graded for noise, so that a BC849 is from the same batch as a BC848 but has passed the noise test. Or, transistors of a quite different part number may be taken from the same batch but have parameters tested to much closer limits than the base type, and these can then be sold at a premium. By this means a transistor manufacturer can maximize the yield from any batch of devices, merely by marking and characterizing them differently.

The other side of this is that cost, and sometimes availability, of a device depends on how tightly it is specified. This is not always the case, as for example in the gain and voltage grades of the BC84... series, which generally have the same cost because they are produced in such high volume that yields of any particular grade can be optimized at will. But the best approach is to use the part with the most relaxed specification that is acceptable—or even none at all if it is for a parameter that does not matter, as for instance may be noise—as this will be the easiest and the cheapest to source.

4.4 JUNCTION FIELD-EFFECT TRANSISTORS

The FET differs from the bipolar device in that current is transported by majority carriers only, whereas in the bipolar carriers of both polarities (majority and minority) are involved. For this reason it was originally known as the "unipolar" transistor. FETs are classified according to the mechanism used by their control terminal as junction FETs (JFETs) or metal-oxide semiconductor FETs (MOSFETs). Fig. 4.27 compares the basic transfer functions of p- and n-channel JFETs and MOSFETs.

The JFET consists of a channel of semiconductor material along which majority carrier current may flow. The current is controlled by a voltage applied to a reverse-biased p-n junction (the gate) formed along the channel. Because the gate is reverse biased, virtually no current flows in it. Therefore, in contrast to the bipolar, which is a low-impedance, current-controlled device, the JFET is a high-impedance, voltage-controlled device. This characteristic is shared with the thermionic valve (vacuum

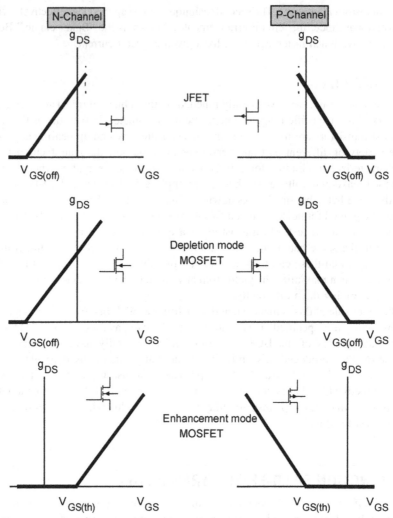

FIGURE 4.27 FET Channel Conductance g_{DS} Versus Gate–Source Voltage V_{GS}.

tube), and venerable circuit designs for one can often be transposed to the other with no change other than operating voltage.

The two ends of the channel are connected to the source and drain terminals. In practice, though the JFET symbol is shown as asymmetrical the channel geometry is usually symmetrical, and it does not matter which terminal is used as the drain and which as the source, since the channel will conduct equally well in either direction. Devices for some applications, notably RF amplifiers, are constructed asymmetrically to optimize interelectrode capacitances and should not be reverse connected.

4.4.1 PINCH-OFF

JFETs work in the depletion mode, which means that current conduction is controlled by depletion of the carriers within some region of the channel. This depletion can be brought about either by increasing the reverse-biased gate voltage or by increasing the drain–source voltage, resulting in a family of output characteristic curves as shown in Fig. 4.28.

There are two distinct regions of operation, depending on applied drain–source voltage. With $V_{GS} = 0$ V, i.e., gate shorted to source, the drain current increases linearly with voltage and the channel acts like a pure resistor, until the pinch-off voltage V_p is reached. At this point the drain current I_D saturates at I_{DSS}, and beyond it current is essentially constant and independent of V_{DS}. As the gate voltage is increased the saturation region extends toward zero and the saturated drain current reduces, until $V_{GS(off)}$, the gate–source cutoff voltage, is reached at which point I_D approaches zero. Because the same physical mechanism is in play, the magnitudes of V_p and $V_{GS(off)}$ are the same, though they are of opposite polarity.

A casual glance through a few JFET data sheets will show that $V_{GS(off)}$ can vary from unit to unit over a very wide range, six-to-one being not uncommon. This is one of the major disadvantages of the device as it greatly complicates bias design, especially for low-voltage applications which cannot afford the extra supply voltage needed to accommodate it.

4.4.2 APPLICATIONS

Use of JFETs incurs a cost penalty. The cheapest general purpose JFET starts at around 10 p in medium quantity, compared with around 2 p for general-purpose

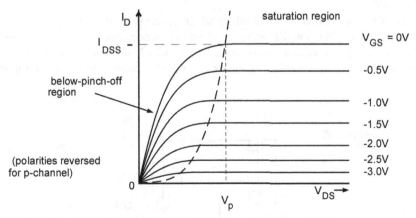

FIGURE 4.28 Typical Family of Output Characteristics for n-channel JFET.

bipolars. Thus JFETs are restricted to applications, which can use their special characteristics. Of these the foremost are as follows:

- analog switches
- RF amplifiers, mixers, and oscillators
- constant current regulators
- high input impedance amplifiers

Analog Switches

The virtue of the FET for use as an analog switch is that its channel is purely resistive. There is no input—output offset voltage, and leakage currents from the control terminal to input or output are very small and often negligible. The input signal can be of either polarity but is limited by the available gate switching voltage and by the gate-channel breakdown rating. Because the JFET operates in depletion mode (Fig. 4.27 explains FET operating modes), the on-state requires that the gate is connected to the source while the off-state requires a gate voltage that is more negative than the source by at least $V_{GS(off)}$ (or more positive, for a p-channel device). This means that the driver supply voltage must be greater by several volts than the expected signal input range. Also, the gate drive circuitry cannot be a straightforward logic output, as it must follow the analog signal during the on-state. A typical switch-plus-driver circuit is as shown in Fig. 4.29.

Because of the complications of the driver circuit it is easier and cheaper to use integrated circuit analog switches, where the FET is integrated along with its driver circuit, than to use a discrete FET and separate driver. This offers the extra benefit that important circuit characteristics such as control signal feedthrough and variation in "on" resistance are already characterized. You should only need to use discrete FET circuits when operating outside the voltage range of readily available IC switches.

RF Circuits

The JFET has been popular as a small-signal RF device (amplifier, mixer, oscillator) for many years. It has low RF noise, good interelectrode capacitance stability and well-defined RF input, and output impedances, which make it easier to design with than bipolars. In addition, its square-law transfer characteristic allows a high

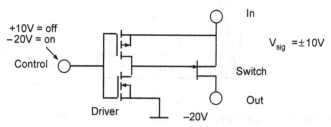

FIGURE 4.29 Analog Switch Configuration.

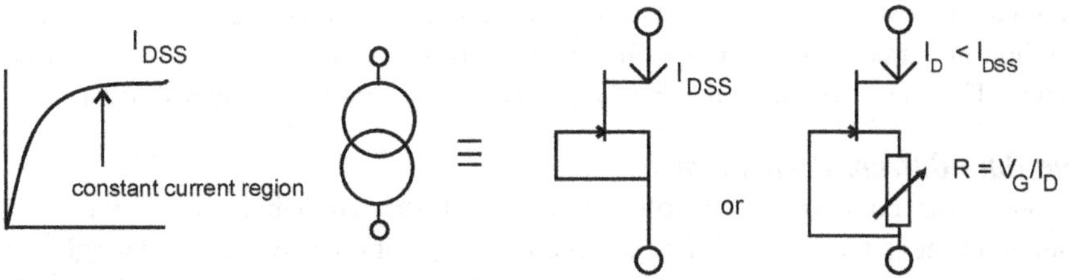

FIGURE 4.30 The Junction Field-Effect Transistor as Current Regulator.

dynamic range. Many devices characterized for HF and VHF use are available at reasonable prices, of the order of 15–20 p for plastic packages.

Current Regulators

A frequent niche application for JFETs is as a one-component current regulator. If the device is connected with gate shorted to source (Fig. 4.30) then once V_{DS} exceeds the pinch-off voltage the drain current will limit to I_{DSS} and remain constant over a wide range of applied voltage. The current can be adjusted downward from I_{DSS} by including a resistor in series with the source, which effectively gives a constant gate bias.

The circuit cannot be used as a precision current source/sink because its output impedance is on the low side for such applications, and the control current is temperature dependent, though like the zener there is a zero temperature coefficient crossover point. Also, the wide variation in pinch-off voltage and I_{DSS} between devices results in a similarly wide variation of current. Even so, it can be useful where the absolute value of current is unimportant, such as in amplifier biasing. It is possible to obtain "current regulator diodes" which are gate–source-shorted JFETs specially characterized for the purpose and available in reasonably tightly selected current bands from 0.2 to 5 mA, though these are relatively expensive, being around 50 p for plastic-packaged types.

4.4.3 HIGH-IMPEDANCE CIRCUITS

The JFET is very useful for the design of high-input impedance amplifiers. Because the gate under normal operating conditions is a reverse-biased junction, the low-frequency input impedance of a JFET front end is limited only by gate leakage and by the resistance of any bias resistor that may be necessary. Gate leakage currents of a few pA at room temperature are readily achievable, although making use of them is another matter—leakage currents due to stray paths across printed circuit boards and connectors are usually the limiting factor. The JFET does not always live up to its promise of high impedance, though.

Firstly, because the current is due to reverse-biased junction leakage, it increases exponentially with temperature in the same way as was seen for the ordinary silicon diode junction. Thus at the maximum commercial temperature limit of 70°C the leakage current is more than 20 times that at room temperature, though it is still better than that of most other devices. At 125°C, the maximum military temperature limit, it is a thousand times worse, and a well-designed bipolar input will have a better performance. This is one reason why JFET input circuits are rare in military designs.

The Gate Current Breakpoint

Secondly the common-mode voltage range is restricted. Obviously if the input voltage exceeds the supply rails then incorrect operation may occur—though you might want to exploit the region between zero V_{GS} and $V_{GS(off)}$ to exceed the supply voltage in one direction or the other. A more subtle mechanism also affects input impedance, which is that gate leakage current is critically dependent on drain–gate voltage and drain current. Gate leakage current is normally characterized as I_{GSS}, i.e., gate current with drain shorted to source. In an operational circuit, an n-channel JFET behaves differently.

n-channel JFETs experience an I_G "breakpoint" above which the gate current rises rapidly with increasing drain–gate voltage. This results from a phenomenon similar to avalanche breakdown but modified by the availability of carriers in the channel due to drain current. These carriers are accelerated by the drain–gate electric field and contribute a leakage current through the gate because of their ionization on impact with the silicon. Consequently, the breakpoint value depends on I_D and is between a third and a half of drain–gate breakdown voltage, as can be seen from Fig. 4.31.

Depressed Z_{in}

The effect of this is that the input impedance of a JFET amplifier will be lower, perhaps by several orders of magnitude, than that calculated from the I_{GSS} figure once the I_G breakpoint is exceeded. This places a limit on the effective common-mode input voltage range. Note that p-channel JFETs, because of their lower carrier mobility, do not exhibit the effect to anything like the same degree.

To overcome the problem the drain–gate bias voltage must be held below the breakpoint. A reduction in drain current will shift the breakpoint knee higher, but the improvement is only marginal. In simple source-follower circuits it may be possible to reduce the drain bias voltage sufficiently as in Fig. 4.32A, but if the full operating range of the FET is required, then a better solution is to add a second FET in cascode with the first (Fig. 4.32B). The drain–gate voltage of the input FET is maintained at a constant low voltage over the whole input range by the source of the cascode FET. The latter is still subject to operation above the breakpoint at the negative end of the input voltage range, but the excess gate current is now added to the operating current with negligible effect.

4.5 MOSFETs

The distinguishing feature of the MOSFET, or metal-oxide semiconductor FET, is that its gate is insulated from the semiconducting channel by a thin layer of oxide. It is not a p-n junction so, unlike the JFET, no gate leakage current flows.

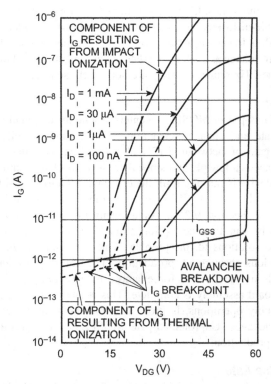

FIGURE 4.31 Plot of Gate Current Versus Drain–Gate Voltage for a Typical Junction Field-Effect Transistor.

Figure courtesy of Siliconix Incorporated.

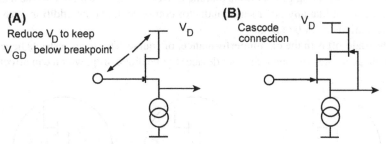

FIGURE 4.32 Reducing the Effect of Gate Current Breakpoint.

4.5.1 LOW-POWER MOSFETs

Low-power MOSFETs have existed for some time, aimed at niche applications similar to JFETs, particularly high-input impedance amplifiers, RF circuits, and analog switches. These devices are planar fabricated, with the source and drain terminals diffused into one side of the substrate. Development of the planar MOS-FET into the double-diffused (DMOS) structure has allowed smaller channel dimensions, which result in lower capacitances and faster switching times. DMOS devices have useable performance as analog switches in the subnanosecond range, and as RF amplifiers into the GHz range, but they are not widely sourced.

Gate Breakdown

The drawback of low-power MOSFETs centers around the extremely thin insulating layer between gate and channel. Generally manufacturers specify a breakdown voltage for the gate of between ±15 and ±40 V. At the same time the gate-channel capacitance is a few pF, and since there is no discharge path for this capacitance when the device is out of circuit the charge needed to exceed the breakdown voltage is very small (around a 100 pico-coulombs) and can easily be generated electrostatically (cf Section 9.2.2). As a result low-power MOSFETs can be destroyed simply by handling them prior to insertion into the circuit.

There are two solutions to this problem. One is to impose rigorous antistatic handling precautions throughout production. The safest procedure is to short together all leads of the device, for example by wrapping a wire link around them, until the board assembly is complete. This is labor-intensive compared with most other assembly operations, is prone to error and is also incompatible with surface-mount packages. Unprotected-gate MOSFETs are therefore not popular for high volume production.

Protection for the Gate

The other solution is to prevent the gate voltage from reaching the breakdown level by incorporating extra protection, in the form of a zener diode rated to just below the breakdown voltage (Fig. 4.33), within the package of the device. Manufacturers may offer versions of the same device with and without gate zener protection. The great advantage of including protection is that the device can now be treated in the assembly process just like any other semiconductor component, and no additional assembly precautions need be taken.

The trade-off is in the circuit performance, of course. The zener restricts negative gate voltage swings (for an n-channel device) to one diode drop, which can affect the

No Protection   Gate-source protection   Gate-body protection

FIGURE 4.33 MOSFET Gate Protection.

design of the gate drive circuit. More importantly, it adds a component of gate leakage current, which exhibits all the properties of diode leakage outlined earlier. In some applications this can nullify the advantage of using a MOSFET.

MOSFET Trade-Offs

The choice facing designers who want to use low-power MOSFETs is fairly simple. Either specify a gate-protected device and accept the performance limitations that this implies, but make yourself popular with the production department. Or, insist on using unprotected devices because you need ultrahigh impedance, and impose extra costs in production, along with possible reliability penalties if the production procedures are not rigorously enforced.

Remember also that the susceptibility of low-power MOSFETs to gate breakdown may not end after they are soldered into circuit. Especially if they are used in high-impedance input circuits, the gate can still be vulnerable if its biasing resistor is in the high MΩ range (or worse still, absent) and stringent handling precautions may need to be observed at the board or equipment level.

4.5.2 VERTICAL METAL-OXIDE SEMICONDUCTOR POWER FIELD-EFFECT TRANSISTORS

The constraint of channel on-resistance (typically tens to hundreds of ohms) and consequent restriction on power handling of the conventional planar MOSFET was removed with the development of the double-diffused vertical MOSFET or VMOS. This range of devices, introduced by International Rectifier in the late 1970s under the HEXFET trademark and subsequently widely second-sourced, can achieve "on" resistances in the milliohm region, and drain–source breakdown voltages up to 500 V and higher.

The basic physical structure of a vertical power MOSFET device is shown in Fig. 4.34. The n-drain drift region is where the main on resistance occurs. This

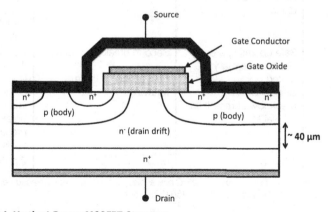

FIGURE 4.34 Vertical Power MOSFET Structure.

region need to be large enough to support a breakdown voltage of ~600 V—hence the dimension shown of about 40 μm.

It is therefore a direct competitor to the power bipolar transistor and outperforms it in many respects. Table 4.2 compares the major differences between these two and their hybrid offspring, the IGBT (see Section 4.6). The necessary trade-offs in switching speed, gain, and power handling that bipolars impose are easier to handle or even absent with MOSFETS.

Table 4.2 Comparison of Power MOSFET, Bipolar, and Insulated Gate Bipolar Transistor (IGBT) Characteristics

	Power MOSFET	Power Bipolar	IGBT
Max voltage rating	Up to 1 kV	Up to 1 kV	Up to 1200 V
Max current rating	Up to 100 A	Up to 500 A	Up to 500 A
Switching speeds	Typically better than 100 ns, independent of temperature	0.3–5 μs typical	0.2–1 μs typical, switching losses rise with temperature
Input characteristics	Voltage, 3–10 V V_{GSth} Gain tempco −0.2%/°C	Current, 20–100 h_{FE} hFE tempco +0.8%/°C	Voltage, 5–8 V V_{GEth} tempco −11 mV/°C
Output characteristics	Resistive, current sharing when paralleled, R_{DSon} tempco 0.7%/°C Inherent body-drain diode	Nonresistive, current hogging when paralleled V_{CEsat} tempco −0.25%/°C No inherent output diode	Hybrid, V_{CEsat} tempco positive at high currents, negative at low currents No inherent output diode
Breakdown	Safe operating area (SOA) is thermally limited Static-induced breakdown precautions advisable	Second breakdown at high V_{CE} Susceptible to thermal runaway	SOA is thermally limited Static-induced breakdown precautions advisable
Unit cost[a]	1.5 p/W	0.75–1 p/W	2 p/W

[a] Average cost per watt of quoted maximum 25°C power dissipation for plastic packaged devices, npn or n-channel, 100 or 600 V voltage rating, 30–150 W power rating.

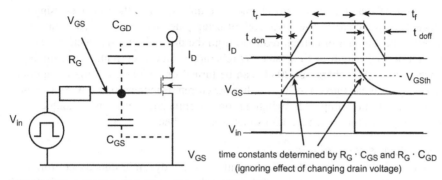

FIGURE 4.35 Gate Switching Capacitance and Waveforms.

The VMOS's features versus bipolar for power switching applications are generally claimed to be the following:

- voltage drive rather than current drive: much less drive power needed, simplified drive circuitry
- increased switching speeds: high efficiency at high frequencies
- resistive output characteristic: good for paralleling devices, although capability is limited due to dissipation at higher currents

Each of these although real, also brings with it disadvantages, and requires some effort to properly realize in practice.

4.5.3 GATE DRIVE IMPEDANCE

Because the VMOS is a majority carrier device, it does not have any of the minority-carrier switching delays associated with bipolars, and its switching speed is governed by how fast the gate can be driven. Since the gate impedance is almost entirely capacitive, the resulting turn-on/turn-off delays depend on the output impedance of the driver circuit (Fig. 4.35).

Thus, although it is possible to drive a VMOS switch directly from a CMOS logic gate, doing so will compromise the achievable switching time. For instance, a typical 74HC-series gate at 5 V has an output drive capability of 4 mA; a typical medium-size VMOS may have an input capacitance of 200 pF. If the switching voltage threshold of the VMOS is 3 V, then the time taken to reach this level is $C \times V/I = 150$ ns. Note that many VMOS devices are characterized for R_{DSon} at $V_{GS} = 10$ V and should not be driven directly from 5 V logic. Worst-case combinations of supply voltage, output level, and VMOS threshold will lead to the VMOS being driven on the knee of its "on" characteristic, with unpredictable and high values of R_{DSon}. Families of "logic-compatible" VMOS are available, characterized at $V_{GS} = 5$ V.

However, this is by no means the end of the story; the actual switching waveforms are complicated by feedback of the drain waveform through the drain–gate (Miller) capacitance C_{GD}. This also has to be charged by the gate drive circuit, along with C_{GS}. Charge time for the Miller capacitance is larger than that for the gate to source capacitance C_{GS} due to the rapidly changing drain voltage during turn-on.

MOSFET manufacturers now characterize this aspect of their devices using the concept of gate charge. The advantage of using gate charge is that you can easily calculate the amount of current required from the drive circuit to switch the device on in a desired length of time, since Q (charge) = time × current. For example, a device with a gate charge of 20 nC can be turned on in 20 μs if 1 ma is supplied to the gate or it can turn on in 20 ns if the gate current is increased to 1 A. So keeping R_G low and I_G high improves switching times significantly. Various techniques can be employed to this end; Fig. 4.36 shows some of them.

Gate–Source Overvoltage

Besides switching times, other factors call for a low dynamic impedance from the gate drive circuit. Excessive V_{GS} will punch through the gate–source oxide layer and cause permanent damage. Transient gate–source overvoltages can be generated by large drain voltage spikes (caused, for instance, by another device connected to the drain, or by induced transients) coupled through the drain–gate capacitance. If the dynamic drive circuit impedance is high—as might be the case if the gate is

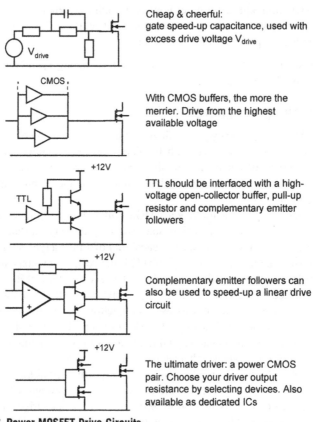

Cheap & cheerful: gate speed-up capacitance, used with excess drive voltage V_{drive}

With CMOS buffers, the more the merrier. Drive from the highest available voltage

TTL should be interfaced with a high-voltage open-collector buffer, pull-up resistor and complementary emitter followers

Complementary emitter followers can also be used to speed-up a linear drive circuit

The ultimate driver: a power CMOS pair. Choose your driver output resistance by selecting devices. Also available as dedicated ICs

FIGURE 4.36 Power MOSFET Drive Circuits.

driven from a pulse transformer—the transient amplitude will only be limited by the potential divider effect of C_{GD} and C_{GS}. A typical ratio of these values is 1:6, so that a 300-V drain transient would be reflected as a 50-V gate transient, which is quite enough to destroy the gate.

The simple precaution if such transients are anticipated (especially if the drain is connected to an external circuit) is to specify a device with an integral zener gate protection diode, or to incorporate one in the external circuit near to the gate—source terminals. VMOS power devices, because of their higher gate capacitance, are inherently less susceptible to handling-induced electrostatic damage than are low-power MOSFETs, but it is nevertheless prudent to specify gate-protected parts if they are available. Otherwise, ensure that the source impedance of the drive circuit is low enough to absorb induced gate transients without them reaching damaging levels (Fig. 4.36).

Source Lead Inductance

The high output di/dt associated with turn-on and turn-off causes a further potential problem for the gate drive. It passes through the source terminal and creates a transient voltage across the inductance of the source lead, which unless the layout is carefully controlled will also appear in series with the gate drive voltage. This acts to reduce V_{GS} and so degrade switching speeds. To minimize this effect, the gate drive circuit should be kept physically separate from the source high-current return right up to the device terminals, so that there is no common impedance between the two (Fig. 4.37).

4.5.4 SWITCHING SPEED

High switching speeds, while desirable for minimizing power losses, have two highly undesirable side effects. They generate significant EMI, and they also generate drain—source overvoltage spikes. EMI is primarily controlled by layout, filtering,

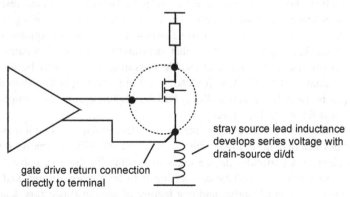

stray source lead inductance develops series voltage with drain-source di/dt

gate drive return connection directly to terminal

FIGURE 4.37 Preventing Source Lead Inductance From Affecting V_G.

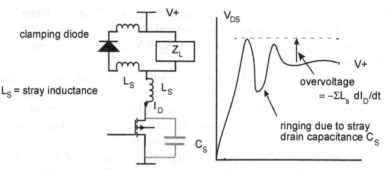

FIGURE 4.38 Drain–Source Turn-Off Transient.

and screening and is the subject of Chapter 8. Here, it is merely noted as a factor opposing the trend to faster switching.

Drain–source overvoltages are generated when current flowing through an inductive load is switched off quickly (cf Section 3.4.4). Even when the main load is noninductive or is clamped, stray inductance can cause significant transients at the current levels and switching speeds offered by VMOS devices (Fig. 4.38). For instance, 20 A being switched across 0.5 μH in 50 ns will create a spike of 200 V. Additionally, the forward recovery of the clamping diode may be insufficient to catch the leading edge of the main inductive transient. As well as ensuring that all connections in the high-*di/dt* area are as short as possible for minimum inductance, a drain–source clamping zener or snubber local to the device is a useful precaution.

4.5.5 ON-STATE RESISTANCE

Finally, it is also wise to remember when running a VMOS device in the on-state at high power that dissipation is thermally limited. This means that I^2R must be restricted to maintain the junction temperature T_j below the maximum permitted in the device data sheet, usually 150°C. The section on thermal management (Section 9.5) outlines how to relate these two factors. However, on-state resistance, R_{DSon}, has a positive temperature coefficient and is normally quoted at a given temperature and V_{GS}. At maximum T_j it will be around 1.8–2 times the quoted value at 25°C so, if the heat sink is sized so that maximum expected dissipation gives maximum permitted temperature, allowable operating current will be about 0.7 times that available at 25°C. Also if steady-state applied V_{GS} is less than that for which R_{DSon} is quoted, the actual R_{DSon} will be higher again. It is dangerous to take data sheet figures at face value.

The advantage of a positive temperature coefficient of R_{DSon} is that it makes paralleling of devices to achieve higher current much easier. With a negative temperature coefficient, if a particular device takes more than its share of current it gets hotter, and this reduces its resistance, causing it to take more current, and so on. This is known as current hogging and is a feature of bipolar transistors, leading to thermal runaway if not dealt with, usually by including a separate emitter resistor for each transistor. The VMOS positive temperature coefficient on the other hand encourages current sharing and prevents thermal runaway.

P-Channel Vertical Metal-Oxide Semiconductor

Although the foregoing discussion has focussed on n-channel VMOS FETs, the same considerations apply to p-channel devices. Many manufacturers offer "complementary" p-channel parts for their more popular n-channel ranges. Because of the differing resistivity of the two base materials there is no such thing as a truly complementary pair. p-type silicon has a much higher resistivity than n-type, so the p-channel device requires a larger active area to achieve the same on-resistance. This reflects in the cost: a p-channel part of the same R_{DSon} and voltage rating will be more expensive than its complementary n-type. Gate threshold voltage, transconductance, and capacitances can be nearly equalized. However, thermal resistance, pulsed and continuous current rating, and SOA are all higher for the p-channel as a result of its larger die. This has the happy consequence that whenever matched operation is needed, the p-channel will have greater operating margins with respect to its thermal ratings.

4.6 INSULATED GATE BIPOLAR TRANSISTORS

A workhorse of the power semiconductor stable is the IGBT. In power electronics there is a constant demand for compact, lightweight, and efficient power supplies, but these demands are not fully satisfied by power bipolars and power MOSFETs. High-current and high-voltage bipolars are available, but their switching speeds are poor. Power MOSFETs have high-speed switching, but high voltage and high current modules are hard to achieve.

The IGBT is a power semiconductor device introduced to overcome these limitations. It reduces the high on-state losses of the MOSFET while maintaining its simple gate drive requirements. It is controlled by the gate voltage as is the power MOSFET, but the output current characteristic is that of a bipolar transistor. These devices combine some desirable features of both the bipolar transistor and of the MOSFET.

4.6.1 INSULATED GATE BIPOLAR TRANSISTOR STRUCTURE

The IGBT combines an insulated gate n-channel MOSFET input with a bipolar pnp output (Fig. 4.39). When comparing the two, the MOSFET and IGBT structures look very similar. Both devices share a similar gate structure and p wells with n+ source contacts. In both devices the n-type material under the p wells is sized in thickness and resistivity to sustain the full voltage rating of the device. The basic difference is the addition of a p+ substrate beneath the n substrate, which creates the output transistor.

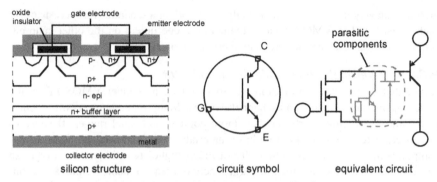

oxide insulator   gate electrode   emitter electrode

p- n+ n+
p+
n- epi
n+ buffer layer
p+
metal

collector electrode

silicon structure

parasitic components

C

G

E

circuit symbol

equivalent circuit

FIGURE 4.39 Insulated Gate Bipolar Transistor Structure and Circuit.

In a power MOSFET most of the conduction losses occur in the n-region, typically 70% in a 500 V device. As shown in the equivalent circuit of Fig. 4.39, the IGBT consists of a pnp driven by an n-channel MOSFET in a pseudo-Darlington configuration. The voltage drop across the IGBT is the sum of two components: a diode drop across the p—n junction and the voltage drop across the driving MOSFET. Thus, unlike the power MOSFET, the on-state voltage drop across an IGBT never goes below a diode threshold. On the other hand, the MOSFET section does not carry the bulk of the output current, the pnp output does; and therefore the output voltage drop at high currents can be simplistically viewed as the voltage drop that would occur across the MOSFET at full current, divided by the gain of the pnp. This is referred to as "conductivity modulation."

The base region of the pnp is not brought out and the emitter—base p-n junction, spanning the entire wafer, cannot be terminated or passivated. This influences the turn-off and reverse blocking behavior of the IGBT and is the cause of its principal disadvantage as discussed shortly.

4.6.2 ADVANTAGES OVER MOSFETS AND BIPOLARS

Compared with the bipolar, the IGBT enjoys the following advantages:

- ease of gate drive, comparable to the MOSFET
- less sensitive to second breakdown limitations
- collector—emitter saturation voltage has a positive temperature coefficient at high currents, making it suitable for paralleling as with the MOSFET.

This has left the power bipolar inhabiting a fairly nondescript, low performance area of application, where only its low cost has saved it from extinction.

And compared with the MOSFET, the IGBT's principal advantage is its lower on-state loss at high currents, since the conductivity modulation of the n-layer greatly increases the current-handling capability of the IGBT for a given die size. This conductivity modulation has been shown to increase the forward conducting current density up to 20 times that of an equivalent MOSFET and 5 times that of a bipolar. Furthermore, as the MOSFET's voltage rating goes up, its inherent reverse diode displays increasing reverse recovery times, which leads to increasing switching losses. The IGBT has no inherent reverse diode, so that an appropriate external one may be selected for the application.

The bipolar output structure also gives the IGBT the edge in respect of maximum output voltage rating. Combined with its suitability for paralleling, this makes it the device of choice for high-voltage, high-current modules. The IGBT technology is certainly the most suitable for breakdown voltages above 1000 V, while the MOSFET is more suitable for device breakdown voltages below 250 V. In between—and of course there is a huge range of applications particularly in the 400–600 V region—the trade-off between the two is more complicated, with MOSFETs tending to be preferred for high-frequency applications and IGBTs for low frequency.

4.6.3 DISADVANTAGES

The Achilles heel of the IGBT is its turn-off characteristic. The biggest limitation to the turn-off speed of an IGBT is the lifetime of the minority carriers in the n-region, which forms the base of the pnp. Since this base is not accessible, external drive circuitry cannot be used to improve the switching time. The charges stored in the base cause the characteristic "tail" in the current waveform of an IGBT at turn-off (Fig. 4.40). When the gate voltage is removed, this excess of minority carriers must be removed before the device will stop conducting completely. As the

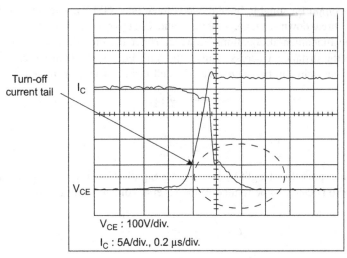

FIGURE 4.40 Turn-Off Waveform of a Typical Insulated Gate Bipolar Transistor, Showing the Tail Current.

Reproduced from International Rectifier published data.

MOSFET channel turns off, electron current ceases, and the output current drops rapidly to a lower value at the start of the tail. The following turn-off speed of the device is determined by the integral bipolar open-base charge decay. This slow switching contributes to a high loss during switch-off, bearing in mind that the loss is the product of V_{CE} and I_C, and hence places a limit on the switching frequency. Even worse, the current tail increases in duration and magnitude at higher temperatures.

The turn-on time is very fast and is determined by the rate of on-voltage saturation of the integral pnp bipolar transistor. IGBT manufacturers are naturally keen to extend the upper frequency limit by reducing the tail, but this can only be done by compromising other parameters, particularly the gain of the pnp transistor and hence the conduction loss in the device.

4.7 WIDE BAND GAP DEVICES

In recent years, the use of wide band gap materials such as silicon carbide (SiC) and gallium nitride (GaN) has become of interest for power semiconductor devices due to their improved conductivity, increased frequency range, and wider thermal range. The term "wide band gap" comes from the basic characteristic of semiconductors where the jump from the valence band to the conduction band is defined in terms of electron volts (eV). For silicon this is about 1.11 eV, however, for GaN devices this is 3.4, and for SiC devices this can be between 2.36 and 3.23.

In a conventional power device such as the vertical power MOSFET as we have seen in Fig. 4.34 in this chapter, the length of the n-drift region needs to be large enough to support a suitable breakdown voltage (say, 600 V). The relationship between the breakdown voltage, mobility, and effectively r_{ds}(on) can be obtained from some fundamental equations. If we consider the electric field gradient with respect to the distance x across a junction using Eq. (4.14), then we can say that the maximum field can be achieved as defined in Eq. (4.15) for the length L.

$$\frac{dE}{dx} = \frac{qN_d}{\varepsilon_{si}} \tag{4.14}$$

$$E_{max} = \frac{qN_dL}{\varepsilon_{si}} \tag{4.15}$$

We can express this length by rearranging Eq. (4.15) into the form shown in Eq. (4.16), and this can also be related to the reverse voltage using Eq. (4.17).

$$L = \frac{E_{max}\varepsilon_{si}}{qN_d} = \sqrt{\frac{2\varepsilon_{si}V_r}{qN_d}} \tag{4.16}$$

$$V_{br} = \frac{\varepsilon_{si}E_c^2}{2qN_d} \tag{4.17}$$

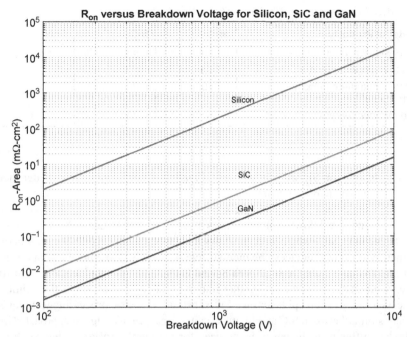

FIGURE 4.41 R_{ds}(on) Resistance Versus Breakdown Voltage for Si, SiC, and GaN Technologies.

This final equation is important as it relates the breakdown voltage to the doping concentration. In the off-state of the device, the electric field must be kept below the critical field E_c and to guarantee a particular breakdown voltage, the depletion region must be as thick as the ratio of the breakdown voltage over E_c. To achieve a high breakdown voltage, Eqs. (4.16) and (4.17) indicate that we need a thick drift region, with low doping concentration.

In the on-state, the resistance will vary approximately as the square of the breakdown voltage, however, practically, E_c will also vary with the level of doping and therefore the resistance will vary at slightly more than the square of the breakdown voltage. These two parameters (breakdown voltage and on resistance) dominate the behavior of the device form a practical perspective, and we can plot the effective limits of performance for a material—also known as the "Silicon Limit." We can also look at alternative materials such as silicon carbide and gallium nitride to observe the differences between them, and this is shown in Fig. 4.41.

As we can see, the different values for critical field and doping concentrations provide significant potential benefits for both silicon carbide and GaN devices over silicon.

As can be seen from Table 4.3, the critical field is an order of magnitude higher for both silicon carbide and GaN material over silicon. The electron saturation velocity is about twice for both SiC and GaN as that for silicon. GaN electron mobility

Table 4.3 Comparison of Electrical Properties

Material Property	Si	SiC	GaN
Critical field (E_c) ($\times 10^7$ V/m)	3	30	35
Electron mobility (cm^2/Vs)	1450	900	2000
Electron saturation velocity ($\times 10^6$ cm/s)	10	22	25
Thermal; conductivity (W/cm^{2}K)	1.5	5	1.3

is a little better than Si, SiC a little worse. Finally, SiC has significantly better thermal conductivity than both GaN and Si.

Taking these parameters into account, the equivalent graph in Fig. 4.41 of on resistance and breakdown voltage shows that in principle GaN should perform even better than SiC, which in turn shows a significant improvement over silicon. In practice, for power devices, the thermal conductivity is also crucial and in this regard SiC has an advantage over both Si and GaN, making it an ideal platform for power semiconductor devices. Another important advantage of SiC is that while it is necessary to grow an epi layer on a SiC substrate, the basic processing techniques are very similar between SiC and Si, making it a conceptually aligned approach. One of the key differences, however, is the need for very high processing temperatures (1500–1700°C).

Typical devices are available from manufacturers such as ROHMor Wolfspeed (now part of Infineon) with voltage ratings of 600, 1200, and even 1700 V, with devices rated to 10 s of A and on resistances of the order of a few 10 s of mΩ. The cost of SiC devices is still relatively high, perhaps an order of magnitude higher that Si devices of a similar current rating, however, usually of higher voltage rating than Si equivalents and lower on resistance.

The device manufacturers are beginning to develop mass market products using SiC and despite the intrinsically higher wafer costs, the prices will continue to drop as more products enter the market, and a wider range of manufacturers develop competitive devices.

An interesting challenge for designers wishing to use SiC or GaN devices in the future will be to develop effective gate drive circuits that will be designed to manage the much higher switching frequencies that are possible with both SiC and certainly with GaN devices. These make passive components smaller, however, the driver circuits will need to be carefully designed to ensure that gate voltages do not have excessive ringing due to inductive effects at these higher switching frequencies. Another problem for the designer is that the gate drive voltages for SiC devices need to be 20 V at the moment, and this means the need for custom-designed gate drivers, as current 12–15 V MOSFET drive circuits will not be adequate for SiC.

Analog Integrated Circuits

5

5.1 The Ideal Op-Amp **211**
   5.1.1 Applications Categories 212
5.2 The Practical Op-Amp **212**
   5.2.1 Offset Voltage 212
      Output Saturation Due to Amplified Offset 213
      Reducing the Effect of Offset 214
      Offset Drift 215
      Circuit Techniques to Remove the Effect of Drift 215
   5.2.2 Bias and Offset Currents 216
      Bias Current Levels 216
      Output Offsets Due to Bias and Offset Currents 216
   5.2.3 Common-Mode Effects 217
      Common-Mode Rejection Ratio 218
      Power Supply Rejection Ratio 218
   5.2.4 Input Voltage Range 219
      Absolute Maximum Input 219
   5.2.5 Output Parameters 219
      Power Rail Voltage 219
      Load Impedance 220
   5.2.6 AC Parameters 222
   5.2.7 Slew Rate and Large-Signal Bandwidth 222
      Slew Rate 223
      Large-Signal Bandwidth 223
      Slewing Distortion 223
   5.2.8 Small-Signal Bandwidth 224
   5.2.9 Settling Time 225
   5.2.10 The Oscillating Amplifier 225
      Ground Coupling 226
      Power Supply Coupling 227
      Output-Stage Instability 227
      Stray Capacitance at the Input 227
      Parasitic Feedback 228

The Circuit Designer's Companion. http://dx.doi.org/10.1016/B978-0-08-101764-7.00005-0

209

The operational amplifier is the basic building block for analog circuits, and progress in op-amp performance is the "litmus test" for linear IC technology in much the same way as progress in memory devices is for digital technology. This chapter will be devoted to op-amps and comparators, with a tailpiece on voltage references. This is not to deny the enormous and ever-widening range of other analog functions that are available, but these are intended for specific niche applications, and little can be generalized about them.

Volumes have already been written about op-amp theory and circuit design and these aspects will not be repeated here. Rather, we shall take a look at the departures from the ideal op-amp parameters that are found in practical devices and survey the trade-offs—including cost and availability, as well as technical factors—that have to be made in real designs. Some instances of anomalous behavior will also be examined.

5.1 THE IDEAL OP-AMP

The following set of characteristics (in no particular order, since they are all unattainable) defines the ideal voltage gain block:

- infinite input impedance, no bias current
- zero output impedance
- arbitrarily large input and output voltage range
- arbitrarily small supply current and/or voltage
- infinite operating bandwidth
- infinite open-loop gain
- zero input offset voltage and current
- zero noise contribution
- absolute insensitivity to temperature, power rail, and common-mode input fluctuations
- zero cost
- off-the-shelf availability in any package
- compatibility between different manufacturers
- perfect reliability

Since none of these features is achievable, you have to select a practical op-amp from the multitude of imperfect types on the market to suit a given application. Some basic examples of trade-offs are as follows:

- A high-frequency ac amplifier will need maximum gain-bandwidth product (GBW) but will not be interested in bias current or offset voltage.

Table 5.1 Parameters for Applications Categories

Category	GBW MHz	Slew Rate V/μs	V_{OS} mV	I_{CC} mA	V_{OS} Drift μV/°C	Noise nV/√Hz	Gain/ Phase Error%
General purpose	1–30	0.5–40	0.5 –20				
Low power	0.05–5	0.03–3	0.5 –20	0.015 –1			
Precision		0.3–10	0.06 –0.5		0.5–4	3–30	
High speed and video	30 –1000	100 –5000	1–25	3–15			0.01 –0.3

- A battery-powered circuit will want the best of all the parameters but at minimum supply current and voltage.
- A consumer design will need to minimize the cost at the expense of technical performance.
- A precision instrumentation amplifier will need minimum input offsets and noise but can sacrifice speed and cheapness.

Device data sheets contain some but not all of the necessary information to make these trade-offs (most crucially, they say nothing about cost and availability, which you must get from the distributor). The functional characteristics often need some interpretation and critical parameters can be hidden or even absent. In general, if a particular parameter you are interested in is not given in the data sheet, it is safest to assume a pessimistic figure. It means that the manufacturer is not prepared to test his devices for that parameter or to certify a minimum or maximum value.

5.1.1 APPLICATIONS CATEGORIES

In fact, although there is a bewildering variety of devices available, op-amps are divided into a few broad categories based on their application, in which the above trade-offs are altered in different directions. Table 5.1 suggests a reasonable range over which you might expect to find a spread of certain critical parameters for op-amps in each category.

5.2 THE PRACTICAL OP-AMP
5.2.1 OFFSET VOLTAGE

Input offset voltage V_{OS} can be defined as that differential dc voltage required between the inverting and noninverting inputs of an amplifier to drive its output to zero. In the perfect amplifier, 0 V in will give 0 V out; practical devices will show offsets ranging from tens of millivolts down to a few microvolts. The offset appears as an error voltage in series with the actual input voltage. Definitions vary, but a "precision" op-amp is usually considered to be one that has a V_{OS} of less than 200 μV and a V_{OS} temperature coefficient (see later) of less than 2 μV/°C. Bipolar input op-amps are the best for very low—offset voltage applications unless you are prepared to limit the bandwidth to a few tens of hertz, in which case the CMOS chopper-stabilized types come into their own. The chopper technique achieves very low values of V_{OS} and drift by repeatedly nulling the amplifier's actual V_{OS} several hundred times a second with the aid of charge storage capacitors.

Offset voltage errors are usually quoted with reference to the amplifier circuit input. The output offset voltage is the input offset times the closed-loop gain. This can have embarrassing consequences particularly in high-gain ac amplifiers where the designer has neglected offset errors because, for performance purposes, they are unimportant. Consider a noninverting ac-coupled amplifier with a gain of 1000 as depicted in Fig. 5.1.

Let us say the circuit is for audio applications and the op-amp is one half of a TL072 selected for low noise and wide bandwidth, running on supply voltages of ±12 V. The TL072 has a maximum quoted V_{OS} of 10 mV. In the circuit shown, this will be amplified by the closed-loop gain to give a dc offset at the output of 10 V—which is far too close to the supply rail to leave any headroom to cope with overloads. In fact, the TL072 is likely to saturate at 9—10 V anyway with ±12-V power rails.

Output Saturation Due to Amplified Offset
The designer may be wanting 2 mV pk—pk ac signals at the input to be amplified up to 2-V pk—pk signals at the output. If the dc conditions are taken for granted then

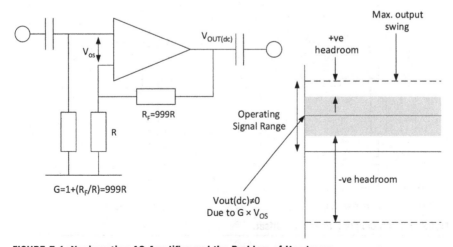

FIGURE 5.1 Noninverting AC Amplifier and the Problem of Headroom.

you might expect at least 20 dB of headroom: ± 1 V output swing with ± 10 V available. But, with a worst-case V_{OS} device virtually no headroom will be available for one polarity of input and 20 V will be available for the other. Unipolar (asymmetrical) clipping will result. The worst outcome is if the design is checked on the bench with a device, which has a much-better-than-worst-case offset, say 1 mV. Then the dc output voltage will only be 1 V, and virtually all the expected headroom will be available. If this design is let through to production then the scene is set for unexpected customer complaints of distortion! An additional problem presents itself if the output coupling capacitor is polarized: the dc output voltage can assume either polarity depending on the polarity of the offset. If this is not recognized it can lead to early failure of the capacitor in some production units.

Reducing the Effect of Offset

The solutions are plentiful. The easiest is to change the feedback to ac coupling, which gives a dc gain of unity so that the output dc voltage offset is the same as the input offset (Fig. 5.2). The inverting configuration is simpler in this respect. The difficulty with this solution is that the time constant $R_f \times C$ can be inordinately long, leading to power-on delays of several seconds.

The second solution is to reduce the gain to a sensible value and cascade gain blocks. For instance, two ac-coupled gain blocks with a gain of 33 each, cascaded, would have the same performance but the offsets would be easily manageable. The bandwidth would also be improved, along with the out-of-band roll-off, if this were necessary. Unfortunately, this solution adds components and therefore cost.

A third solution is to use an amplifier with a better V_{OS} specification. This will either involve a trade-off in gain bandwidth, power consumption or other parameters, or cost. For instance, in the above example, AD's OP-227G with a maximum

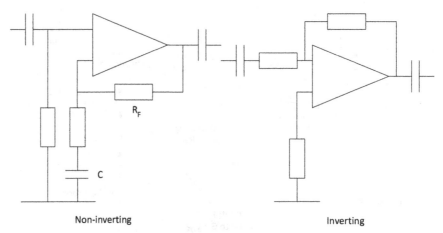

Non-inverting   Inverting

FIGURE 5.2 AC Coupling to Reduce Offset.

offset of 180 µV might be a suitable candidate, though it is noticeably more expensive. The overall cost might work out the same though, given the reduction in components over the second solution.

Offset Drift

Offset voltage drift is closely related to initial offset voltage and is a measure of how V_{OS} changes with temperature and time. Most manufacturers will specify drift with temperature, but only those offering precision devices will specify drift over time. Present technology for standard devices allows temperature coefficients of between 5 and 40 µV/°C, with 10 µV/°C being typical. For bipolar inputs, the magnitude of drift is directly related to the initial offset at room temperature. A rule of thumb is 3.3 µV/°C for each millivolt of initial offset. This drift has to be added to the worst-case offset voltage when calculating offset effects and can be significant when operating over a wide temperature range.

Early MOS-input op-amps suffered from poor offset voltage performance due to gate threshold voltage shifts with time, temperature, and applied gate voltage. New processes, particularly developments in silicon gate technology, have overcome these problems and CMOS op-amps (Texas Instruments' LinCMOS range for instance) can achieve bipolar-level V_{OS} figures with extremely good drift, 1−2 µV/°C being quoted.

Circuit Techniques to Remove the Effect of Drift

Microprocessor control has allowed new analog techniques to be developed and one of these is the nulling of input amplifier offsets, as in Fig. 5.3. With this technique the initial circuit offsets can be calibrated out of the system by applying a zero input, storing the resultant input value (which is the sum of the offsets) in nonvolatile memory and subsequently subtracting this from real-time input values. With this technique, only offset drifts, not absolute offset values, are important. Alternatively, for the cost of a few extra components—analog switches and interfacing—the

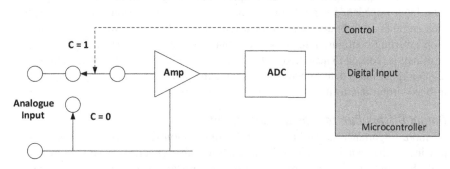

True Input Value = [Input(C = 1)] − [Input(C = 0)]

FIGURE 5.3 Offset Nulling With a Microcontroller.

nulling can be done repetitively in real time, and even the drift can be subtracted out. (This is the microprocessor equivalent of the chopper op-amps discussed earlier.)

5.2.2 BIAS AND OFFSET CURRENTS

Input bias current is the average dc current required by the inputs of the amplifier to establish correct bias conditions in the first stage. Input offset current is the difference in the bias current requirements of the two input terminals. A bipolar input stage requires a bias current, which is directly related to the current flowing in the collector circuit, divided by the transistor gain. FET-input (or biFET) op-amps on the other hand do not require a bias current as such, and their input currents are determined only by leakage and the need for input protection.

Bias Current Levels

Input bias currents of bipolar devices range from a few microamps down to a few nanoamps, with most industry-standard devices offering better than 0.5 μA. There is a well-established trade-off between bias current and speed; high speeds require higher first-stage collector currents to charge the internal node capacitance faster, which in turn requires higher bias currents. Precision bipolar op-amps achieve less than 20 nA while some devices using current nulling techniques can boast picoamp levels. Junction field-effect transistor (JFET) and CMOS devices routinely achieve input currents of a few picoamps or tens of picoamps at 25°C, but because this is almost entirely reverse-bias junction leakage it increases exponentially with temperature (see Section 4.1.3). Industry-standard JFET op-amps are therefore no better than bipolar ones at high temperatures, though precision JFET and CMOS still show nanoamp levels at the 125°C extreme. Note that even the 25°C figure for JFETs can be misleading, because it is quoted at 25°C *junction* temperature: many JFET op-amps take a fairly high supply current and warm up significantly in operation, so that the junction temperature is actually several degrees or tens of degrees higher than ambient.

The significance of input bias and offset currents is twofold: they determine the steady-state input impedance of the amplifier, and they result in added voltage offsets. Input impedance is rarely quoted as a parameter on op-amp data sheets since bias currents are a better measure of actual effects. It is irrelevant for the closed-loop inverting configuration, since the actual impedance seen at the op-amp input terminals is reduced to near zero by feedback. The input impedance of the noninverting configuration is determined by the change in input voltage divided by the change in bias current due to it.

Output Offsets Due to Bias and Offset Currents

Of more importance is the bias current's contribution to offsets. The bias current flowing in the source resistance R_S at each terminal generates a voltage in series with the input; if the bias currents and source resistances were equal, the voltages would cancel out and no extra offset would be added (Fig. 5.4).

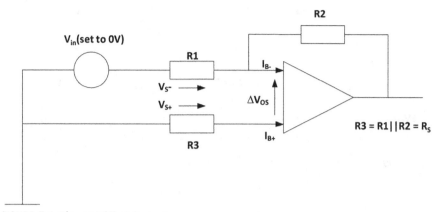

FIGURE 5.4 Bias and Offset Currents.

Ideal situation: I_B, R_S equal at $+$ and $-$ terminals, so $V_{S+} = V_{S-} = I_B \times R_S$ and $\Delta V_{OS} = 0$

Bad design: R_S not equal at $+$ and $-$ terminals so, neglecting I_{OS}, $V_{S+} = I_B \times R_3$, $V_{S-} = I_B \times R_1//R_2$ and $\Delta V_{OS} = I_B \times (R_1//R_2 - R_3)$

Practical op-amp: I_{B-} differs from I_{B+} by I_{OS}, R_S equal at both terminals, so $V_{S+} = I_{B+} \times R_S$, $V_{S-} = (I_B + I_{OS}) \times R_S$ and $\Delta V_{OS} = I_{OS} \times R_S$

As it is, the offset current generates an effective offset voltage given by $I_{OS} \times R_S$ (with a temperature coefficient determined by both), which adds to, or subtracts from, the inherent offset voltage V_{OS} of the op-amp. Clearly, whichever dominates the output depends on the magnitude of R_S. Higher values demand an op-amp with lower bias and offset currents. For instance, the current and voltage offsets generated by a 741's input circuit are equal when $R_S = 33$ KΩ (typical $V_{OS} = 1$ mV, $I_{OS} = 30$ nA). The same value for the TL081 JFET op-amp is 1000 MΩ ($V_{OS} = 5$ mV, $I_{OS} = 5$ pA).

I_B itself does not contribute to offset *provided* that the source resistances are equal at each terminal. If they are not then the offset contribution is $I_B \times \Delta R_S$. Since I_B can be an order of magnitude higher than I_{OS} for bipolar op-amps, it pays to equalize R_S: this is the function of R_3 in the circuit above. R_3 can be omitted or changed in value if current offset is not calculated to be a problem. Apart from the disadvantage of an extra component, R_3 is also an extra source of noise (generated by the noise component of I_B), which can weigh heavily against it in low-noise circuits.

5.2.3 COMMON-MODE EFFECTS

Two factors, which because they do not appear in op-amp circuit theory can be overlooked until late in the design, are common-mode rejection ratio (CMRR) and power supply rejection ratio (PSRR). Fig. 5.5 shows these schematically. Related to these is common-mode input voltage range.

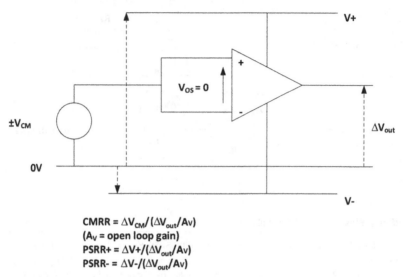

$$CMRR = \Delta V_{CM}/(\Delta V_{out}/A_V)$$
$$(A_V = \text{open loop gain})$$
$$PSRR+ = \Delta V+/(\Delta V_{out}/A_V)$$
$$PSRR- = \Delta V-/(\Delta V_{out}/A_V)$$

FIGURE 5.5 Common-Mode Power and Supply Rejection Ratio.

Common-Mode Rejection Ratio

An ideal op-amp will not produce an output when both inputs, ignoring offsets, are at the same (common mode) potential throughout the input range. In practice, gain differences between the two inputs, and variations in offset with common-mode voltage, combine to produce an error at the output as the common-mode voltage varies. This error is referred to the input (that is, divided by the gain) to produce an equivalent input common-mode error voltage. The ratio of this voltage to the actual common-mode input voltage is the CMRR, usually expressed in decibels. For example, a CMRR of 80 dB would give an equivalent input voltage error of $100\ \mu V$ for every 1 V change at both $+$ and $-$ inputs together. The inverting amplifier configuration is inherently immune to common-mode errors since the inputs stay at a constant level, whereas the noninverting and differential circuits are susceptible.

CMRR is not necessarily a constant. It will vary with common-mode input level and temperature, and always worsens with increasing frequency. Individual manufacturers may specify an average or a worst-case value and will always specify it at dc.

Power Supply Rejection Ratio

PSRR is similar to CMRR but relates to error voltages referred to the input as a result of changes in the power rail voltages. As before, a PSRR of 80 dB with a rail voltage change of 1 V would result in an equivalent input error of $100\ \mu V$. Again, PSRR worsens with increasing frequency and may be only 20–30 dB in the tens-to-hundreds of kilohertz range, so that high-frequency noise on the power rails is easily reflected on the output. There may also be a difference of several tens of decibels between the PSRRs of the positive and negative supply rails, due to the difference in internal biasing arrangements. For this reason, it is unwise to expect equal but antiphase power rail signals, such as mains frequency ripple, to cancel each other out.

5.2.4 INPUT VOLTAGE RANGE

Common-mode input voltage range is usually defined as the range of input voltages over which the quoted CMRR is met. Errors quickly increase as it is exceeded. The input range may or may not include the negative supply rail, depending on the type of input. The popular LM324 range and its derivatives have a pnp emitter—coupled pair at the input, which allows operation down to slightly below the negative rail. The CMOS-input devices from Texas, National, STM, and Intersil also allow operation down to the negative rail. Some of these op-amps stop a few volts short of the positive rail, as they are optimized for operation from a single positive supply, but there are also some devices available that include both rails within their input range, known unsurprisingly as "rail-to-rail" input op-amps. Conventional bipolar devices of the 741 type, designed for ±15 V rails, cannot swing to within less than 2 V of each rail, and biFET types are even more restricted.

Absolute Maximum Input

The common-mode operating input voltage is normally different from the absolute maximum input voltage range, which is usually equal to the supply voltage. If you exceed the maximum input voltage without current limiting then you are likely to destroy the device; this can quite easily happen inadvertently, apart from circuits connected to external inputs, if for instance a large-value capacitor is discharged directly through the input. Even if current is limited to a safe value, overvoltages on the input can lead to unpredictable behavior. Latch-up, where the IC locks itself into a quasistable state and may draw large currents from the power supply, leading to burnout, is one possibility. Another is that the sign of the inputs may change, so that the inverting input suddenly becomes noninverting. (This was a well-known fault on early devices such as the 709.) These problems most frequently arise with capacitive coupling direct to one or other input, or when power rails to different parts of the circuit are turned on or off at different times. The safe way to guard against them is to include a reasonable amount of resistance at each input, directly in series with the input pin.

5.2.5 OUTPUT PARAMETERS

Two factors constrain the output voltage available from an op-amp: the power rail voltage and the load impedance.

Power Rail Voltage

It should be obvious that the output cannot swing to a greater value than either power rail. Unfortunately it is often easy to overlook this fact, particularly as the power connections are frequently omitted from circuit diagrams, and with different quad op-amp packages being supplied from different rails it is hard to keep track of which device is powered from what voltage. More seriously, with unregulated supplies the actual voltage may be noticeably less than the nominal. The required output must be calculated for the worst-case supply voltage.

Historically, most op-amps could not swing their output right up to either supply rail. The profusion of CMOS-output devices have dealt with this limitation, as have many of the types intended for single-supply operation, which have a current sink at the output and can reach within a few tens of millivolts of the negative (or ground) supply terminal. Other conventional bipolar and biFET parts cannot swing to within less than 2 V of either rail. The classic output stage (Fig. 5.6) is a complementary emitter follower pair, which gives low output impedance, but the output available in either direction is limited by $(V_{DR(min)} + V_{BE})$. Depending on the detailed design of the output, the swing may or may not be symmetrical in either polarity. This fact is disguised in some data sheets where the maximum peak-to-peak output voltage swing is quoted, rather than the maximum output voltage relative to the supply terminals.

Load Impedance

Output also depends on the circuit load impedance. This may again seem obvious, but there is an erroneous belief that because feedback reduces the output impedance of an op-amp in proportion to the ratio of open- to closed-loop gains, it should be capable of driving very low−load resistors. Well, of course, to an extent it is, but Ohm's law is not so easily flouted, and a low output resistance can only be driven to a low output voltage swing, depending entirely on the current drive capability of the output stage. The maximum output current that can be obtained from most devices is limited by package dissipation considerations to about $\pm10\,\text{mA}$. In

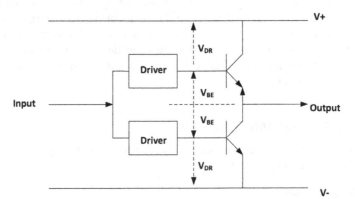

FIGURE 5.6 Output Voltage Swing Restrictions.

some cases, the output current spec is given as a particular output voltage swing when driving a stated value of load, typically 2–10 kΩ. The "rail-to-rail" op-amps with CMOS output will in fact only give a full rail-to-rail swing if they are driven into an open circuit; any output load, including, of course, the feedback resistor, reduces the total available swing in proportion to the ratio of output resistance to load resistance (Fig. 5.7).

If you want more output current it is quite to buffer the output with an external complementary emitter follower or something similar, provided that feedback is taken from the final output. Take care with short-circuit protection when doing this (or else do not be surprised if you have to keep replacing transistors), and also bear in mind that you have changed the high-frequency response of the combination and the closed-loop circuit may now be unstable.

Some single-supply op-amps are not designed both to source and sink current and, when used with split supplies, may have some crossover distortion as the output signal passes through the mid-supply value.

Output current protection is universally provided in op-amps to prevent damage when driving a short circuit. This does not work in the reverse direction, that is, when the output voltage is forced outside either supply rail by a fault condition. In this case, there will be one or two forward-biased diode junctions to the power rail, and current will flow through these limited only by the fault source impedance. Preventative measures for circuits where this is likely are dealt with in Section 6.2.3.

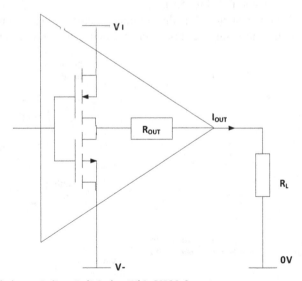

FIGURE 5.7 Limits on Rail-to-Rail Swing With CMOS Outputs.

5.2.6 AC PARAMETERS

The performance of an op-amp at high frequency is described by a motley collection of parameters, each of which refers to slightly different operating conditions. They are as follows:

- large-signal bandwidth, or full-power response: the maximum frequency, at unity closed-loop gain, for which a sinusoidal input signal will produce full output at rated load without exceeding a given distortion level. This bandwidth figure is normally determined by the slew-rate performance.
- small-signal or unity-gain bandwidth, or GBW: the frequency at which the open-loop gain falls to unity (0 dB). The "small-signal" label means that the output voltage swing is small enough that slew-rate limitations do not apply.
- slew rate: the maximum rate of change of output voltage for a large input step change, quoted in volts per microsecond.
- settling time: elapsed time from the application of a step input change to the point at which the output has entered and remained within a specified error band about the final steady-state value.

These parameters are illustrated in Fig. 5.8.

5.2.7 SLEW RATE AND LARGE-SIGNAL BANDWIDTH

These two specifications are intimately related. All conventional voltage feedback op-amps can be modeled by a transconductance gain block (with a gain of g_m) driving a transimpedance amplifier (with an amplifier gain of $-A$) with capacitive feedback (capacitor value C_c) (Fig. 5.9).

The compensation capacitor C_c is the dominant factor setting the op-amp's frequency response. It is necessary because a feedback circuit would be unstable if the

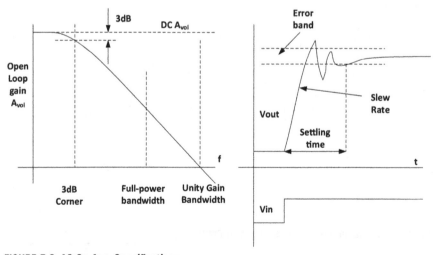

FIGURE 5.8 AC Op-Amp Specifications.

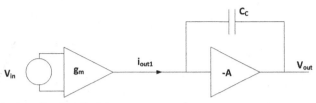

FIGURE 5.9 Op-Amp Slewing Model.

gain block's high-frequency response was not limited. Digital designers avoid capacitors in the signal path because they slow the response time, but this is the price for freedom from unwanted oscillations when working with linear circuits.

Slew Rate

The exact value of the price is measured by the slew rate. From the above circuit, you can see that the rate of change of V_{out} is determined entirely by i_{out1} and C_c (remember $dV/dt = I/C$). As an example, the 741's input section current source can supply 20 µA and its compensation capacitor is 30 pF, so its maximum slew rate is 0.67 V/µs. Op-amp designers have the freedom to set both these parameters within certain limits, and this is what distinguishes a fast, high supply–current device from a slow, low supply–current one. "Programmable" devices such as the LM4250 or LM346 make the trade-off more obvious by putting it in the circuit designer's hands.

If i_{out1} can be increased without affecting the transconductance g_m, then slew rate can be improved without a corresponding reduction in stability. This is one of the major virtues of the biFET range of op-amps. The JFET-input stage can be run at high currents for a low g_m relative to the bipolar and so can provide an order of magnitude or more increase in slew rate.

Large-Signal Bandwidth

Slew-rate limitations on dV_{out}/dt can be equated to the maximum rate of change of a sine wave output. The time derivative of a sine wave is

$$\frac{d}{dt}[V_p \sin \omega t] = \omega V_p \cos \omega t \tag{5.1}$$

where $\omega = 2\pi f$.

This has a maximum value of $2\pi f \times V_p$, which relates frequency directly to peak output voltage. If V_p is equated to the maximum dc output swing then f_{max} can be inferred from the slew rate and is equal to the large-signal or full-power bandwidth,

$$2\pi \times f_{\text{max}} = \text{slewrate}/V_p \tag{5.2}$$

Slewing Distortion

Operating an op-amp above the slew-rate limit will cause slewing distortion on the output. In the limit the output will be a triangle wave (Fig. 5.10) as it alternately

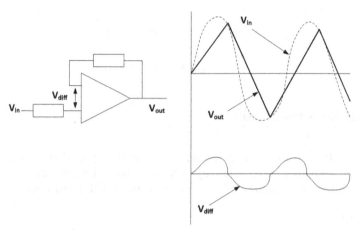

FIGURE 5.10 Slewing Distortion.

switches between positive and negative slewing, which will decrease in amplitude as the frequency is raised further. If the positive and negative slew rates differ, there will be asymmetrical distortion on the output. This can generate an unexpected effect equivalent to a dc offset voltage, due to rectification of the asymmetrical feedback waveform or overloading of the input stage by large distortion signals at the summing junction. Also, slewing is not always linear from start to finish but may exhibit a fast rise for the first part of the change followed by a reversion to the expected rate for the latter part.

5.2.8 SMALL-SIGNAL BANDWIDTH

The op-amp frequency response shown in Fig. 5.8A exhibits the same characteristic as a simple low-pass RC filter. The 3-dB frequency or corner frequency is that point at which the open-loop gain has dropped by 3 dB from its dc value. It is set by the compensation capacitor C_c and is in the low hertz or tens of hertz range for most devices. The gain then "rolls off" at a constant rate of 20 dB per decade (a ten-times increase in frequency produces a 10-fold gain reduction) until at some higher frequency the gain has dropped to 1. This frequency therefore represents the unity-gain bandwidth of the part, also called the small-signal bandwidth.

The fact of a constant roll-off means that it is possible to speak of a constant "gain-bandwidth product" for a device. The LM324's op-amps for instance have a typical unity-gain bandwidth of 1 MHz, so if you wanted to use them at this frequency you could only use them as voltage followers—and small-signal ones at that, since large output swings would be slew-rate limited. A gain of 10 would be achievable up to 100 kHz, a gain of 100 up to 10 kHz and so on (but see the comments on open-loop gain later). This gain-bandwidth trade-off is illustrated in Fig. 5.11. On the other hand, many more recent devices have unity-gain bandwidths

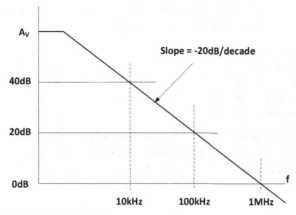

FIGURE 5.11 Gain-Bandwidth Roll-Off.

of 5−30 MHz and can therefore offer reasonable gains up to the MHz region. Anything with a GBW of more than 30 MHz is justifiably offered as a "high-speed" device.

5.2.9 SETTLING TIME

When an op-amp is faced with a step input, as compared with a linear function such as a sinusoid or triangle wave, the step takes some time to propagate to the output. This time includes the delay to the onset of output slewing, the slewing time, recovery from slew-limited overload, and settling to within a given output error. Students of feedback theory will know that a feedback-controlled system's response to a step input exhibits some degree of overshoot (Fig. 5.8B) or undershoot depending on its damping factor. Op-amps are no different. For circuits whose output must slew rapidly to a precise value, particularly analog-to-digital converters and sample-and-hold buffers, the settling time is an important parameter.

Op-amps specifically intended for such applications include settling time parameters in their specifications. Most general-purpose ones do not, although a graph of output pulse response is often presented from which it can be inferred. When present, settling time is usually specified for unity gain, relatively low impedance levels, and low or no capacitive loading. Because it is determined by a combination of closed-loop amplifier characteristics, both linear and nonlinear, it cannot be directly predicted from the open-loop specs of slew rate and bandwidth, although it is reasonable to assume that an amplifier that performs well in these respects will also have a fast settling time.

5.2.10 THE OSCILLATING AMPLIFIER

Just about every analog designer has been bugged by the problem of the feedback amplifier that oscillates (and its converse, the oscillator that does not) at some time or other. There are really only a few fundamental causes of unwanted oscillations, they are all curable, and they can be listed as follows:

- feedback-loop instability
- incorrect grounding
- power supply coupling
- output-stage instability
- parasitic coupling

The most important clue in tracking down instability is the frequency of oscillation. If this is near the unity-gain bandwidth of the device then you are most probably suffering feedback-induced instability. This can be checked by temporarily increasing the closed-loop gain. If feedback is the problem, then the oscillation should stop or at least decrease in frequency. If it does not, look elsewhere.

Feedback-loop instability is caused by too much feedback at or near the unity-gain frequency, where the op-amp's phase margin is approaching a critical value. (Many books on feedback circuit theory deal with the question of stability, gain and phase margin, using tools such as the Bode plot and the Nyquist diagram, so this is not covered here.)

Ground Coupling

Ground loops or other types of incorrect grounding cause coupling from output back to input of the circuit via a common impedance in its grounded segment. This effect has been covered in Chapter 1, but the circuit topology is repeated here, in Fig. 5.12. If the resulting feedback sense gives an output component in-phase with the input then positive feedback occurs, and if this overrides the intended negative feedback you will have oscillation. The frequency will depend on the phase contribution of the common impedance, which will normally be inductive and can vary over a wide range.

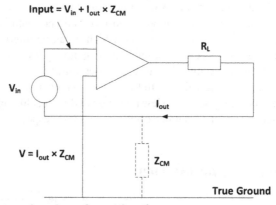

FIGURE 5.12 Common Impedance Ground Coupling.

Power Supply Coupling

Power supplies should be properly bypassed to avoid similar coupling through the common-mode power supply impedance. PSRR falls with frequency, and typical $0.01-0.1$ µF decoupling capacitors may resonate with the parasitic inductance of long power leads in the mehahertz region, so these problems usually show up in the $1-10$ MHz range. Using $1-10$ µF tantalum capacitors for power rail bypassing will drop the resonant frequency and stray circuit Q to the level at which problems are unlikely (compare Fig. 3.19 for capacitor resonances).

Output-Stage Instability

Localized output-stage instability is most common when the device is driving a capacitive load. This can create output oscillations in the high-MHz range, which are generally cured by good power rail decoupling close to the power supply pins, with the decoupling ground point close to the return point of the load impedance, or by including a low-value series resistor in the output within the feedback loop.

Capacitive loads also cause a phase lag in the output voltage by acting in combination with the op-amp's open-loop output resistance (Fig. 5.13). This increased phase shift reduces the phase margin of a feedback circuit. A typical capacitive load, often invisible to the designer because it is not treated as a component, is a length of coaxial cable. Until the length starts to approach a quarter wavelength at the frequency of interest, coax looks like a capacitor: for instance, 10 m of the popular RG58C/U 50-Ω type will be about 1000 pF. The capacitance can be decoupled from the output with a low-value series resistor, and high-frequency feedback provided by a small direct feedback capacitor C_F compensates for the phase lag caused by C_L.

Stray Capaoitance at the Input

A further phase lag is introduced by the stray capacitance C_S at the op-amp's inverting input. With normal layout practice, this is of the order of $3-5$ pF, which becomes

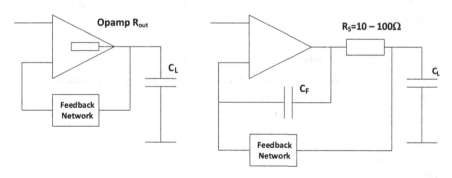

Phase lag @ freq f = tan^{-1}(f/fc) degrees
Where fc=1/2π × R$_{out}$ × C$_L$

Isolate an large value of C$_L$ with R$_S$
and C$_F$ typically 20pF

FIGURE 5.13 Output Capacitive Loading.

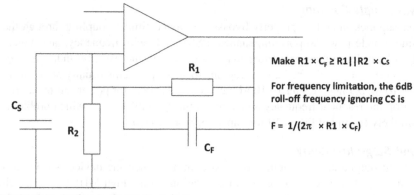

Make R1 × C$_F$ ≥ R1||R2 × C$_S$

For frequency limitation, the 6dB
roll-off frequency ignoring CS is

F = 1/(2π × R1 × C$_F$)

FIGURE 5.14 Adding Feedback Capacitance.

significant when high-value feedback resistors are used, as is common with MOS-
and JFET-input amplifiers. The roll-off frequency due to this capacitance is determined by the feedback network impedance as seen from the inverting input. The
small-value direct feedback capacitance C_F of Fig. 5.14 can be added to combat
this roll-off, by roughly equating time constants in the feedback loop and across
the input. In fact this technique is recommended for all low-frequency circuits as
with it you can restrict loop bandwidth to the minimum necessary, thereby cutting
down on noise, interference susceptibility, and response instability.

Parasitic Feedback
Finally in the catalog of instability sources, remember to watch out for parasitic
coupling mechanisms, especially from the output to the noninverting input. Any
coupling here creates unwanted positive feedback. Layout is the most important factor: keep all feedback and input components close to the amplifier, separate input
and output components, keep all pc tracks short and direct, and use a ground plane
and/or shield tracks for sensitive circuits.

5.2.11 OPEN-LOOP GAIN
One of the major features of the classical feedback equation, which is used in almost
all op-amp design:

$$A_{CL} = \frac{A_{OL}}{(1 + A_{OL}\beta)} \tag{5.3}$$

where β is the feedback factor, A_{OL} is the open-loop gain, A_{CL} is the closed-loop
gain.

If you assume a very high A_{OL} then the closed-loop gain is almost entirely determined by β, the feedback factor. This is set by external (passive) components and can
therefore be very tightly defined. Op-amps always offer a very high dc open-loop

gain (80 dB as a minimum, usually 100–120 dB), and this can easily tempt the designer into ignoring the effect of A_{OL} entirely.

Sagging A_{OL}

A_{OL} does, in fact, change quite markedly with both frequency and temperature. We have already seen (Fig. 5.11) that the ac A_{OL} rolls off at a constant rate, usually 20 dB/decade, and this determines the gain that can be achieved for any given bandwidth. In fact when the frequency starts approaching the maximum bandwidth the excess gain available becomes progressively lower, and this affects the validity of the high-A_{OL} approximation. If your circuit has a requirement for precise gain then you need to evaluate the actual gain that will be achieved.

As an example, take $\beta = 0.01$ (for a gain of 100) and $A_{OL} = 10^5$ (100 dB) at dc. The actual gain, from the feedback equation, is

$$A_{CL} = \frac{10^5}{(1 + 10^5 \times 0.01)} = 99.9 \tag{5.4}$$

Now raise the frequency to the point at which it is a decade below the maximum expected bandwidth at this gain. This will have reduced A_{OL} to 10 times the closed-loop gain or 1000. The actual gain is now:

$$A_{CL} = \frac{1000}{(1 + 1000 \times 0.01)} = 90.9 \tag{5.5}$$

which shows a 10% gain reduction at one-tenth the desired bandwidth!

A_{OL} also changes with temperature. The data sheet will not always tell you how much, but it is common for it to halve when going from the low temperature extreme to the high extreme. If your circuit is sensitive to changes in closed-loop gain, it would be wise to check whether the likely changes it will experience in A_{OL} are acceptable and if not, either reduce the closed-loop gain to give more gain margin, or find an op-amp with a higher value for A_{OL}.

5.2.12 NOISE

A perfect amplifier with perfect components would be capable of amplifying an infinitely small signal to, say, 10 V p–p with perfect resolution. The imperfection that prevents it from doing so is called noise. The noise contribution of the amplifier circuit places a lower limit on the resolution of the desired signal, and you will need to account for it when working with low-level (submillivolt) signals or when the signal-to-noise ratio needs to be high, as in precision amplifiers and audio or video circuits.

There are three noise sources that you need to consider:

- amplifier-generated noise
- thermal noise
- electromagnetic interference

The third of these is either electromagnetically coupled into the circuit conductors at RF or by common-mode mechanisms at lower frequencies. It can be minimized by good layout and shielding and by keeping the operating bandwidth low and is mentioned here only to warn you to keep it in mind when thinking about noise. Chapter 8 discusses electromagnetic compatibility in more depth.

Definitions

Two important definitions for noise calculations are the mean value and the root mean square (RMS) value. For any signal we can calculate the mean value using the integration over a time period as shown by:

$$\overline{v_n} = \frac{1}{T} \int_0^T v_n(t)dt = 0 \tag{5.6}$$

And the RMS is the square root of the integration of the square of the instantaneous values as given by:

$$\overline{v_{n(RMS)}^2} = \frac{1}{T} \int_0^T v_n(t)^2 dt \neq 0 \tag{5.7}$$

Using these definitions, we can make some important statements about noise signals which are as follows:

- The instantaneous value of a noise signal is undetermined.
- To characterize a noise signal, the *mean* value, *mean square*, and *RMS value* are used.
- The standard deviation is equal to the RMS; the variance is equal to the mean square value (literally the RMS^2).
- The mean square value is a measure for the normalized noise power of the signal.

If we consider RMS noise, then this can be considered as a voltage or current noise. The power dissipation by a 1-Ω resistor with a dc voltage of x V applied across it is equivalent to a noise source with an RMS voltage of x V_{RMS}. In a similar manner we can say that the power dissipation by a 1-Ω resistor with a dc current of y A applied across it is equivalent to a noise source with an RMS current of y V_{RMS}.

5.2.13 CALCULATING THE EFFECT OF NOISE IN A CIRCUIT

Typically, a circuit contains many noise generators. For circuit analysis we usually sum all noise sources into a single one (either at the output or the input of the circuit) and treat the circuit as noiseless.

Summing j noise sources is done by adding their mean square values as shown by:

$$V_{n(RMS)}^2 = V_{n1(RMS)}^2 + V_{n2(RMS)}^2 \cdots + V_{nj(RMS)}^2 \tag{5.8}$$

This is only valid provided that the noise sources are independent of each other (uncorrelated). For circuit analysis this is usually the case.

Using this approach we can therefore calculate the effective total noise power in a circuit, by summing the individual noise powers. For example, if we have two noise sources in a circuit with values of RMS noise voltage of 10 and 20 μV, respectively. The total noise power can be calculated:

$$V^2_{\text{total (RMS)}} = (10 \ \mu V)^2 + (20 \ \mu V)^2 = 500(\mu V)^2 \tag{5.9}$$

This results in an RMS noise voltage of the square root of 500 $(\mu V)^2$, which is 22.36 μV. At this point it is worth considering precisely what we have calculated, and in fact we have calculated that for two components with uncorrelated noise of 20 and 10 μV, the combined overall noise voltage is 22.36 μV, which is the RMS of the noise signal.

Power Spectral Density of Noise

We have considered the mathematical nature of noise signals using RMS calculations; however, it is important to note that the noise is always spread out over the frequency spectrum. We therefore must consider not only the RMS noise but also the way that the noise is distributed over the spectrum and how we can possibly mitigate the effects of noise as a result on our design. The way that we think about noise is that the overall spectrum is divided into 1-Hz bandwidth "slices" so the noise power *density* is the mean square noise within that 1-Hz bandwidth.

As a result, if we consider a total noise power in terms of voltage squared, then the power spectral density (PSD) in our 1-Hz bandwidth is therefore defined in terms of volts squared per hertz (V^2/Hz). In a similar manner, if we have a noise *voltage*, then this is in terms of the square root of the PSD, and therefore we can define the equivalent voltage noise spectral density in terms of V/$\sqrt{\text{Hz}}$.

If we calculate the autocorrelation function of the time domain noise function, we can express the noise spectral density in the frequency domain using the integration of the individual spectral harmonics as defined by:

$$\overline{v^2_{n(\text{RMS})}} = \int_0^\infty S(f)df \tag{5.10}$$

where $S(f)$ is the autocorrelation function. It is beyond the scope of this book to go much further into the description of how this is derived, however, it can be calculated using the Wiener–Khinchin theorem, which states that the spectral density is the Fourier transform of the autocorrelation function of a random signal (further details can be found in many signal processing textbooks).

Types of Noise

We have two main types of noise to consider in practice—thermal noise and flicker noise. White noise has a flat (or constant) spectral density, i.e., $S(f)$ is a constant and is produced by thermal noise generators (or Johnson, Boltzmann). Examples of sources of such noise are resistors, bipolar transistors, diodes, and MOSFET transistors. While, of course, the noise may be not exactly white up to infinity, we can assume the noise is effectively white into the terahertz range. Flicker noise has a different

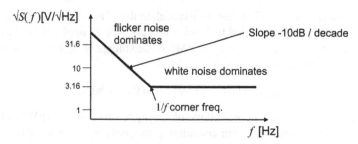

FIGURE 5.15 Typical Noise Spectral Density of a MOS Transistor.

characteristic where the noise spectral density is proportional to $1/f$ (where f is the frequency). Flicker noise is particularly important in MOS transistors, especially at low frequencies as we shall see. It is important to remember that MOS transistors have both flicker noise and white noise.

If we take an example of a MOS transistor and plot its noise spectral density as in Fig. 5.15, then we can see that the flicker noise will dominate the behavior at low frequencies, but that the white noise continues to very high frequencies.

If we consider a white noise source, in contrast, if we define a white noise source with a power spectral density (PSD) of 10 $\mu V^2/Hz$, this means that the level of noise power will be 10 μV^2 *across the whole spectrum*. We can see this graphically if we plot the PSD for a single white noise source as shown in Fig. 5.16.

Thermal Noise

The other two sources, such as the dc offset and bias error components discussed earlier, are conventionally referred to the op-amp input. Thermal, or "Johnson" noise is generated in the resistive component of any circuit impedance by thermal agitation of the electrons. All resistors around the input circuit contribute this. It is given by

$$e_n = \sqrt{4kTBR} \tag{5.11}$$

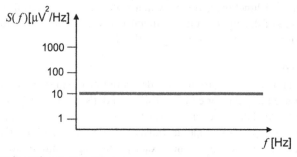

FIGURE 5.16 White Noise Source Example.

where e_n = RMS value of noise voltage, k = Boltzmann's constant, 1.38×10^{-23} J/K, T = absolute temperature, B = bandwidth in which the noise is measured, R = circuit resistance.

As a rule of thumb, it is easier to remember that the noise contribution of a 1-kΩ resistor at room temperature (298K) in a 1-Hz bandwidth is 4 nV RMS. The noise is proportional to the square root of bandwidth and resistance, so a 100-kΩ resistor in 1 Hz, or a 1-kΩ resistor in 100 Hz, will generate 40 nV. Noise is a statistical process. To convert the RMS noise to peak-to-peak, multiply by 6.6 for a probability of less than 0.1% that a peak will exceed the calculated limit, or 5 for a probability of less than 1%.

As we have seen, thermal noise is a form of white noise—uniform across the whole spectrum, and it is present in resistors and, as we have seen, has the following basic equation that describes the noise power in V^2/Hz.

If we consider this in more detail, it is clear that it is *not* dependent on frequency, but *is* dependent on the resistance and temperature, so this explains why the characteristic of the noise is "white" i.e., uniform across the frequency range. As a designer, it is important to understand the impact that this will have on a circuit's performance, and we can illustrate that with an example. Consider a 100-kΩ resistor, where the temperature is room temperature approximately (27°C), then we can calculate the RMS noise power (S) using:

$$S = 4kTR \ \ V^2/Hz$$
$$S = 4 \times 1.38 \times 10^{-23} \times (273 + 27) \times 100 \times 10^3 \ \ V^2/Hz \quad (5.12)$$
$$S = 1.66 \times 10^{-15} \ \ V^2/Hz$$

And from this we can calculate the noise voltage by simply taking the square root of the noise power as given in:

$$N = \sqrt{4kTR} \ \ V/\sqrt{Hz}$$
$$N = \sqrt{1.66 \times 10^{-15}} \ \ V/\sqrt{Hz} = 4.07 \times 10^{-8} \quad (5.13)$$

If we look at this in more detail, we can see that the noise voltage RMS value is of the order of 40 nV/$\sqrt{Hz}$, and so as we have seen previously in Fig. 5.16, this means that across the whole spectrum there will be 40 nV (RMS) of noise voltage.

So, what are the implications of this for our circuit designer? Consider the situation where we have a circuit that is designed to provide amplification for a signal that will eventually go into an analog-to-digital converter. If we consider the case where the converter is 16 bits, then we can state that the number of quantization levels for an N-bit converter is 2^N and the resolution is given by $V_{FS}/(2^N - 1)$; (where V_{FS}: full scale voltage). This is equivalent to the smallest increment level (or step size) q of that converter.

If we have a converter of 16 bits, and a voltage supply of 3.3 V, giving a $V_{FS} = 3.3$ V, then we can estimate the resolution as being of the order of 50 μV, and so the RMS noise from this single resistor will be of the order of 0.16% of the resolution of this circuit.

We can see how this can quickly increase if we start to build circuits using multiple components, and if we take a simple example of a potential divider, ideal amplifier, and an RC low-pass filter, we can investigate the noise behavior of the complete circuit. Now, as we have already calculated the noise for the 100-kΩ resistor, we can use that in our calculations. As the amplifier is ideal, we do not need to consider its noise performance (although in practice, of course, that would also need to be included in the calculation). Finally, we can assume that ideal capacitors have negligible thermal noise, although again in practice, there will probably be a figure due to parasitic elements that needs to be included, although this will be very small.

As we have already calculated, we have the thermal noise power in each resistor of the potential divider where $R_1 = R_2 = 100$ k, and this was calculated to be 1.66×10^{-15} V^2/Hz. As $R_1 = R_2$ we know that the voltage gain (A) of the potential divider is 0.5, and therefore the combined noise power contribution will be multiplied by the square of the voltage gain to give the effective contribution due to the potential divider resistors.

So, given that the noise in R_1 is 1.66×10^{-15} V^2/Hz, the noise after the potential divider *contributed from R_1* will be

$$S_o = (0.5)^2 \times 1.66 \times 10^{-15} = 4.14 \times 10^{-16} \text{ V}^2/\text{Hz} \qquad (5.14)$$

And there will be the same contribution from R_2. The filter resistor is only 100 Ω, however, it will also make a contribution to the overall noise in the system.

$$S = 4 \times 1.38 \times 10^{-23} \times (273 + 27) \times 100 \text{ V}^2/\text{Hz}$$
$$S = 1.66 \times 10^{-18} \text{ V}^2/\text{Hz} \qquad (5.15)$$

As the amplifier has been considered to be ideal, and the capacitor is also ideal (with no noise contribution), we can therefore calculate the overall noise of the system as follows:

$$S_{\text{total}} = S_{R_1} + S_{R_2} + S_{LP}$$
$$S = 4.14 \times 10^{-16} + 4.14 \times 10^{-16} + 1.66 \times 10^{-18} \text{ V}^2/\text{Hz} \qquad (5.16)$$
$$S = 8.30 \times 10^{-16} \text{ V}^2/\text{Hz}$$

We can therefore calculate the noise voltage:

$$N = \sqrt{S} \text{ V}/\sqrt{\text{Hz}}$$
$$N = 2.88 \times 10^{-8} \text{ V}/\sqrt{\text{Hz}} \qquad (5.17)$$
$$N = 28.8 \text{ nV}/\sqrt{\text{Hz}}$$

This is an interesting result as it shows that even though the overall noise contribution of an individual resistor is 40 nV/$\sqrt{\text{Hz}}$, due to the gain of the circuit, the overall noise on the output will be less than a single resistor, even though we have three resistors in the circuit. It is also an illustration that we need to be careful in assumptions of noise, and that it will quickly become difficult to correctly predict the noise contribution for anything other than the simplest circuits. Thus far we have assumed that the amplifier is noiseless, but in practice this will often be the largest noise source in the circuit.

Amplifier Noise

Amplifier noise is what you will find specified in the data sheet (sometimes; where it is not specified it can be two to four times worse than an equivalent low-noise part). It is characterized as a voltage source in series with one input, and a current source in parallel with each input, with the amplifier itself being considered noiseless. The values are specified at unity bandwidth, as RMS nanovolts or nanoamps per root-hertz; alternatively they may be specified over a given bandwidth. Because you need to add together all noise contributions, it is usually easiest to calculate them at unity bandwidth and then multiply the overall result by the square root of the bandwidth. This assumes a constant noise spectral density over the bandwidth of interest, which is true for resistors but may not be for the op-amp (see later). Noise, being statistical, is added on a root-mean-square basis. So the general noise model for an op-amp circuit is as shown in Fig. 5.17.

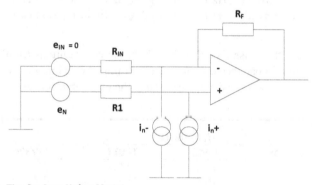

FIGURE 5.17 The Op-Amp Noise Model.

Contributor	Cause	Output Voltage Contribution
R_{IN}	Thermal noise	$\sqrt{(4kTR_{IN})} \times A_V \times \sqrt{B} = N(R_{IN})$
R_1	Thermal noise	$\sqrt{(4kTR_1)} \times (A_V + 1) \times \sqrt{B} = N(R_1)$
R_F	Thermal noise	$\sqrt{(4kTR_F)} \times \sqrt{B} = N(R_F)$
i_{n-}	Amplifier current noise	$i_{n-} \times R_F \times \sqrt{B} = N(i_{n-})$
i_{n+}	Amplifier current noise	$i_{n+} \times R_1 \times (A_V + 1) \times \sqrt{B} = N(i_{n+})$
e_n	Amplifier voltage noise	$e_n \times (A_V + 1) \times \sqrt{B} = N(e_n)$

Total output noise $= \sqrt{[N(R_{IN})^2 + N(R_1)^2 + N(R_F)^2 + N(i_{n-})^2 + N(i_{n+})^2 + N(e_n)^2]}$

When the noise is added in RMS fashion, if any noise source is less than a third of another it can be neglected with an error of less than 5%. This is a useful feature to remember with complex circuits where it is difficult to account accurately for all generator resistances.

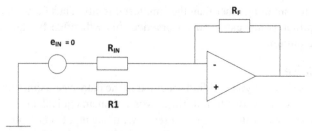

FIGURE 5.18 Standard Op-Amp Inverting Amplifier.

As an example of how to apply the noise model, let us examine the trade-offs between a high-impedance and a low-impedance circuit for different op-amps. The circuit is the standard inverting configuration with R_1 sized according to the principle laid out earlier for minimization of bias current errors (R_3 in Fig. 5.4). R_{IN} is the sum of generator output impedance and amplifier input resistor (Fig. 5.18). The op-amps chosen have the following noise characteristics (at 1 kHz):

OP27: $e_n = 3$ nV/$\sqrt{\text{Hz}}$ $i_n = 0.4$ pA/$\sqrt{\text{Hz}}$ (low-noise precision bipolar)
TL071: $e_n = 18$ nV/$\sqrt{\text{Hz}}$ $i_n = 0.01$ pA/$\sqrt{\text{Hz}}$ (low-noise biFET)
LMV324: $e_n = 39$ nV/$\sqrt{\text{Hz}}$ $i_n = 0.21$ pA/$\sqrt{\text{Hz}}$ (industry-standard low-voltage bipolar)

Working from the noise model of Fig. 5.15, the contributions (in nV/$\sqrt{\text{Hz}}$) are tabulated for a low-impedance circuit and a high-impedance circuit, with the major contributor in each case shown emphasized and the negligible contributors shown in brackets:

Low impedance, $R_{IN} = 200\ \Omega$, $R_1 = 180\ \Omega$, $R_F = 2$ KΩ

Noise contributor	OP27	TL071	LMV324
$N(R_{IN})$	17.9	17.9	17.9
$N(R_1)$	18.7	18.7	18.7
$N(R_F)$	(5.6)	(5.6)	(5.6)
$N(i_n-)$	(0.8)	(0.02)	(0.42)
$N(i_n+)$	(0.79)	(0.02)	(0.46)
$N(e_n)$	**33**	**198**	**429**
Total noise voltage	41.9	200	430

High impedance, $R_{IN} = 200$ KΩ, $R_1 = 180$ KΩ, $R_F = 2$ MΩ

Noise contributor	OP27	TL071	LMV324
$N(R_{IN})$	565	565	565
$N(R_1)$	590	**590**	**590**
$N(R_F)$	178	178	178
$N(i_n-)$	**800**	(20)	420
$N(i_n+)$	792	(19.8)	460
$N(e_n)$	(33)	198	429
Total noise voltage	1402	836	1127

Some further rules of thumb follow from this example:

- high-impedance circuits are noisy
- in low-impedance circuits, op-amp voltage noise will be the dominant factor
- in high-impedance circuits, one or other of resistor noise or op-amp current noise will dominate: use a biFET or CMOS device and delete R_1
- do not expect a low-noise voltage op-amp to give you any advantage in a high-impedance circuit

Noise Bandwidth

Deciding the actual noise bandwidth is not always simple. The bandwidth used in the noise calculations is a notional "brick-wall" value, which assumes infinite attenuation above the cut-off frequency. This of course is not achievable in practice, and the circuit bandwidth has to be adjusted to reflect this fact. For a single-pole response with a cut-off frequency f_c and a roll-off of 6 dB/octave, the noise bandwidth is $1.57f_c$. For a cascade of single-pole filters the ratio of the noise bandwidth to cut-off frequency decreases.

For more complex circuits it is usually enough to make some approximation to the actual bandwidth. If the low-frequency cut-off is more than a decade below the high-frequency one then it can be neglected with little error, and the noise bandwidth can be taken as from dc to the high-frequency cut-off. The exception to this is in very low−frequency and dc applications (below a few tens of hertz), because at some point the op-amp noise contribution starts to rise with decreasing frequency. This region is known as $1/f$ or "flicker" noise. All op-amps show this characteristic, but the point at which the noise starts to rise (the $1/f$ noise corner) can be reduced from a few hundred hertz to below 10 Hz by careful design of the device.

Modeling and Simulation of Noise

Luckily, we have the option to carry out noise analysis using simulations, and we have two choices to consider when we do that. The first option is to add random noise sources that have the correct RMS value of noise and complete time domain simulations, and while this is very accurate, it is also extremely time consuming to complete. The second option is to work in the frequency domain, and in this way the noise analysis in a simulators works like a small-signal frequency analysis (ac) by sweeping the frequency and plotting the output. In the noise analysis, however, unlike the conventional frequency analysis, the sum of the individual noise contributions over the frequency range specified is calculated.

This is particularly useful in situations such as the example we have just considered, where there is a frequency dependence in the characteristic, and as such it would be a laborious task to calculate the overall frequency behavior, whereas in a simulator, a single analysis will give that response.

If we implement the circuit in a circuit simulator (in this case Saber) then the noise voltage spectral response can be simulated.

As we know from the circuit in our previous example (potential divider and simple amplifier, plus low-pass filter), there is a low-pass filter on the output, and the noise is subject to the same filtering as the signal, so it is a useful check to see that the noise response exhibits the same low-pass response. If we compare the overall noise at a low frequency, we can see that the simulator predicts a value for the noise voltage RMS of 28.818 nV/$\sqrt{\text{Hz}}$, which is entirely consistent with our calculated value.

5.2.14 SUPPLY CURRENT AND VOLTAGE

Circuit diagrams often leave out the supply connections to op-amp packages, for the very good reason that they create extra clutter, and the purpose of a circuit diagram is to communicate information as clearly as it can. When a single supply or a dual-rail supply is used throughout a circuit then confusion is unlikely, but with several different voltage levels in use it becomes difficult to work out exactly which op-amp is supplied by what voltage, and it is then better practice to show supplies to each package.

Supply Voltage

By far the largest number of recent op-amp introductions is aimed at low-power, single-supply applications where the circuit is battery operated. The lithium battery voltage of 3 V is a major driving force in this trend. Although "low power" and "single supply" are independent parameters, they usually coexist as circuit specifications. A few years ago, op-amps were associated with ±15 V supplies, which shrank to ±5 V then to just +5 V; now, nominal +3 V supplies are common, with surprisingly little sacrifice in device performance. But low-power, low-voltage devices are not as forgiving of system design shortcomings because they have less input range to accommodate poorly behaved input signals, less head room to deal with dynamic range requirements, and less output drive capacity. System design decisions should still favor higher voltage rails where these are possible.

Supply Current

One of the disbenefits of not showing supply connections is that it is easy to forget about supply current (I_S). Data sheets will normally give typical and maximum figures for I_S at a specified voltage and no load. If the supply voltages are the same in the circuit as on the data sheet, and if none of the outputs are required to deliver any significant current, then it is reasonable just to add the maximum figures for all the devices in circuit to arrive at a worst-case power consumption. At other supply voltages, you will have to make some estimate of the true supply current, and some data sheets include a graph of typical I_S versus supply voltage to aid in this. Also, note that I_S varies with temperature, usually increasing with cold.

When an op-amp output drives a load, be it resistive, capacitive or inductive, the current needed to do so is drawn from one supply rail or the other, depending on the polarity of the output. In the worst case of a short-circuit load, I_S is limited by the device's output current limiting. It is quite possible for the load currents to dominate power supply drain. With typical quiescent I_S figures of a milliamp or so, you only need an output load resistance of 10 kΩ being driven with a ±10 V swing to double the actual current consumption of the circuit. When calculating worst-case load currents in these circumstances, you need to know not only the maximum output swing into resistive loads but also the current that may be needed to drive capacitive loads.

I_S Versus Speed and Dissipation

Op-amp supply current is usually a trade-off against speed. You can find devices that are spec'd at 10 μA I_S, but such a part can only offer a slew rate of 0.03 V/μs. Conversely, fast devices require more current, often up to 10 mA. At these levels, package dissipation rears its head. An op-amp run at ±15 V with 10 mA I_S is dissipating 300 mW. With a thermal resistance of 100−150°C/W (the data sheet will give you the exact value) its junction temperature will be 30−45°C above ambient (see Section 9.5.1), and this is before it drives any load! This could well prevent the use of the part at high ambient temperatures and will also affect other parameters that are temperature sensitive. With such a device, make sure you know what its operating temperature will be before getting deeply involved in performance calculations.

5.2.15 TEMPERATURE RATINGS

And so we come naturally to the question of over what temperature range can you use a particular device. Analog ICs historically have been marketed for three distinct sectors, with three specified temperature ranges:

*   Commercial: 0 to +70°C
*   Industrial: −40 to +85°C (occasionally −25 to +85°C)
*   Military: −55 to +125°C

The picture is nowadays slightly blurred with the introduction of parts for the automotive market, which may be spec'd over −40 to +125°C, and with some Japanese suppliers (predominantly in the digital rather than analog area) offering nonstandard ratings such as −20 to +75°C.

If you are designing equipment for the typical commercial environment of 0−50°C then you are not going to worry much about device temperature ratings: just about every IC ever made will operate within this range. At the other extreme, if you are designing for military use then you will be buying military-qualified components, paying the earth for them, and this book will be of little use to you. But the question quite frequently arises, what parts should I use when my ambient temperature range goes a few degrees below 0 or above 70°C?

In theory, you should use industrial temperature−rated devices. Unhappily, there are three good reasons why you might not:

*   the part you want to use may not be available in the industrial range;
*   if it is available, it may be too expensive;
*   even if it is listed as available, it may actually prove to be on a long lead-time or otherwise hard to get.

So the question resolves itself into: can I use commercial parts outside their specified temperature range? And the answer is: maybe. No IC manufacturer will give you a guarantee that the part will operate outside the temperature range that he specifies. But the fact is, most parts will, and there are two main factors that limit such use, namely specifications and reliability.

Specification Validity

The manufacturer will specify temperature-sensitive parameters (which is most of them) either at a nominal temperature (25°C) or over the temperature range. These specifications have bite, in that if the part fails to meet them the customer is entitled to return it and ask for a replacement. So the manufacturer will test the parts at the specification limits. However, he is not responsible for what happens outside the temperature range, and it is more than likely that some parameters will drift out of their specification when the temperature limits are overstepped. Very often these parameters are unimportant in the application, such as offset voltage in an ac amplifier. Therefore you can, with care, design a circuit with wider tolerances than that would be needed for the published figures and trust that these will be sufficient for wide temperature range abuse.

It is of course a risky approach, and two extra risks are that some parameters may change much more outside the specified temperature range than they do within it, and that you may successfully test a sample of manufacturer A's product, but manufacturer B's nominally identical parts behave quite differently. We shall comment on this again in Section 5.2.15.

Package Reliability

The second factor is reliability. The reliability of any semiconductor device worsens with increasing temperature; a temperature rise of 10°C halves the expected lifetime. So operating ICs at high temperatures is to be avoided wherever possible, but there is no magic cut-off at 70 or 85°C. The maximum junction temperature should always be observed, but this is usually in the region of 100–150°C.

At low temperatures the problem is included moisture. Molded plastic packages allow some moisture to creep along the lead-to-plastic interface (this is worse at high temperatures and humidities), and this can accumulate over the surface of the chip, where it is a long-term corrosive influence. When the operating temperature dips below 0°C the moisture freezes, and the resulting change in conductivity and volume can give sudden changes in parameters, which are well outside the drift specifications. The effect is very much less with "hermetic" packages using a glass–ceramic–metal seal, and in fact progress in plastic packages has advanced to the point where included moisture is not as serious a problem as it used to be. Other board-related problems arise when equipment is used below 0°C due to condensation of airborne moisture on the cold printed circuit board surface, as ambient temperature rises.

5.2.16 COST AND AVAILABILITY

The subtitle to this section could be, why use industry standards? Basically, the application of op-amps (along with virtually all other components) follows the 80/20 rule beloved of management consultants: 80% of applications can be met with 20% of the available types. These devices, because of their popularity, become "industry standards" and are sourced by several manufacturers. Their costs are low and their availability is high. The majority of other parts are too specialized to fulfill more than a handful of applications and they are only produced by one or perhaps two manufacturers. Because they are only made in small quantities their cost is high, and they can sometimes be out of stock for months.

When to Use Industry Standards

The virtue of selecting industry standards is that the parts are well established, unlikely in the extreme to run into sourcing problems or be withdrawn (the humble 741 has been around for over 30 years!) and, because of the competition between manufacturers, they will remain cheap. If they will do the job, use them in preference to a sole-sourced device. For companies with many different designs of product, keeping the variety of component parts low and reusing them in new designs has the benefit of increasing the total purchase of any given part. This potentially reduces its price further.

Another hidden advantage of older, more established devices is that their quirks and idiosyncrasies are well known to the suppliers' applications support engineers, and you are less likely to run into unusual effects that are peculiar to your usage and that take days of design time to resolve.

But nothing comes for free: the negative aspect of multisourcing is that many parameters go unspecified for cheap devices, and this leaves open the possibility that different manufacturers' nominally identical parts can differ substantially in those parameters that are omitted from the common spec. If you have designed and tested a circuit with manufacturer A's devices, and they happen to be quite fast, you will be heading for production problems when your purchasing manager buys a few thousand of manufacturer B's devices that are slower. For instance, TI's data sheet for the LM324 gives a typical slew rate of 0.5 V/μs at 5-V supply; but National, who could fairly be said to have invented the part, do not mention slew rate at all in theirs.

To deal with this, design the circuit from the outset to be insensitive to those parameters that are badly specified, unspecified or (worse) specified differently in the data sheets of each manufacturer. Or, look for a more tightly specified part.

When Not to Use Industry Standards

Within the last few years there has been a countertrend to the imperative for multisourcing, and the use of industry standards. Hundreds of new types have been introduced, and many of them are much better than their predecessors. They not only minimize the trade-offs between speed, power, precision and cost but are also more fully specified. You can select a part by function and application—for instance, a DAC buffer or 75-Ω cable driver—rather than by comparing technologies, or by looking at a particular specification such as gain-bandwidth product. Following the manufacturer's selection guides on the basis of application will often lead quickly to the most suitable part.

Selecting more application-specific ICs in this way steers the design process away from industry standards. But there are a number of reasons why alternate sourcing has become less of a necessity, despite its advantages given above. The average product life cycle—sometimes months rather than years—is much shorter than the lifetime of a good op-amp. In addition, qualifying multiple sources is a task that many designers do not have time or expertise to do fully. Finally, for highly competitive products, you will have to choose parts that give your design the edge (even if they are proprietary) and for which there may be nothing comparable in performance, cost, or functionality.

Quad or Dual Packages

Comparing prices, the LM324 does offer, in fact, the lowest cost-per-op-amp (5 p). This points up another factor to bear in mind when selecting devices: choose a quad or dual package in preference to a single device, when your circuit uses several gain stages. This reduces both unit cost and production cost. Such parts often have quiescent supply currents only slightly greater than a single-channel device, but with better offset, temperature tracking of drift, matching, and other specs. The disadvantages are inflexibility in supply voltage and pc layout, and possible thermal, power rail, or RF interaction between gain blocks on a single substrate.

Some parts are available only in dual or quad configurations because single-channel versions would not have enough applications. Conversely, highest speed op-amps, with bandwidths above several hundred megahertz, are often available in singles only, because of internal cross talk. However, pinouts in multichannel configurations are less standardized than the basic single-channel unit, so substitutes are harder to find.

5.2.17 CURRENT-FEEDBACK OP-AMPS

There are also op-amps that use current-feedback topology instead of the more familiar voltage feedback. Voltage feedback is the classic, well-understood mechanism, which we have been discussing all the way through this section so far. In current feedback, the error signal is a current flowing into the inverting input; the input buffer's low impedance, in contrast to a voltage amplifier's high input impedance, allows large currents to flow into it with negligible voltage offset. This current is the slewing current, and slew rate is a function of the feedback resistor and change in output voltage. Therefore, the current-feedback amplifier has nearly constant output transition times, regardless of amplitude.

A very small change in current at the inverting input will cause a large change in output voltage. Instead of open-loop voltage gain, the current-feedback op-amp is characterized by current gain or "transimpedance" Z_S. As long as $Z_S \gg R_F$, the feedback resistor, the steady-state (nonslewing) current at the inverting input is small, and it is still possible to use the usual op-amp assumptions as initial approximations for circuit analysis, i.e., the differential voltage between the inputs is negligible, as is the differential current (Fig. 5.20).

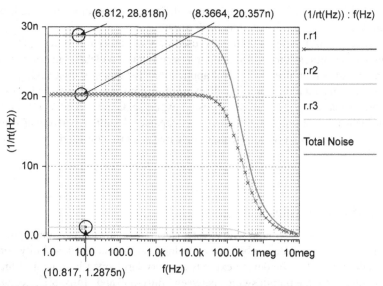

FIGURE 5.19 Noise Voltage Spectral Response Using a Simulated Noise Analysis.

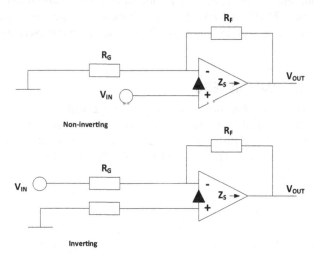

FIGURE 5.20 The Current-Feedback Circuit.

In performance, current feedback generally offers higher slew rate for a given power consumption than voltage feedback, and voltage feedback offers you flexibility in selecting a feedback resistor, two high-impedance inputs, and better dc specifications. With a current-feedback op-amp, you first set the desired *bandwidth* via the feedback resistor, and then the gain is set according to the usual resistive ratios. This means that the wider the bandwidth, the lower will be the operating impedances.

If R_F is doubled, the bandwidth will be halved. The circuit becomes *less* stable when capacitance is added across the feedback resistor.

Current-feedback devices tend to be used only at higher frequencies, for applications such as professional video and high-performance wideband instrumentation. The same part can be used in several applications for quantity cost savings, using only as much bandwidth as needed. They are less common in lower end consumer applications because they need more design expertise. Current feedback is no "better" or "worse" than voltage, which is also capable of similar performance in the right design, but it does provide an alternative that is worth considering in the appropriate application.

5.3 COMPARATORS

A comparator is just an op-amp with a faster slew rate, and with its output optimized for switching. It is intended to be used open-loop, so that feedback stability considerations do not apply. The device exploits the very large open-loop gain of the op-amp circuit so that the output swings between "fully-on" and "fully-off", depending on the polarity of the differential input voltage, and there should be no stable state in between. Input-referred and open-loop parameters—offsets, bias currents, temperature drift, noise, CMRR, and PSRR, supply current and open-loop gain—are all specified in the same way as op-amps. Output and ac parameters are specified differently.

5.3.1 OUTPUT PARAMETERS

The most frequent use of a comparator is to interface with logic circuitry, so the output circuit is designed to facilitate this. Two configurations are common: the open collector, and the totem pole (Fig. 5.21). The open-collector type requires a pull-up resistor externally, while the totem pole does not. Both types interface

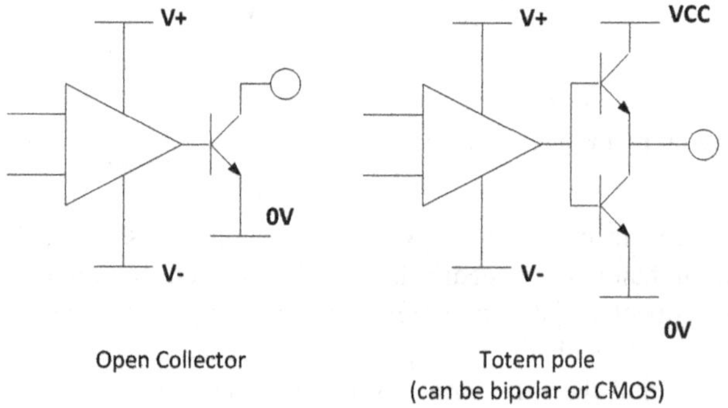

Open Collector

Totem pole
(can be bipolar or CMOS)

FIGURE 5.21 Comparator Outputs.

readily to the classical LSTTL logic input, which requires a higher pull-down current than is needed to pull it up. The CMOS input, which only takes a small current at the transition due to its input capacitance, is even easier. The output is specified either in terms of its saturation voltage, sink current, leakage current, and maximum collector voltage for the open-collector type or in terms of high- and low-level output voltages at specified load currents for the totem-pole type.

Because the totem-pole type is invariably aimed at logic applications, it is always specified for 3.3 or 5 V output levels. The open-collector type, which includes the highly popular LM339/393 and its derivatives, is more flexible since any output voltage can be obtained simply by pulling up to the required rail, which can be separate from the analog supply rails.

5.3.2 AC PARAMETERS

Because the comparator is used as a switch, the only ac parameter that is specified is the response time. This is the time between an input step function and the point at which the output crosses a defined threshold. It includes the propagation delay through the IC and the slewing rate of the output. Outside of the device itself, two factors have a large effect on the response time:

- the input overdrive
- the output load impedance.

Overdrive

For the specifications, an input step function is applied, which forces the differential input voltage from one polarity to the other. The overdrive, as in Fig. 5.22, is the final steady-state differential voltage. Usually, the step amplitude is held constant, and its offset is varied to give different overdrive values. The greater the overdrive, the more current is available from the differential input stage to propagate the change of state through to the output, although beyond a certain point there is no gain to be had from increasing it. Small overdrives can lead to surprisingly long response times, and you should check the data sheet carefully to see if the device is being specified in similar fashion to how your circuit will drive it.

The specification test assumes that the step function has a much shorter rise time than the response to be measured. Response time specs are virtually meaningless

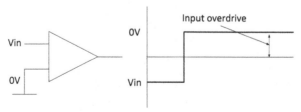

FIGURE 5.22 Comparator Overdrive.

when the comparator is driven by slow rise time analog signals. We shall discuss this more fully under the heading of hysteresis.

Load Impedance

The output load resistance R_L (for open-collector types) and capacitance C_L have a major influence on the output slewing rate. The capacitance includes the device output capacitance, circuit strays and the input capacitance of the driven circuit (this last is usually the most significant). The slewing rate is determined by the current that is available to charge and discharge the capacitance, following the rule $dV/dt = I/C$. For the negative-going transition this current is supplied by the output sink transistor and is in the region of 10–50 mA, assuring a fast edge, but the current available to charge the positive transition is supplied by the pull-up device or resistor and may be an order of magnitude lower. The choice of output resistor directly affects the positive-going rise time (Fig. 5.19) and the power dissipation of the circuit (Fig. 5.23).

The Advantages of the Active Low

On this latter point, it is worth remembering that if you expect low duty–cycle pulses at the output, want low power drain and a fast leading edge and have a choice of logic polarity, that the preferable configuration is to use an active low output as in Fig. 5.24A. The signal is normally off so that power drain is low, and the leading edge transition depends on the output transistor rather than the pull-up. If a fast trailing edge is also needed, the pull-up can be reduced in value without significantly affecting power drain if the duty cycle is low. It is easy and cheap to provide a logic inverter if you really need positive-going pulses.

Pulse Timing Error

Continuing this train of thought, you can see that it is quite easy for the pulse timing to be affected by the output rise- and fall-times. This is quite often the source of unexpected errors in circuits, which convert analog levels into pulse widths for timing

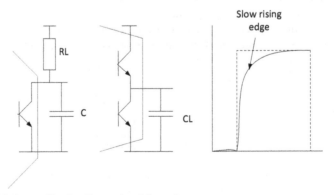

FIGURE 5.23 Output Slewing Versus Load Capacitance.

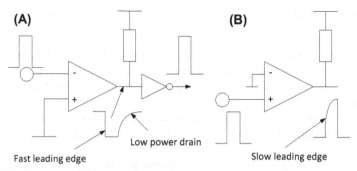

FIGURE 5.24 Comparator Output Configurations.

(A) Preferred configuration (B) poor configuration.

measurement. Because the pulse rising edge is slowed to a greater extent than the falling edge, the point at which it crosses the following logic gate's switching threshold is different, so that rising and falling analog inputs result in different switching points. This effect is demonstrated in Fig. 5.25. The problem is generally more visible with CMOS-input-level gates than it is with TTL-input-level ones, as TTL's switching threshold is closer to 0 V whereas the CMOS threshold is ill-defined, being anywhere between 0.3 and 0.7 times its supply rail. The difference can amount to a microsecond or more in low-power circuits.

5.3.3 OP-AMPS AS COMPARATORS (AND VICE VERSA)

You may often be faced with a circuit full of multiple op-amp packages and the need for a single comparator. Rather than invest in an extra package for the comparator

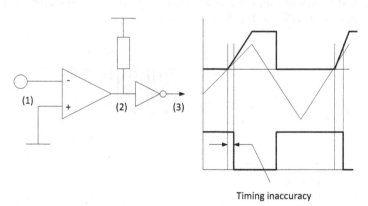

FIGURE 5.25 Timing Error Through Pull-Up Delay.

function, it is quite to use a spare op-amp as a comparator with the following provisos:

- The response time and output slew rate are adequate. Typical cheap op-amp slew rates of 0.5 V/μs will traverse the logic "gray area" from 0.8 to 2 V in about 3 μs; this is too slow for some logic functions. Faster op-amps make better comparators.
- In some op-amps, recovery from the saturated state can take some time, causing appreciable delays before the output starts to slew. This is hardly ever specified on data sheets.
- The output voltage swing and drive current are adequate and correct for the intended load. Clearly you cannot drive 5-V logic directly from an op-amp output that swings to within 2 V of ±15 V supply rails. Some form of interface clamping is needed; this could take the form of a feedback zener arrangement so that the output is not allowed to saturate, which confers the additional benefit of reduced response time. Drive current is not a problem with CMOS inputs.

It is also possible, if you have to, to use a comparator as an op-amp. (In most cases: some totem-pole outputs cannot be operated in the linear mode without drawing destructively large supply currents.) It was never designed for this, and will be hideously unstable unless you slug the feedback circuitry with large capacitors, in which case it will be slow. Also, of course, it is not characterized for the purpose, so for some parameters you are dealing with an unknown quantity. Unless the application is completely noncritical it is best to design op-amp circuits with op-amps.

5.3.4 HYSTERESIS AND OSCILLATIONS

When the analog input signal is changing relatively slowly, the comparator may spend appreciable time in the linear mode while the output swings from one saturation point to the other. This is dangerous. As the input crosses the linear-gain region the device suddenly becomes a very high—gain open-loop amplifier. Only a small fraction of stray positive feedback is needed for the open-loop amplifier to become a high-frequency oscillator (Fig. 5.26).

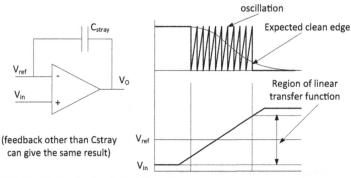

FIGURE 5.26 Oscillation During Output Transitions.

The frequency of oscillation is determined by the phase shift introduced by the stray feedback and is generally of the same order as the equivalent unity-gain bandwidth. This is not specified for comparators, but for typical industry-standard devices is several of MHz. The term "relatively slowly" as used above means relative to the period of the oscillation, so that any traverse of the linear region, which takes longer than a few hundred nanoseconds must be regarded as slow: this of course applies to a very large proportion of analog input signals!

The Subtle Effects of Edge Oscillation

This oscillation can be particularly troublesome if you are interfacing to fast logic circuits, especially when connecting to a clock input. It can be hard to spot on an oscilloscope, as you will probably have the time base set low for the analog signal frequency, but the oscillations appear to the digital input as multiple edges and are treated as such: so for instance a clock counter might advance several counts when it appears to have had only one edge, or a positive-going clock input might erroneously trigger on a negative-going edge.

Even when you do not have to contend with high-speed logic circuits, the oscillation generated by the comparator can be an unexpected and unwelcome source of RF interference.

Minimize Stray Feedback

The preferred solution to this problem is to reduce the stray feedback path to a minimum so that the comparator remains stable even when crossing the linear region. This is achieved by following three golden rules:

- keep the input drive impedance low;
- minimize stray feedback capacitance by careful layout; and
- avoid introducing other spurious feedback paths, again by careful layout and grounding.

The lower the input impedance, the more feedback capacitance is needed to generate enough phase shift for instability. For instance, 2 pF and 10 kΩ gives a pole frequency of 8 MHz, a perfectly respectable oscillation frequency for many high-speed comparators. It is hard to reduce stray capacitance much below 2 pF, so the moral is, keep the drive impedance below 10 kΩ, and preferably an order of magnitude lower.

Minimum stray capacitance from output to input should always be the layout designer's aim; follow the rules quoted in Section 5.2.10 for high-frequency op-amp stability. Most IC packages help you in this regard by not putting the output pin close to the noninverting input pin. Do not look this particular gift horse in the mouth by running the output track straight back past the inputs! Guarding the inputs (see Section 2.4.1) can be useful. And, again as with op-amp circuits, do not introduce ground-loop or common-mode feedback paths by incorrect layout.

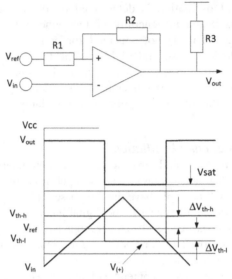

FIGURE 5.27 Hysteresis.

Ignoring input and output leakage currents

V_{out} (H) $= V_{cc} - (V_{cc} - V_{ref})(R_3/[R_1 + R_2 + R_3])$

V_{out} (L) $= V_{sat}$

$V_{th-h} = \alpha \times V_{cc} + (1 + \alpha) \times V_{ref}$ where $\alpha = R_1/(R_1 + R_2 + R_3)$

$\Delta V_{th-h} = \alpha \times (V_{cc} + V_{ref})$

$V_{th-l} = \beta \times V_{sat} + (1 - \beta) \times V_{ref}$ where $\beta = R_1/(R_1 + R_2)$

$\Delta V_{th-l} = \beta \times (V_{sat} - V_{ref})$

A common simplification is that $R_3 << R_1 + R_2$ so that $\alpha = \beta$, and that V_{ref} is half of V_{cc} and $V_{sat} = 0$, in which case ΔV_{th} (the total hysteresis band) reduces to $\beta \times V_{cc}$.

Hysteresis

Another approach to the problem of unwanted oscillation is to kill it with hysteresis. This approach is used when the above methods fail or cannot be applied, and you can also use it as a legitimate circuit technique in its own right, as in the well-known Schmitt trigger. Hysteresis is the application of deliberate positive feedback to propel the output speedily and predictably through the linear region. The principle of hysteresis is shown in Fig. 5.27.

Note that although this looks superficially like the classic inverting op-amp configuration, feedback is applied to the *non*inverting input and is therefore operating in the positive sense. Note also that the application of hysteresis modifies the switching threshold in both directions, and that it is modified differently in either direction by the presence of R_3. This resistor is shown in the circuit of Fig. 5.27 to emphasize that it must be included in calculating hysteresis; we have assumed that the comparator is the open-collector type. If the output is the totem-pole type, then R_3 is omitted but the output levels and impedance must be taken into consideration. These values directly affect the switching threshold and can cause surprisingly large inaccuracies.

Because hysteresis deliberately alters the switching threshold, it cannot be indiscriminately applied to all comparator circuits to clean up their oscillatory tendencies, nor should it. The techniques outlined previously should be the first priority. But it is not always possible to keep drive impedances low and where high impedance is necessary, hysteresis is a valuable tool. If the minimum input dV/dt is predictable, you can also apply a judicious amount of ac hysteresis (by substituting a capacitor for R_2), which will prevent oscillation without affecting the dc threshold: but beware of slow-moving inputs, or you will simply end up with a longer time-constant oscillator!

5.3.5 INPUT VOLTAGE LIMITS

When an op-amp is operating in closed loop, the differential voltage at its inputs is theoretically zero. If it is not then the feedback loop is open, either by design or because of one or another form of overloading. Comparators on the other hand are intended for open-loop operation, and their differential input voltage is never expected to be zero.

Data sheets specify the maximum voltage range of differential input signals and this should not be overlooked. If it is exceeded, too much current through the breakdown of the input transistor base–emitter junctions (or MOS gates) can degrade the input offset and bias current parameters. Most of the industry-standard LM339 derivatives have a differential limit equal to the supply rail limit, but some comparators have quite restricted differential input ranges. For instance the fast NE529 has a differential input restriction of ±5 V, with a common mode of ±6 V. These two quantities interact: both inputs at +4 V will satisfy the common-mode limit, but if one is left at +4 V the other cannot be taken below −1 V because the differential voltage is then greater than 5 V.

Even if the normal operating differential range is kept within limits, it is possible for abnormal conditions (such as cycling of separate power rails) to breach the limit. If this is at all likely, and if the condition cannot be prevented, at the very least include some input current protection resistance. You can calculate the required values from the expected or possible overvoltage divided by the absolute maximum input current, or from the power dissipation, which is always quoted on device data sheets.

Comparator Parameters Versus Input Voltage

Also, while considering large differential input voltages, remember that unexpected things can happen to the comparator even when the limits are not exceeded. Response time is usually specified for a common-mode voltage of zero and may degrade when the common-mode limits are approached; this applies equally to bias currents. Some data sheets show curves of input bias current, which have

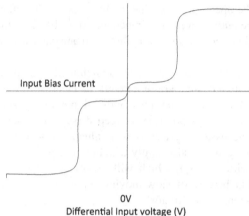

Differential Input voltage (V)

FIGURE 5.28 Input Bias Current Steps.

step changes (Fig. 5.28) at certain differential input voltages, due to internal dc feedback. Notice these and make sure your circuit can cope!

In multichannel packages, some comparators may remain unused. Never leave unused inputs open, as that device could oscillate on its own, which would then be coupled into the other devices in the same package. If both inputs are grounded, the unpredictable offset voltage will mean that the output voltage, and hence unit supply current, will vary. The safest course is to ground one input and supply the other from another fixed voltage within its differential and common-mode limit (which might include the supply rail), so that the device is always saturated.

5.3.6 COMPARATOR SOURCING

Exactly the same comments about sourcing apply to comparators as were made earlier about op-amps (see Section 5.2.15). Like the LM324 op-amp, the most popular and cheapest part per comparator is the quad LM339, with its dual counterpart the LM393 not far behind.

5.4 VOLTAGE REFERENCES

The need for a stable reference voltage is found in power supplies, measurement instrumentation, DAC/ADC systems and calibration standards. Two techniques exist to provide such references, one based on the precision zener diode and the other on the band-gap voltage of silicon.

5.4.1 ZENER REFERENCES

We have already discussed the operation of the basic zener diode (Section 4.1.7). To produce a reference from a zener, it must be temperature compensated, fed from a constant current and buffered. Temperature compensation is achieved by selecting a low-tempco zener voltage in the range 5.5–7 V and mating it with a silicon diode so that the voltage tempcos cancel. The combination is driven from a constant current generator and buffered to give a constant output voltage regardless of load.

Since surface breakdown increases noise and degrades stability, a precision zener is usually fabricated below the surface of the IC, which contains its support circuitry, but this gives a greater spread of tempco and absolute voltage. The overall reference must therefore allow adjustment of these parameters, normally by laser wafer trimming. Such references can offer long-term stability of 50 ppm/year and absolute accuracy of 0.1% with ±10 ppm/°C tempco. Better performance is obtained if the reference can be stabilized with an on-chip heater, as in the LM399 for example. This takes a comparatively large power drain and has a warm-up time measured in seconds but offers sub-ppm tempcos.

5.4.2 BAND-GAP REFERENCES

A significant disadvantage of the zener reference is that its output voltage is set at around 6.9 V and it therefore needs a comparatively high supply voltage. A competing type of reference overcomes this and other problems, notably cost and supply current, and has become extremely widespread since its invention by Robert Widlar in 1971. The fundamental circuit is shown in Fig. 5.29. In this circuit I_1 and I_2 differ by a fixed ratio and V_{ref} is given (neglecting base currents) by

$$V_{\text{ref}} = V_{BE3} + I_2 \times R_2 = V_{BE3} + (V_{BE1} - V_{BE2})R_2/R_1 \tag{5.18}$$

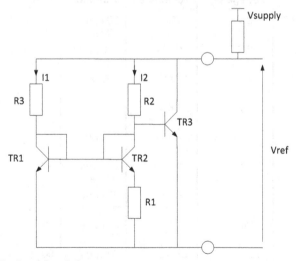

FIGURE 5.29 The Band-Gap Reference.

The temperature coefficient of the second term can be arranged by suitable selection of I_1, R_1, and R_2 to cancel that of the V_{BE3} term. This turns out to occur when V_{ref} is in the neighborhood of 1.2 V, which is equivalent to the "band-gap" voltage of a silicon junction at 0°K.

Such a band-gap reference, relying only on matched transistors, is easily integrated along with biasing, buffer, and amplifier circuitry to give a complete reference in a single package. It is capable of a lower minimum operating current and a sharper knee than any zener. As well as the unprocessed band-gap voltage of 1.2 V (actual voltage depends on detailed internal design and process variations and varies between 1.205 and 1.26 V) devices are available with trimmed outputs of 2.5, 5, and 10 V, principally for use in digital-to-analog/analog-to-digital conversion circuits. Other voltages are available, and there are several adjustable parts offered as well.

Costs and Interchangeability

There is an obvious trade-off between initial voltage tolerance and tempco on the one hand, and cost and availability on the other, since the manufacturer has to accept a lower yield and longer test and trim time for the closer tolerances. Initial voltage can be trimmed exactly with a potentiometer, but this method adds both parts and production cost, which will offset the higher cost of a tighter tolerance part. Trimming the reference voltage can also worsen the reference temperature coefficient in some configurations, and there is the extra tempco of the trimming components to include. Table 5.2 shows a sample of typical two-terminal 1.2-V references,

Table 5.2 Some Voltage References

Type	Output Voltage (V)	Tolerance	Tempco	Min. Current (μA)	Cost £, 25+
MAX6520EUR-T	1.2	±1%	20 ppm/°C typ	50	1.29
LM4041B-1.2	1.225	±0.2%	100 ppm/°C	45	0.97
ICL8069DCZR	1.23	±1.6%	100 ppm/°C	50	0.78
ICL8069CCZR	1.23	±1.6%	50 ppm/°C	50	1.27
LM385Z-1.2	1.235	±2%	20 ppm/°C avg	10	0.30
LM385Z-1.2	1.235	±1%	20 ppm/°C avg	10	0.55
LT1004CZ-1.2	1.235	±4 mV	20 ppm/°C	10	1.68
ZRA124A01	1.24	±1%	30 ppm/°C	50	0.67
ZRA125F02	1.25	±2%	30 ppm/°C	50	0.55

including their tolerance, tempco, minimum operating current and cost. Most of these are available in different grades, corresponding to tighter or looser tolerances and tempcos.

Although it would appear from this table that there is a wide choice of types offering much the same performance, not all of these are directly interchangeable. The minor differences in regulation voltage may catch you out if you have designed a circuit for a given voltage tolerance and subsequently want to change to a different type. The preferable solution is to allow as wide a tolerance as possible in the first place. Also, there are variations in the allowable or required capacitive loading. Some parts require a decoupling capacitor of $0.1-1$ μF across them, others require that such a capacitor is *not* included. The parts are mostly supplied in the TO-92 package or the small outline SOT23, but not all pinouts are the same. Again, check before specifying alternatives.

5.4.3 REFERENCE SPECIFICATIONS

Line and Load Regulation

Line regulation is the change in output voltage due to a specified change in input voltage, normally quoted in microvolts per volt. Load regulation is a similar change due to a change in load current, expressed either in percent for a given current change or as a dynamic resistance in ohms. It should, but does not always, include self-heating effects due to dissipation change.

Output Voltage Tolerance

This is the deviation from nominal output voltage. It is quoted at a given temperature and input voltage or current, and the nominal voltage will differ under other conditions. Generally it is expressed as a percentage figure, but the asymmetry of device yields may persuade a manufacturer to quote upper and lower bounds and the nominal figure may not be in the middle of them. In your circuit design, it is best to ignore the nominal voltage and work everything out for upper and lower limits.

Output Voltage Temperature Coefficient

This is the output voltage change due to an ambient temperature difference, usually from 25°C. Because neither band-gap nor zener references exhibit a straight line voltage–temperature curve (see Fig. 5.26) manufacturers choose different ways to express their temperature coefficients, sometimes as an average value across the range in ppm/°C, sometimes as different values at a series of spot temperatures, and sometimes as a worst-case error band in mV. To evaluate different manufacturers' references properly you need to correct for these differences in specification (Fig. 5.30).

Long-Term Stability

Usually expressed in ppm/1000 h or in microvolts change from the nominal voltage, this is a difficult specification to verify and so is often quoted as a typical figure based on characterization of a sample. It is rarely specified on the cheaper

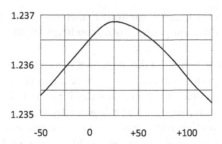

FIGURE 5.30 Typical Band-Gap Reference Temperature Characteristics.

components. Zeners tend to stabilize after a couple of years, so for ultraprecision applications the practice of burning in zener references at high temperatures to speed up the settling process is sometimes followed.

Settling Time
This is the time taken for the output to settle within a specified error band after application of power. It is typically in the tens-to-hundreds of microseconds region, and is normally only of interest if you are concerned about the dynamic performance of the reference circuit—for instance, if the application has to wake up rapidly from "sleep" mode. It does not include any long-term effects due to thermal shifts, but of course these do occur, more noticeably at higher operating currents.

Minimum Supply Current
The regulation of a two-terminal device is not maintained below a certain minimum current. Typical values for band-gap references are $50-100\ \mu A$, with $10\ \mu A$ being available although some earlier devices are much higher than this. The very low useable operating currents combined with low dynamic resistance at these currents make band-gap devices very much preferable to zener types for low-power circuitry. The maximum operating current is usually based on the point at which the device goes outside its regulation specification, but may also be determined by allowable power dissipation.

5.5 CIRCUIT MODELING
Virtually every op-amp supplier provides Spice models, which are a very useful approximation of device performance. There are two opposing criteria for such a model. It should use the fewest internal elements to ease computing, but it should also give an accurate representation of the device as a "black box." You can use these models as a necessary (but not entirely sufficient) step in the design process. Models cannot capture a device's every sensitivity to supply variations or temperature and load changes. Dynamic performance such as slew rate and overshoot are especially difficult to model, and peculiarities such as behavior at or beyond the common-mode limits will be entirely absent.

The circuit design must be characterized for the entire range of performance characteristics that an off-the-shelf part might show, but generally available Spice models use typical rather than worst-case specs.

Even a perfect model would not capture what is just as critical in high-performance analog design: your physical circuit that surrounds the part. Just a few picofarads of circuit-board capacitance will change the frequency response, for example. Common impedances in the power or ground circuits (see Section 1.1) can affect stability and power supply rejection. Conductive residues on the circuit provide a leakage path between IC pins. No model of itself will detail your circuit layout strays or ground topology.

This does not mean you should not use manufacturer supplied models. Use them for initial assessments of your circuit, to about ±20% accuracy. At the same time, recognize that the model itself is neither perfect nor does it include the subtleties of your design. Check with the supplier to understand which modeled parameters are typical, which are worst case, which are at room temperature, and other similar limitations and simplifications. The typically short development timescale, and the project manager champing at the bit, may constrain your ability to experiment and tempt you to go straight from the model to the final layout. But if there is any critical performance issue that you know is not covered by the model, be prepared for a few design iterations, and do not be afraid to breadboard the design if possible.

Some modern simulators such as Saber have libraries of characterized parts, which have been built using much improved behavioral models, however, the limitation is always the accuracy of the source measured data used, and the limitations of using a small sample of parts to obtain that data.

The best advice for effective use of simulation is to use it to evaluate the initial design to the limits of accuracy defined earlier, and use advanced techniques such as Monte Carlo (statistical analysis) to establish the range of operation of the circuit, and see how well the component tolerance work across a wide range of scenarios, not just attempting to fine-tune the nominal case.

Most suppliers offer evaluation boards and suggested circuit-board layout drawings, especially for high-performance or complex parts. An evaluation board shows you what the part can do in a reference design. The layout can serve as a starting point for your own implementation, so you will not waste time discovering mistakes the application engineers have already made and dealt with. The first question an applications support engineer asks when a designer calls with a problem such as oscillation in high-frequency current-feedback circuits is, "Did you use the evaluation board layout?"

Digital Circuits

CHAPTER OUTLINE

The Circuit Designer's Companion. http://dx.doi.org/10.1016/B978-0-08-101764-7.00006-2
Copyright © 2017 Elsevier Ltd. All rights reserved.

259

The great success story of digital electronics is due to one simple fact: information can be reduced to a stream of binary data, which can be represented as one of two discrete voltage levels. These data can be manipulated and processed at will, and the quantity of information you can process depends only on the speed at which you can do it. The infinite variability of analogue voltage levels is replaced by two dimensions of quantization, in voltage and time. In theory, all voltage levels below a given threshold represent binary 0, and all levels above the threshold represent binary 1. Again in theory, time is divided into discrete units by a reference clock, and the boundary between each unit marks the transition from 1 bit of data to the next.

By this means, the unpredictability and variability of analogue, or linear, electronic phenomena is factored out of the design process. (It is replaced by another kind of unpredictability and variability, which of complex software phenomena, but that is not the subject of this book.) Voltage drifts, component tolerances, offsets, and impedance inaccuracies become instantly irrelevant. At the same time, programmability allows a single piece of hardware to perform widely different tasks, including ones that perhaps were not even envisaged when it was designed and built. To incorporate such programming flexibility into analogue hardware would be impossible.

The millions of successful digital designs worldwide testify to these advantages. At the same time, much to the relief of those who would bewail the apparent soullessness of the digital universe, analogue phenomena have not been completely banished. They have merely changed their appearance. Ohm's law still holds, and the grand laws of electromagnetic field theory still maintain their grip on digital electrons, even tightening it as designer's strive for ever greater speed. Variability finds its way into the gap between 1 and 0 and into the spaces between one clock period and the next. It is the digital designer's task to understand it and deal with it.

6.1 LOGIC INTEGRATED CIRCUITS

The interfaces between logic (integrated circuits) ICs including signal, clock, and power supply lines must be considered to achieve a reliable digital design. This applies whether the devices concerned are microprocessors, their support chips, application-specific ICs (ASICs), programmable logic arrays or field programmable gate arrays (FPGAs), or simple "glue" logic.

6.1.1 NOISE IMMUNITY AND THRESHOLDS

A logic input can take any value of voltage, nominally from one supply rail to the other, although due to transmission line effects (Section 1.3), the actual voltage can exceed either supply rail on transitions. Each input is designed so that any voltage below one level, conventionally V_{IL}, is regarded as a logic "0" and any voltage above another level, V_{IH}, is regarded as logic "1" (Fig. 6.1).

These levels are characterized for each logic or microprocessor family, and worst-case values of V_{IL} and V_{IH} can be found on any data sheet. Note that, as with any hardware-determined parameter, they may vary with temperature and you should ensure that the values you use are guaranteed across the device's temperature range. They are also a function of supply voltage. If all ICs are fed from the same supply, this is not a problem; but it becomes more significant if you are interfacing logic circuits, which may be fed from different supply rails.

The significance of the band between V_{IL} and V_{IH} is that the input logic state (and therefore the output state) is undefined, while the voltage is in this band. Therefore transitions between logic states must happen as rapidly as possible, and no decisions must be taken while the input is in transit or for a given period (the "settling time") thereafter. This is why clocked, or synchronous, circuits are normally more reliable than unclocked or asynchronous ones for complex logic operations: the state of the clock determines when logic decisions are taken, and it is arranged that all data transitions occur when the clock is inactive.

Susceptibility to Noise

Provided that all signals to logic inputs, whether from other logic outputs or from interfaces to other circuits, lie outside the $V_{IL} - V_{IH}$ band when they are active, then in theory no misinterpretation of the input should occur. The difference between a "low" output logic level (V_{OL}) and V_{IL}, or that between a "high" output level (V_{OH}) and V_{IH}, is the noise immunity (expressed in volts) of the logic interface (Fig. 6.2). Notice that noise immunity is not a property of any particular device but of the interface between devices. The noise immunity of a family of devices (such as LVT or HCMOS) only refers to interfaces between devices of the same family.

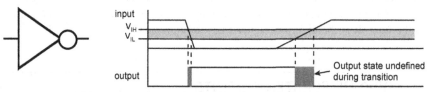

FIGURE 6.1 Transitions Through Logic Thresholds.

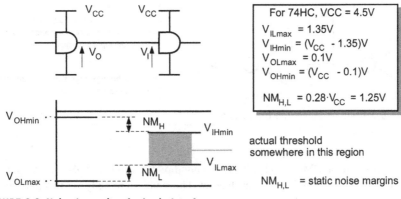

For 74HC, VCC = 4.5V

$V_{ILmax} = 1.35V$
$V_{IHmin} = (V_{CC} - 1.35)V$
$V_{OLmax} = 0.1V$
$V_{OHmin} = (V_{CC} - 0.1)V$

$NM_{H,L} = 0.28 \cdot V_{CC} = 1.25V$

actual threshold
somewhere in this region

$NM_{H,L}$ = static noise margins

FIGURE 6.2 Noise Immunity of a Logic Interface.

Current Immunity

The noise immunity value gives meaning to the ability of the interface to withstand externally coupled noise without corruption of the perceived logic level. So, for instance, the HCMOS–LSTTL interface can tolerate a variation of 2.4 V in the high state or 0.47 V in the low state. These are worst-case values, and the actual circuit could tolerate somewhat more before a change of state occurred. But the voltage difference is only part of the story. When noise is coupled into an interface, the impedance of the interface is just as important, since this determines what voltage will be developed by a given induced interference current. The impedance is normally defined by the output driver (as long as transmission line effects are neglected) and the effective noise current threshold of the interface, given by the noise immunity voltage divided by the driver output impedance, gives a truer picture of the actual noise immunity of a given combination.

The metal-gate 4000B CMOS logic family has a high output impedance at 5 V compared with the other families, so that its current immunity is significantly worse. However, as the supply voltage increases so its output impedance goes down, and the combined effect means that its immunity at 15 V V_{CC} is about 10 times better than at 5 V. It is inherently insensitive to low voltage inductively coupled noise but shows poor rejection of capacitively coupled noise. For general-purpose 5 V applications the 74HC family is preferred. It is also true that a microcontroller's high output resistance means that it does not compare favorably with standard logic.

Use of a Pull-Up

Note that the figures for high-state and low-state immunities are often different because of the differing drive impedances and voltage thresholds in the two states. A negative immunity value indicates that, if nothing further is done, this particular interface combination will be unreliable *by design*. For instance, the 2.7 V minimum high output of the venerable LS-TTL family is less than the required 3.15 V

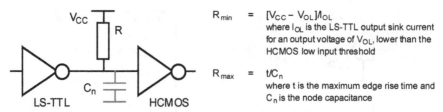

R_{min} = $[V_{CC} - V_{OL}]/I_{OL}$
where I_{OL} is the LS-TTL output sink current for an output voltage of V_{OL}, lower than the HCMOS low input threshold

R_{max} = t/C_n
where t is the maximum edge rise time and C_n is the node capacitance

FIGURE 6.3 Logic Interface Pull-Up Resistor.

minimum V_{IH} for HCMOS so LS-TTL driving directly into HCMOS is in danger of incorrectly transmitting the logic high level. The standard remedy for this particular situation (if you are still using LS-TTL!) is a pull-up resistor to V_{CC} to ensure a higher output from the LS-TTL (Fig. 6.3). The minimum resistor value is a function of the driver output capability, and the maximum value depends on permissible timing constraints. Alternatively, use the HCTMOS family, whose inputs are characterized especially for driving from LS-TTL levels.

Dynamic Noise Immunity

The static noise margins as discussed above apply until the interference approaches the operating speed of the devices. When very fast interference is present, higher amplitude is necessary to induce upset. The dynamic noise margin is measured by applying an interference pulse of known magnitude and increasing its width until the device just begins to switch. This yields a plot of noise margin versus pulse width such as shown in Fig. 6.4. The high level and low level dynamic noise margins may be different.

You may often be forced to interface different logic families. Typically, a 3.3 V microprocessor may need to drive 5 V buffers or vice versa, or you may not be able to source a particular part in the same family as the rest of the system, or you may need to change families to optimize speed/power product. You can normally expect logic interfaces of the same family to be compatible, but whenever different families

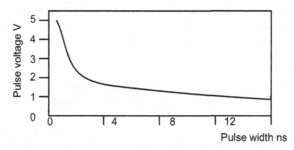

FIGURE 6.4 Dynamic Noise Immunity of 74HC Series Devices.

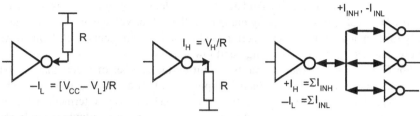

FIGURE 6.5 Logic Output Loading.

or a custom interface are used you have to check the logic threshold aspects of each one. The voltage level conversion issue is very common, to the extent that there are families of devices such as the 74LVT series, which are characterized for an input range of $2.0V_{IH}$ and $0.8V_{IL}$, but can still operate from a 5 V rail, or vice versa, and can accept 5 V-swing inputs while being operated from a 3.3 V rail.

6.1.2 FAN-OUT AND LOADING

The output voltage levels that are used to fix noise immunity thresholds are not absolute. They depend as usual on temperature but more importantly on the output current that the driver is required to source or sink. This in turn depends on the type of loading that each output sees (Fig. 6.5).

Any driver has an output voltage versus current characteristic, which saturates at some level of loading (Fig. 6.6). The characteristic is tailored so that at a given load current, the output voltage V_{OH} or V_{OL} is equivalent to the input threshold voltage (V_{IH} or V_{IL}) plus the noise immunity for that particular logic family. This load current then corresponds to the sum of the input currents for a given number N of standard gates of that family, and N is called the "fan-out": that number of standard gates that the output can drive and still keep the interface within the noise threshold limits. The fan-out is normally specified against each output of a device for other devices of

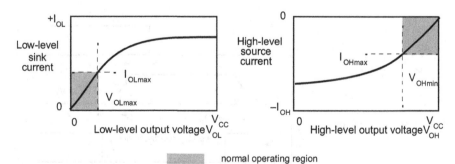

normal operating region

FIGURE 6.6 Logic Driver Output Characteristics.

the same family, but it can be calculated for other logic family interfaces or indeed for any DC load current simply by comparing the output voltage and current capability for each logic state with the current and voltage requirements of each input. As before, fan-out figures for logic high and low may differ.

For CMOS families there is rarely a limit on the number of inputs that can be driven at DC by one output. When low-power parts are required to drive high-power types then fan-out may be insufficient and you have to use intermediate buffer devices. The low drive capability of many microprocessor bus outputs severely restricts the number of other components that may be placed on the bus without interposing additional bus buffers—which accounts for the phenomenal popularity of the 74XX244 type of octal bus driver!

Dynamic Loading

The DC load current taken by the input side of the interface is only part of the total load. Indeed, for CMOS-input logic ICs, it is negligible and has no significant effect on fan-out calculations. But every input has an associated capacitance to ground, and the charging or discharging of this capacitance limits the speed at which the node can operate. Typical logic IC input capacitances are 5−10 pF; and these must be summed for all connected inputs, together with an allowance for interconnection capacitance, which is layout dependent but is again typically 5 pF, to reach the total load capacitance facing the driver.

Driver dynamic output current ability is rarely specified on data sheets but some manufacturers give application guidance. For instance, the 74HC range can offer typically ± 40 mA for standard devices and ± 60 mA for buffers at 4.5 V supply. This current slews the interface node capacitance C_n (Fig. 6.7) that you have just calculated and you have to ensure that slewing from logic 0 to the logic 1 threshold, or vice versa, is accomplished before the input data level is required to be valid. As an example, a 100 pF capacitance slewing from 0 to 3 V with 40 mA drive will take 7.5 ns, and this time (plus a safety factor) has to be added onto the other specified propagation delays to ensure adequate timing margin. If the figures do not add up, you will need to add extra buffer devices (which add their own propagation delays), reduce the load, reduce the operating frequency, or go to a faster logic family.

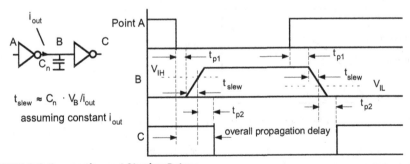

FIGURE 6.7 Propagation and Slewing Delays.

If you choose to run CMOS devices with a high load capacitance and accept that the edges will be slower, then be aware that this also reduces the reliability of the device because of the higher transient currents that the output drivers are handling.

6.1.3 INDUCED SWITCHING NOISE

This phenomenon is more colloquially known as "ground bounce." We are not talking here about external noise signals but about noise that is induced on the supply rails by the switching action of each logic gate in the circuit.

As each gate changes state, a current pulse is taken from the supply pins because of the different device currents required in each state, the external loading, the transient caused by charging or discharging the node capacitance, and the conduction overlap in the totem pole output stage. All these effects are present in all logic families to some extent, although CMOS types suffer little from the first two. In most cases, the node capacitance charging current dominates, more so with higher-speed circuits. The capacitance C_n must be charged with a current of

$$I = C_n \times \frac{dV}{dt} \tag{6.1}$$

Thus a 74AC-series gate with a dV/dt of around 1.6 V/ns will require a 50 mA current pulse when charging a 30 pF node capacitance. Fig. 6.8 shows the current paths. The significance of the supply current spike is that it causes a disturbance in the supply voltage and also in the groundline because of the inductive impedance of the lines. A pulse with a di/dt of 50 mA/ns through a track inductance of 20 nH (one inch of track) will generate a voltage pulse of 1 V peak, which is approaching the noise margin of fast logic. Supply voltage spikes are not too much of a problem as the logic high level noise immunity is usually good, and they can be attenuated by proper decoupling, as the next section shows. Groundline disturbances are more threatening. Pulses on a high-impedance groundline can easily exceed the noise

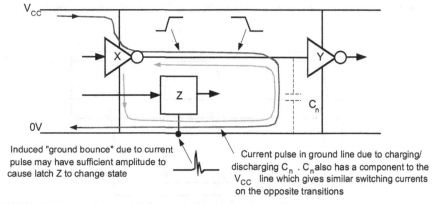

Induced "ground bounce" due to current pulse may have sufficient amplitude to cause latch Z to change state

Current pulse in ground line due to charging/discharging C_n. C_n also has a component to the V_{CC} line which gives similar switching currents on the opposite transitions

FIGURE 6.8 Induced Ground Noise Due to Switching Currents.

threshold and cause spurious switching of innocent gates. Only if a good, low-inductance ground plane system is maintained, as discussed in Chapter 1 and again in Section 2.2.4, can this problem be minimized.

Synchronous Switching

The supply pin pulse current is magnified in synchronous systems when several gates switch simultaneously. A typical example is an octal bus buffer or latch whose data change from #FF$_H$ to #00$_H$. If all outputs are heavily loaded, as may be the case when the device is driving a large data bus, a formidable current pulse—exceeding an amp in fast systems—will pass through the ground pin. Worse, if 7 bits of an octal latch change simultaneously, the induced ground bounce may corrupt the state of the eighth bit. You need to ensure that such devices are grounded to their loads with a very low-inductance ground system, preferably a true ground plane.

Ground noise on a microprocessor board can easily be observed by hooking a wide-bandwidth oscilloscope to the groundline—you can connect the scope probe tip and its ground together and still see the noise, since the magnetic fields due to the ground currents will induce a signal in the loop formed by the probe leads. What you see is a regular series of narrow, ringing pulses spaced at the clock period. The amplitude of each pulse varies because the sum of the data transitions is random, but the timing does not. Such noise (Fig. 6.9) is impossible to remove entirely.

6.1.4 DECOUPLING

No matter how good the V_{CC} and ground connections are, you cannot eliminate all line inductance. Except on the smallest boards, track distance will introduce an impedance that will create switching noise from the transient currents discussed in the last section. This is the reason for decoupling (Fig. 6.10).

Distance

The purpose of a decoupling capacitor is to maintain a low dynamic impedance from the individual IC supply voltage to ground. This minimizes the local supply voltage droop when a fast current pulse is taken from it. The word "decoupling" means

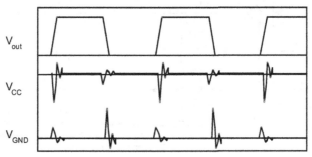

FIGURE 6.9 Switching Noise on Power and Ground Rails.

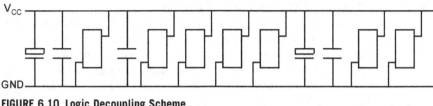

FIGURE 6.10 Logic Decoupling Scheme.

isolating the local circuit from the supply impedance. Bearing in mind the speed of the current pulses just discussed, it should be clear that the capacitor must be located close to the circuit it is decoupling. "Close" in this context means less than half an inch for fast logic, such as 74AC or ECL, especially when high-current devices such as bus drivers are involved, extending to several inches for low-current, slow devices such as 4000B-series CMOS.

If the decoupling current path between IC and capacitor is too long, the track inductance in conjunction with the capacitor forms a high-Q LC tuned circuit, and the ringing it generates will have a worse effect than no decoupling at all.

Capacitor Type and Value

The crucial factor for high-speed logic decoupling is lead inductance rather than absolute value. Fig. 3.19, showing the impedance of different capacitor types, is instructive in this regard. Minimum lead inductance offers a low impedance to fast pulses. Small chip capacitors are preferred, and the smaller the better, since this minimizes the package inductance. 0805, 0603, or even 0402 size is acceptable.

You can calculate the value if you want to by matching the transient current demand to the acceptable power rail voltage droop. Take, for example, a 74HC octal buffer each of whose outputs when switching takes a transient current of 50 mA for 6 ns (calculated from Eq. 6.1). The total peak current demand is then 0.4 A.

The acceptable voltage droop is perhaps 0.4 V (equivalent to the worst system noise margin). Assume that the local decoupling capacitor supplies all of the current to hold the droop to this level, which is reasonable if other decoupling capacitors on the board are isolated by track inductance. Then the minimum capacitor value is as follows:

$$C = \frac{It}{V} = 0.4 \times 6 \times \frac{10^{-9}}{0.4} = 6 \, \text{nF} \tag{6.2}$$

On the other hand, the actual value is noncritical, especially as the variables in the above calculation tend to be somewhat vague, and you will prefer to use the same component in all decoupling positions for ease of production. Values between 10 and 100 nF are recommended, a good compromise being 22 nF, which has both low self-inductance and respectable reservoir capacitance. It also tends to be cheaper than the higher values, particularly in the low-performance Z5U or Y5V ceramic grades, which are adequate for this purpose.

Capacitors Under the Integrated Circuit Package

Very high-speed and high-current logic ICs push the location requirement of the decoupling capacitor to its limits: it has to be right next to the supply pins. In fact, the inductance of the lead-out wires within the chip package becomes significant, and this has meant that locating power and ground pins in the middle of, or all around, the package rather than at opposite corners is necessary for high-performance large scale ICs. For such devices it is necessary to locate the decoupling capacitors underneath the chip, on the opposite side of the board. The leads to the capacitors are then limited to the vias between the device pads, the planes, and the capacitor pads. This is easily achievable with surface mount construction but of course not if you are restricted to through hole.

In fact the power and ground planes themselves (see Section 2.2.4) are more effective at reducing the high frequency noise than are the decoupling capacitors because their associated capacitance has no significant inductance. The closer together the planes are the higher is this capacitance. You still need discrete decoupling capacitors for midfrequency decoupling, but their positioning becomes less important, as long as they are still located close to the relevant IC pins.

Low-Frequency Decoupling

You also need to decouple the supply rail against lower-frequency ripple due to varying logic load currents, as distinct from transient switching edges. The frequency of these ripple components is in the megahertz range and lower, so that widely distributed capacitance and low self-inductance are less important. Typically, they can be dealt with by a few tantalum electrolytics of $1-2$ μF placed around the board, particularly where there are several devices that can turn on together and produce a significant drain from the power supply, such as burst refresh in dynamic RAM. Additionally, a single large capacitor of $10-47$ μF at the power entry to the board is recommended to cope with frequency components in the kHz range.

Under normal circumstances, logic circuits are inherently insensitive to ripple on the supply lines. The exception is when they are faced with slow edges; if the ripple is at a substantially higher frequency than the edge and is modulating the signal, then as the signal passes through the transition region the logic element may undergo spurious switching (Fig. 6.11).

The safest way to deal with slow edges is to apply hysteresis with a Schmitt trigger logic input, as described in Section 6.2.2.

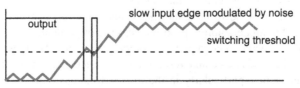

FIGURE 6.11 Spurious Switching on Slow Edges due to Ripple.

Guidelines

The minimum requirements for good decoupling are as follows:

- one 22 µF bulk capacitor per board
- one 1 µF tantalum capacitor per 10 packages of SSI/MSI (small-scale integration/medium-scale integration) logic or memory
- one 1 µF tantalum capacitor per 2–3 LSI (large-scale integration) packages
- one 10–100 nF ceramic multilayer capacitor for each supply pin of an LSI package with multiple supply pins
- one 10–100 nF ceramic multilayer capacitor for each octal IC or for each MSI package
- one 10–100 nF ceramic multilayer capacitor per four packages of SSI logic.

When in doubt, calculate the requirements for individual power/speed-hungry devices to make sure you have enough capacitors, and that they are in the right places.

6.1.5 UNUSED GATE INPUTS

Frequently you will have spare gates or latches in a package left over or will not be using all the inputs of a multiinput gate or latch. All such unused logic inputs must be tied to a fixed voltage, either high or low, and should never be left floating. Noise immunity of floating inputs is poor, so you should not float spare inputs of used gates, and especially not preset/clear inputs of latches or flip-flops, which are very sensitive to spikes. Fig. 6.12 illustrates the options.

You must connect *all* unused CMOS inputs either to V_{CC} or ground. Floating any input is inadmissible, whether its gate is used or not. This is because the CMOS input has a very high impedance and consequently can float to any voltage if unconnected, and this voltage could be within the threshold switchover region of the gate. At this point both the P-channel and N-channel input transistors are conducting, which results in excessive current drain through the package. Due to the high gain of buffered gates, it is possible for a gate to oscillate, resulting in even higher current drain.

CMOS inputs can be connected directly to either rail; a protection resistor is unnecessary as long as the supply is not expected to carry noise spikes that would exceed the maximum input voltage.

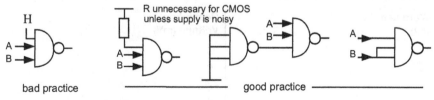

FIGURE 6.12 Connecting Unused Inputs.

6.2 INTERFACING

6.2.1 MIXING ANALOGUE AND DIGITAL

The two main problems that designers who have to integrate analogue and digital circuits on the same (printed circuit board) pcb face are

- preventing digital switching noise from contaminating the analogue signal, and
- interfacing the wide range of analogue input voltages to the digital circuit.

Generating analogue outputs from digital signals is not usually a problem, whereas generating digital inputs from analogue signals is.

Ground Noise

The high-frequency switching noise that was discussed in Section 6.1.3 must be kept out of analogue circuits at all costs. An analogue-to-digital interface quantizes a variable analogue signal into a digital word, and the number of bits in the word determines the resolution that can be achieved of the signal. Assuming a full-scale voltage range of 0—10 V, which is typical of many analogue to digital converters (ADCs), Table 6.1 shows the voltage levels that correspond to 1 bit change in the digital word.

You can see that the more resolution is demanded of the interface, the smaller the voltage change that will cause 1 bit change. 8 bits is regarded as commonplace in ADC circuits, 12 bits as reasonably high resolution (0.025%), and 16 bits as precision.

The significance of these diminishing voltage levels is that any noise that is coupled into the analogue input will cause unwanted fluctuation of the digital value. For a 12-bit converter, a 1-bit uncertainty will be given by noise of 2.4 mV at the converter input; for a 16-bit converter, this reduces to 150 μV. By contrast, the switching noise on the digital groundline is normally tens of millivolts and frequently hundreds of millivolts peak amplitude. If this noise were coupled into the converter input—and it is hard to keep ground noise out of the input—you would be unable to use a converter of greater precision than 8—10 bits.

Table 6.1 Analogue to Digital Converter Resolution Voltage for Different Word Lengths, 10 V Full-Scale

Word Length (bit)	Resolution Voltage (mV)
8	39
10	10
12	2.4
14	0.6
16	0.15

Filtering

One partial solution to this problem is to filter the bandwidth of the analogue signal to well below that of the noise so that the effective noise signal is reduced. For slowly varying analogue signals this works reasonably well, especially if the noise injection occurs at the input of the signal-processing amplifier so that bandwidth limitation has maximum effect. Filtering is in any case good practice to minimize susceptibility to external noise.

Filtering the input amplifier is no use if the noise is injected into the ADC itself. For fast ADCs and wide-bandwidth analogue signals you cannot take this approach anyway and the only available solution is to prevent the injection of digital noise at its source.

Segregation

The basic rule to follow when designing an analogue-to-digital interface is to segregate the circuits, including grounds, completely. This means that

- separate analogue and digital grounds should be established, connected only at one point
- the analogue and digital sections of the circuit should be physically separated, with no digital tracks traversing the analogue section or vice versa. This will minimize cross talk between the circuits.

It should be appreciated that no grounding scheme that establishes a multiplicity of different grounds can ever be optimum because there will always be circuits that need to communicate signals across different ground areas. These signals are then particularly exposed to the nuances of both internal and external interference, or indeed may be the source of it. You should always strive to make such circuits low-risk in terms of their bandwidth and sensitivity or else keep a single ground system for all circuits (both digital and analogue) and take extreme care in its layout so that ground noise from one noisy part of the system does not circulate in another sensitive part.

Single-Board Systems

The appropriate grounding schemes for single-board and multi board systems are shown in Fig. 6.13. If your system has a single ADC, perhaps with a multiplexer to select from several analogue inputs, then the connection between analogue and digital grounds can be made at this ADC as in Fig. 6.13A. This scheme requires that the analogue and digital power supply returns are not linked together anywhere else, so that two separate power supply circuits are needed. The analogue and digital grounds must be treated as entirely separate tracks, despite being nominally at the same potential; unavoidable noise currents circulating in the digital ground will then not couple into the "clean" analogue ground. The digital ground should be of gridded or ground plane construction, whereas the analogue section may benefit from a single-point grounding system or may have a separate ground plane of its own. On no account should you extend the digital ground plane over the analogue

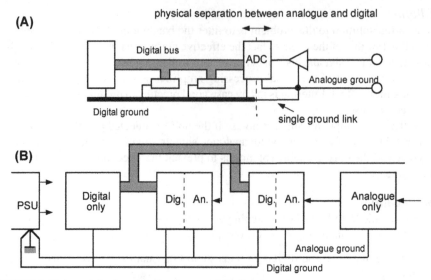

(A)

physical separation between analogue and digital

Digital bus

ADC

Analogue ground

Digital ground

single ground link

(B)

PSU

Digital only

Dig. An.

Dig. An.

Analogue only

Analogue ground

Digital ground

FIGURE 6.13 Layout for Separate Analogue and Digital Grounds.

(A) Single-Board (B) Multiboard. *ADC*, analogue to digital converter.

section of the board, since there will then be capacitive coupling from one ground plane to another.

Multiboard Systems

When your system consists of several boards, some entirely digital, some entirely analogue, and some a mixture of the two, with an external power supply, then you cannot make the connection between digital and analogue grounds at the ADC. There may be several ADCs in the one system. Instead, make the link at the power supply (Fig. 6.13B) and run separate analogue and digital grounds to each board that requires them. Digital-only boards should be located physically closer to the power supply to minimize the radiating loop area or length.

6.2.2 GENERATING DIGITAL LEVELS FROM ANALOGUE INPUTS

The first rule when you want to use a varying analogue voltage to generate an on/off digital signal—as distinct from an analogue-to-digital conversion—is to always use either a comparator or a Schmitt trigger gate. *Never* feed an analogue signal straight into an ordinary TTL or CMOS gate input.

The reason is that ordinary gates do not have well-defined input voltage switching thresholds. Not only that but they are also very critical of slow rise time inputs. Very few analogue input signals have the slew rate, typically faster than 5 V/μs, required to produce a clean output from an ordinary logic gate. The result of applying a slow analogue voltage to a logic gate is shown in Fig. 6.14.

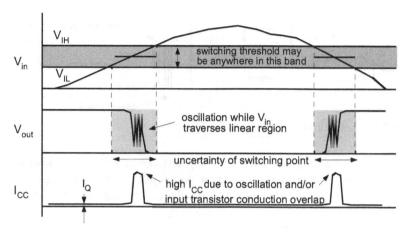

FIGURE 6.14 The Effect of a Slow Input to a Logic Gate.

A Schmitt trigger gate, or a comparator with hysteresis (see Section 5.3.4), will get over the slow rise time problem. A Schmitt trigger gate has the same output characteristics as an ordinary gate, but it includes input hysteresis to ensure a fast transition. The threshold levels of typical Schmitt devices, such as the 74HC14, are specified within wide tolerances and so do not overcome the variability of the actual switching point. When the analogue levels corresponding to high and low states can be kept above V_{IH} and below V_{IL}, respectively, a Schmitt is adequate. For more precision you will need to use a comparator with an accurately specified reference voltage.

Secondly, if the analogue supply rail range is greater than the logic supply, interfacing the analogue signal straight to the logic input will threaten the gate with damage. This is possible even if the normal signal range is within the logic supply range; abnormal conditions such as turn-on or turn-off may exceed the rails. This, of course, is also a problem with Schmitt trigger gates. Normally, the inputs are protected by clamp diodes to the supply and ground rails, but the current through these must be limited to a safe level so a resistor in series with the input is essential. More positive steps to limit the input voltage, such as running the analogue section from the same supply voltage as the logic (heeding the earlier advice about separate digital and analogue ground rails), are to be preferred.

Debouncing Switch Inputs

On the face of it, switch inputs to digital circuitry must be the easiest of interfaces. All you should need are an input port or gate, a pull-up resistor, and a single pole switch (Fig. 6.15). This circuit, though it undoubtedly works, is prone to a serious problem because of the electromechanical nature of the switch and the speed of logic devices.

When a switch contact operates, the current flow is not cleanly initiated or interrupted. As the contacts come together or part, the instantaneous contact resistance

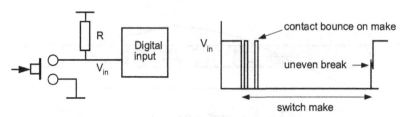

FIGURE 6.15 Contact Bounce.

varies due to contamination, and the mating surfaces may "bounce" apart a few times due to the springiness of the material. As a result the switching edge is irregular and may easily consist of several discrete edges, extending over a period of typically 1 ms. You can verify this behavior simply by observing the input waveform of Fig. 6.15 on a storage scope.

Of course, the digital input responds very fast to each crossing of the switching threshold, and consequently the port or gate sees several transitions each time the switch is operated, before it settles to a steady-state 1 or 0. This may not be a problem for level-sensitive inputs, but it undoubtedly is for edge-sensitive ones such as counter or latch clock inputs. Mistriggering of counter circuits that are fed from a switch input is commonly caused by this phenomenon.

The simple solution to contact bounce is to filter the logic input with an RC network (Fig. 6.16A). The RC time constant must be significantly longer than the bounce period to effectively attenuate the contact noise. This has the extra advantage of protecting against induced impulsive or RF interference, but it requires additional discrete components and demands that the logic input must be a Schmitt trigger type, since the input rise time has been deliberately slowed.

If the switch input may change state quickly, an RC time constant that is sufficiently long to cure the bounce will slow the response to the switch unacceptably. This can be overcome in two ways: the R–S latch, Fig. 6.16B, which requires a changeover rather than single-throw switch, or a software- or hardware-implemented delay. Fig. 6.16C shows the hardware delay, which uses a continuously clocked shift register and OR gate to effectively "window out" the bounce. The delay can be adjusted to suit the bounce period. These two solutions are most suited to realization with semicustom logic arrays or ASICs, where the overhead of the extra logic is low.

6.2.3 PROTECTION AGAINST EXTERNALLY APPLIED OVERVOLTAGES

Logic inputs and outputs that are taken off-board will be subject at some time in the life of the system to an overvoltage. Your philosophy in this respect should be, if it can happen, it will. Overvoltages can be applied by misconnection of the board or of external equipment or can be due to static buildup. The latter is a particular threat to CMOS inputs with their high impedance, but the effect of a large static discharge can also be disastrous for other logic families.

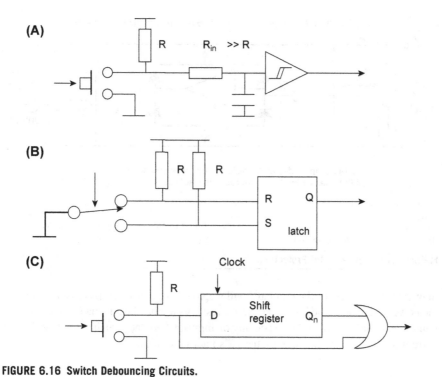

FIGURE 6.16 Switch Debouncing Circuits.

There are three major consequences of an overvoltage on a logic signal line:

- immediate damage of the device due to rupture of the track metallization or destruction of the silicon;
- progressive degradation of device characteristics when the overvoltage does not have enough energy to destroy it immediately;
- latch-up, where damage may be caused by excessive power supply current following a transient overvoltage.

Modern logic families include some protection both at their inputs and outputs in the form of clamping diodes to the supply lines, but these diodes are limited in their current handling ability, and therefore the potential fault current that can be applied due to an overvoltage must be limited. This is best achieved by the methods of Fig. 6.17.

The external clamp diodes are used to take the lion's share of the incoming overload current and divert it to the V_{CC} or 0 V rails; the resistors shown dotted are needed if the IC's internal diodes would otherwise still take too much current because of the ratio of their forward voltage to the external diodes' forward voltages. The power rail takes the excess incoming current and must therefore be of a low enough impedance for its voltage to be substantially unaffected by this current being dumped into it. This

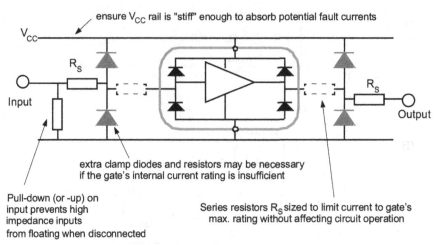

ensure V_{CC} rail is "stiff" enough to absorb potential fault currents

extra clamp diodes and resistors may be necessary
if the gate's internal current rating is insufficient

Pull-down (or -up) on
input prevents high
impedance inputs
from floating when disconnected

Series resistors R_S sized to limit current to gate's
max. rating without affecting circuit operation

FIGURE 6.17 Logic Fate I/O Protection.

may call for a review of the regulator philosophy or for extra clamps to be applied on the power rail local to the interfaces. Series resistors R_S may be adequate in themselves without external clamp diodes, especially on inputs, where they can be used to limit the current to what can be handled by the IC's internal diodes.

6.2.4 ISOLATION

Even if you take precautions against input/output abuse, it is not good practice to take logic signals directly into or out of equipment. As well as facing the threat of overload on individual lines, you also have to extend the ground and/or supply rails outside the equipment to provide a signal return path. These then act as antennas both to radiate ground noise out of the equipment and to conduct external interference back into it. It is very much safer to keep power rails within the bounds of the equipment case.

A common technique to achieve this is to electrically isolate all signal lines entering or leaving the equipment. As well as guarding against interference, this eliminates problems from ground loops and ground differentials. Digital signals lend themselves to the use of optocouplers. An optocoupler is basically an LED chip integrated in the same package as a light-sensitive device such as a photodiode or phototransistor, the two components being electrically separate but optically coupled. A typical isolation scheme using such devices is shown in Fig. 6.18.

One optocoupler is needed per digital channel. Optocouplers can be sourced as single, dual, or quad packages; and the price for commercial grade units can vary depending mainly on the required speed and level of integration from 25 p to £5 per channel. Clearly, in cost- or space-sensitive applications the number of isolated channels should be minimized. This tends to mean that isolation is more common in industrial products than consumer ones.

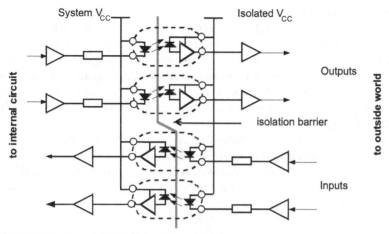

FIGURE 6.18 Interface Isolation Using Optocouplers.

Optocoupler Trade-offs

There are a number of quite complex trade-offs to make when you use opto-isolation. Factors to consider are as follows:

Speed of the interface: cheap couplers with standard transistor outputs have switching times of 2–5 µs so are limited to data rates of around 100 kbits/s maximum. High-speed devices with data rates of 10 Mbits/s are available but cost over £5 per channel.

Power consumption: standard transistor output types offer a current transfer ratio (CTR) typically between 10% and 80%. This is the ratio of LED input current to transistor output current in the on-state. Thus for a required output current of 1 mA with a CTR of 20%, 5 mA would be needed through the LED. In addition, CTR degrades with time, and you should include an extra safety margin of between 20% and 50%, depending on expected lifetime and operating current, to ensure end-of-life circuit reliability. Reducing the operating current reduces the speed of the interface. Darlington-output optocouplers are available with CTRs of 200%–500%, but these unfortunately have turn-off times of around 100 µs so are only useful for low-speed purposes. Optocoupler drive current can be a significant fraction of the overall power requirement, especially on the isolated side.

Support circuitry: a simple phototransistor or photodarlington output needs several passive components plus a buffer gate to interface it correctly to logic levels. Alternatively, you can get optocouplers that have logic-compatible inputs and outputs, especially the faster ones, but at a significantly higher cost. Low-current LED drive requirements can be met directly by a logic gate and series limiting resistor, whereas if you are using a cheaper opto with higher LED current, you will need an extra buffer.

Coupling Capacitance

Although an optocoupler breaks the electrical connection at DC, with an isolation voltage measured in kilovolt, there is still some residual coupling capacitance, which reduces the isolation at high frequencies. The specification figures of 0.5–2 pF are increased somewhat by stray wiring capacitance, which is layout dependent. Input and output pins are invariably on opposite sides of the package. There is no point in designing in an optocoupler for isolation if you then run the output tracks back alongside the input tracks!

The coupling capacitance of individual channels, multiplied by the number of channels in the system, means that a significant level of high frequency ground noise may still be coupled out of an isolated system or fast rise time transients or RFI may still be coupled in. (This is another argument for minimizing the number of channels.) In addition, high common mode dV/dt signals can be coupled directly into the photodiode or transistor input through this capacitance and cause false switching. This effect is reduced by incorporating an electrostatic screen across the optical path and connecting it to the output ground pin, and some optocouplers are available with this screen included. Common mode transient immunity can vary from worse than 100 V/µs to better than 5 kV/µs (for the expensive devices).

Alternatives to Optocouplers

Two alternatives to optocouplers for isolating digital signals are relays and pulse transformers. The relay is a well-established device and is a good choice if its restrictions of size, weight, speed, power consumption, and electromechanical nature are acceptable. Pulse transformers are most useful for passing wide bandwidth, high-speed digital data for which optocouplers are too slow or too expensive. They can also be designed for good immunity to high dV/dt interference. The data must be coded or modulated to remove any DC component. This requires an overhead of a few gates and a latch per channel, but this overhead may be acceptable, especially if you are already using semicustom silicon, and may easily be outweighed by the attractions of high-speed and low-power consumption.

6.2.5 CLASSIC DATA INTERFACE STANDARDS

When you want to connect logic signals from one piece of equipment to another, it is not sufficient to use standard logic devices and make direct gate-to-gate connections, even if they are isolated from the main system. Standard logic is not suited to driving long lines; line terminations are unspecified and noise immunity is low, so that reflections and interference would give unacceptably high data corruption. External logic interfaces must be specially designed for the purpose.

At the same time, it is essential that there is some commonality of interface between different manufacturers' equipment. This allows the user to connect, say, a computer from manufacturer A to a printer from manufacturer B without worrying about electrical compatibility. There is therefore a need for a standard definition for electrical interface signals.

This need has been recognized for many years, and there are a wide variety of data interchange standards available. The logic of the marketplace has dictated that only a small number of these are dominant. This section will consider the two main commercial ones: EIA-232F and EIA-422. EIA-232F is an update of the popular RS-232C standard published in 1969, to bring it into line with the international CCITT V.24 and V.28 and ISO IS2110 standards. EIA-422 is the same as the earlier RS-422 standard. The prefix changes are cosmetic, purely to identify the source of the standards as the EIA.

EIA-232F

The boom in data communications has led to many products, which make interface conformity claims by quoting "RS-232" in their specifications. Some of these claims are in fact quite spurious, and discerning users will regard interface conformity as an indicator of product quality and test it early on in their evaluation. The major characteristics of the specification are given in Table 6.2. As well as specifying the electrical parameters, EIA-232F also defines the mechanical connections and pin configuration and the functional description of each data circuit.

By modern standards the performance of EIA-232F is primitive. It was originally designed to link data terminal equipment (DTE) to modems, known as data communications equipment (DCE). It was also used for data terminal-to-mainframe interfaces. These early applications were relatively low speed, less than 20 kbaud, and used cables shorter than 50 ft. Applications that call for such limited capability are now abundant, hence the standard's great popularity. Its new revision recognizes this by replacing the phrase "data communication equipment" with "data circuit-terminating equipment," also abbreviated to DCE. It does not clarify exactly what is a DTE and what is a DCE, and since many applications are simple DTE (computer) to DTE (terminal or printer) connections, it is often open to debate as to what is at which end of the interface. Although a point-to-point connection provides the correct pin terminations for DTE-to-DCE, a useful extra gadget is a cable known as a "null modem" (Fig. 6.19), which creates a DTE-to-DTE connection. The common sight of an installation technician crouched over a 9-way connector swapping pins 2 and 3, to make one end's receiver listen to the other end's driver, has yet to disappear.

EIA-232F's transmission distance is limited by its unbalanced design and restricted drive current. The unbalanced design is very susceptible to external noise pickup and to ground shifts between the driver and receiver. The limited drive current means that the slew rate must be kept slow enough to prevent the cable becoming a transmission line, and this puts a limit on the fastest data rate that can be accommodated. Maximum cable length, originally fixed at 50 ft, is now restricted by a requirement for maximum load capacitance (including receiver input) for each circuit of 2500 pF. As the line length increases so does its capacitance, requiring more current to maintain the same transition time. The graph of Fig. 6.20 shows the drive current versus load capacitance required to maintain the 4% transition time relationship at different data rates. In practice, the line length is limited to

Table 6.2 Major Electrical Characteristics of EIA-232F, EIA-422, and EIA-485

Interface	EIA-232F	EIA-422	EIA-485
Line type	Unbalanced, point to point	Balanced, differential, multidrop (one driver per bus)	Balanced, differential, multiple drivers per bus (half duplex)
Line impedance	Not applicable	100 Ω	120 Ω
Max. line length	Load dependent, typically 15 m depending on capacitance	L ≈ 105/B m B = bit rate, kB/s	Max. recommended 1200 m, depending on attenuation
Max. data rate	20 kB/s	10 MB/s	10 MB/s
Driver			
Output voltage	±5 to ±15 V loaded with 3 −7 kΩ +V = logic 0, −V = logic 1	±10 V max differential unloaded, ±2 V min loaded with 100 Ω	±6 V max differential unloaded, ±1.5 V min loaded with 54 Ω
Short circuit current	500 mA max	150 mA max	150 mA to gnd, 250 mA to −7 or +12 V
Rise time	4% of unit interval (1 ms max) 30 V/μs max. slew rate	10% of unit interval (min 20 ns)	30% of unit interval
Output with power off	>300 Ω output resistance	±100 μA max leakage	>12 kΩ output resistance
Receiver			
Sensitivity	±3 V max thresholds	±200 mV	±200 mV
Input impedance	3–7 kΩ, <2500 pF	4 kΩ min	12 kΩ
Common mode range	Not applicable	±7 V	+12 to −7 V

3 m or less for data rates more than 20 kb/s. Most drivers can handle the higher transmission rates over such a short length without drawing excessive supply current.

Note that there are several common "enhancements" that are not permitted by strict adherence to the standard. EIA-232F makes no provision for tristating the driver output, so multiple driver access to one line is not possible. Similarly, paralleling receivers is not allowed unless the combined input impedance is held between 3 and 7 kΩ. It does not consider electrically isolated interfaces: no specification is offered for isolation requirements, despite their desirability. It does not specify the

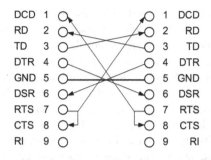

DB9S connectors

FIGURE 6.19 The Null Modem.

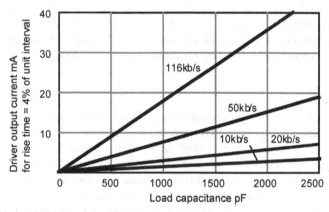

FIGURE 6.20 EIA-232F Transmit Driver Output Current Versus C_L.

communication data format. The usual "one start bit, eight data bits, two stop bits" format is not part of the standard, just its most common application. It is not directly compatible with another common single-ended standard, EIA-423, although such connections will usually work. In addition, you cannot legitimately run EIA-232F off a ± 5 V supply rail—the minimum driver *output* voltage is specified as ± 5 V, loaded with 3–7 $k\Omega$ and with an output impedance of 300 Ω.

The standard calls for slew rate limiting to 30 V/μs maximum. Although you can do this with an output capacitor, which operates in conjunction with the output transistor's current limit while it is slewing, this will increase the dissipation and reduce the maximum possible cable length. It is preferable to use a driver, which has on-chip slew rate limiting, requiring no external capacitors and making the slew rate independent of cable length.

EIA-422

Many data communications applications now require data rates in the megabaud region, for which EIA-232F is inadequate. This need is fulfilled by the EIA-422 standard, which is an electrical specification for drivers and receivers for use in a balanced or differential, point-to-point or multidrop high-speed interface using twisted pair cable. Table 6.1 summarizes the EIA-422 specification in comparison with EIA-232F. One driver and up to 10 receivers are allowed. The maximum data rate is specified as 10 Mbaud, with a trade-off against cable length; maximum cable length at 100 kbaud is 4000 ft. Note that unlike EIA-232F, EIA-422 does not specify functional or mechanical parameters of the interface. These are included in other standards, which incorporate it, notably EIA-449 and EIA-530.

EIA-422 achieves its high-speed and long-distance capabilities by specifying a balanced and terminated design. The balanced design reduces sensitivity to external common mode noise and allows a ground differential of up to a few volts to exist between the driver and one or more of the receivers without affecting the receiver's thresholds. A cable termination, together with increased driver current, allows fast slew rates, which in turn allows high data rates. If the cable is not terminated, serious ringing on the edges occurs, which may cause spurious switching in the receiver. The specified termination of 100 Ω is closely matched to the characteristic impedance of typical twisted pair cables. Only one termination is used, at the receiver at the far end of the cable.

Interface Design

By far the easiest way to realize either EIA-232F or EIA-422 interfaces is to use one of the many specially tailored driver and receiver chip sets that are available. The more common ones, such as the 1488 driver/1489 receiver for EIA-232F or the 26LS31 driver/26LS32 receiver for EIA-422, are available competitively from many sources and in low-power CMOS versions. You can also obtain combined driver/receiver parts so that a small interface can be handled with one IC. Because the 9-pin implementation of EIA-232F is so common, a single package 3-transmitter plus 5-receiver part is also widely sourced. The high-voltage requirement of EIA-232F, typically ±12 V supplies, is addressed by some suppliers who offer on-chip DC-to-DC converters from the +5 V rail.

Fig. 6.21 suggests typical interface circuits for the two standards. Note the inclusion of power supply isolating diodes, to protect the rest of the circuit against the inevitable overvoltages that will come its way. You can also construct an interface, particularly the simpler EIA-232F, using standard components such as op-amps, comparators, CMOS buffer devices, or discrete components if you are prepared to spend some time characterizing the circuit against the requirements of the standard and against expected overload conditions. This may turn out to be marginally cheaper in component cost, but its overall worth is somewhat questionable.

6.2.6 HIGH PERFORMANCE DATA INTERFACE STANDARDS

This section briefly reviews some of the newer data interface standards that have grown up for high-speed purposes around particular applications and have subsequently become more widely entrenched.

(A)

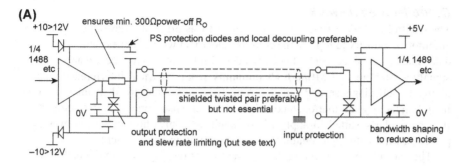

(B)

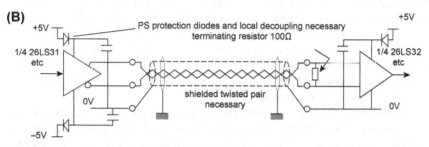

FIGURE 6.21 Typical EIA-232F and EIA-422 Interface Circuits.

(A) EIA-232F (B) EIA-422.

EIA-485

EIA-485 shares many similarities with EIA-422 and is widely used as the basis for in-house and industrial data communication systems. For instance, one variant of the SCSI interface (HVD-SCSI: high voltage differential–small computer systems interface) uses 485 as the basis for its electrical specification. 485-compliant devices can be used in 422 systems, though the reverse is not necessarily true. The principal difference is that 485 allows multiple transmitters on the same line, driving up to 32 unit loads, with half-duplex (bidirectional) communication. One Unit Load is defined as a steady-state load allowing 1 mA of current under a maximum common mode voltage of 12 V or 0.8 mA at -7 V. ULs may consist of drivers or receivers and failsafe resistors (see below) but do not include the termination resistors. The bidirectional communication means that 485 drivers must allow for line contention and for driving a line that is terminated at each end with 120 Ω. The two specifications are compared in Table 6.2.

One further problem that arises in a half-duplex system is that there will be periods when no transmitters are driving the line, so that it becomes high impedance, and it is desirable for the receivers to remain in a fixed state in this situation. This means that a differential voltage of more than 200 mV should be provided by a suitable passive circuit that complies with both the termination impedance requirements and the unit load constraints. A network designed to do this is called a "failsafe" network.

Controller Area Network

The Controller Area Network (CAN) standard was originally developed within the automotive industry to replace the complex electrical wiring harness with a two-wire data bus. It has since been standardized in ISO 11898. The specification allows signaling rates up to 1 MB/s, high immunity from electrical interference, and an ability to self-diagnose and repair errors. It is now widespread in many sectors, including factory automation, medical, marine, aerospace, and of course automotive. It is particularly suited to applications requiring many short messages in a short period of time with high reliability in noisy operating environments.

The ISO 11898 architecture defines the lowest two layers of the OSI/ISO seven layer model, that is, the data link layer and the physical layer. The communication protocol is carrier sense multiple access, with collision detection and arbitration on message priority (CSMA/CD+AMP). The first version of CAN was defined in ISO 11519 and allowed applications up to 125 kB/s with an 11-bit message identifier. The 1 MB/s ISO 11898:1993 version is standard CAN 2.0 A, also with an 11-bit identifier, while Extended CAN 2.0 B is provided in a 1995 amendment to the standard and provides a 29-bit identifier.

The physical CAN bus is a single twisted pair, shielded, or unshielded, terminated at each end with 120 Ω. Balanced differential signaling is used. Nodes may be added or removed at any time, even while the network is operating. Unpowered nodes should not disturb the bus, so transceivers should be configured so that their pins are in a high-impedance state with the power off. The standard specification allows a maximum cable length of 40 m with up to 30 nodes and a maximum stub length (from the bus to the node) of 0.3 m. Longer stub and line lengths can be implemented, with a trade-off in signaling rates. The recessive (quiescent) state is for both bus lines to be biased equally to approximately 2.5 V relative to ground; in the dominant state, one line (CANH) is taken positive by 1 V, while the other (CANL) is taken negative by the same amount, giving a 2 V differential signal. The required common mode voltage range is from -2 to $+7$ V, i.e., ± 4.5 V about the quiescent state.

Universal Serial Bus

The Universal Serial Bus (USB) is a cable bus that supports data exchange between a host computer and a wide range of simultaneously accessible peripherals. The attached peripherals share USB bandwidth through a host scheduled, token-based protocol. The bus allows peripherals to be attached, configured, used, and detached while the host and other peripherals are in operation. There is only one host in any USB system. The USB interface to the host computer system is referred to as the Host Controller, which may be implemented in a combination of hardware, firmware, or software.

USB devices are either hubs, which act as wiring concentrators and provide additional attachment points to the bus, or system functions such as mice, storage devices or data sources, or outputs. A root hub is integrated within the host system to provide one or more attachment points.

The USB transfers signal and power over a four-wire point-to-point cable. A differential input receiver must be used to accept the USB data signal. The receiver has an input sensitivity of at least 200 mV when both differential data inputs are within the common mode range of 0.8–2.5 V. A differential output driver drives the USB data signal with a static output swing in its low state of <0.3 V with a 1.5 kΩ load to 3.6 V and in its high state of >2.8 V with a 15 kΩ load to ground. A full-speed USB connection is made through a shielded, twisted pair cable with a characteristic impedance (Z_0) of 90 Ω 15% and a maximum one-way delay of 26 ns. The impedance of each of the drivers must be between 28 and 44 Ω. The detailed specification controls the rise and fall times of the output drivers for a range of load capacitances.

In version 1.1, there are two data rates as follows:

- the full-speed signaling bit rate is 12 Mb/s;
- a limited capability low-speed signaling mode is also defined at 1.5 Mb/s.

Both modes can be supported in the same USB bus by automatic dynamic mode switching between transfers. The low-speed mode is defined to support a limited number of low-bandwidth devices, such as mice. To provide guaranteed input voltage levels and proper termination impedance, biased terminations are used at each end of the cable. The terminations also allow detection of attachment at each port and differentiate between full-speed and low-speed devices. The USB 2.0 specification adds a high-speed data rate of 480 MB/s between compliant devices using the same cable as 1.1, with both source and load terminations of 45 Ω.

The most recent specification, USB 3.0, has a data rate of up to 5 Gbit/s, with an achievable data rate of 3.2 Gbit/s being seen as reasonable. The first devices commercially available to support the new USB 3.0 specification were on the market in 2010.

The cable also carries supply wires, nominally +5 V, on each segment to deliver power to devices. Cable segments of variable lengths, up to several meters, are possible. The specification defines connectors, and the cable has four conductors: a twisted signal pair of standard gauge and a power pair in a range of permitted gauges.

The clock is transmitted, encoded along with the differential data. The clock encoding scheme is nonreturn-to-zero with bit stuffing to ensure adequate transitions. A SYNC field precedes each packet to allow the receiver(s) to synchronize their bit recovery clocks.

Ethernet

Ethernet is a well-established specification for serial data transmission. It was first published in 1980 by a multivendor consortium that created the DEC-Intel-Xerox (DIX) standard. In 1985 Ethernet was standardized in IEEE 802.3, since when it has been extended a number of times. "Classic" Ethernet operates at a data transmission rate of 10 Mbit/s. Since the 1990s, Ethernet has developed in the following areas:

- Transmission media
- Data transmission rates

- Fast Ethernet at 100 Mbit/s (1995)
- Gigabit Ethernet at 1 Gbit/s (1999)
- Network topologies.

Nowadays Ethernet is the most widespread networking technology in the world in commercial information technology systems and is also gaining importance in industrial automation. All network users have the same rights under Ethernet. Any user can exchange data of any size with another user at any time, and any network device that is transmitting is heard by all other users. Each Ethernet user filters the data packets that are intended for it out from the stream, ignoring all the others.

In the standard Ethernet, all the network users share one collision domain. Network access is controlled by the CSMA/CD procedure (Carrier Sense Multiple Access with Collision Detection). Before transmitting data, a network user first checks whether the network is free (carrier sense). If so, it starts to transmit data. At the same time it checks whether other users have also begun to transmit (collision detection). If that is the case, a collision occurs. All the network users concerned now stop their transmission, wait for a period of time determined according to a randomizing principle, and then start transmission again. The result of this is that the time required to transmit data packets depends heavily on the network loading and cannot be determined in advance. The more collisions occur, the slower the entire network becomes.

This lack of determinism can be overcome by a variant of the basic approach known as *switched* Ethernet. This refers to a network in which each Ethernet user is assigned a port in a switch, which analyses all the data packets as they arrive, directing them on to the appropriate port. Switches separate former collision domains into individual point-to-point connections between the network components and the relevant user equipment. Preventing collisions makes the full network bandwidth available to each point-to-point connection. The second pair of conductors in the four-wire Ethernet cable, which otherwise is needed for collision detection, can now be used for transmission, so providing a significant increase in data transfer rate.

The Ethernet interface at each user is defined according to Fig. 6.22. It is usual to find structured twisted pair local area network wiring already integrated within a building, and the cabling characteristics are given in IEC 11801 and related standards (see Table 1.8); hence the 10BaseT and 100BaseT variants are the most popular of the Ethernet implementations, and the appropriate MAU/MDI using the RJ45 connector are included in most types of computer. The maximum lengths are set by signal timing limitations in the Fast Ethernet implementation, and an Ethernet system implementation relies on correct integration of cable lengths, types, and terminations.

In contrast to the coaxial versions of Ethernet, which may be connected in multidrop, each segment of twisted pair or fiber route is a point-to-point connection between hosts; this means that a network system that is more than simply two hosts requires a number of hubs or switches, which integrate the connections to each user. A hub will simply pass through the Ethernet traffic between its ports without controlling it in any way, but a switch does control the traffic, separating packets to their destination ports.

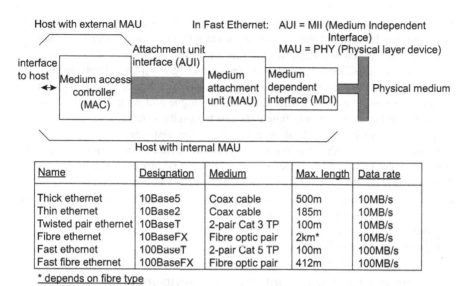

FIGURE 6.22 Ethernet Interface and Media.

Name	Designation	Medium	Max. length	Data rate
Thick ethernet	10Base5	Coax cable	500m	10MB/s
Thin ethernet	10Base2	Coax cable	185m	10MB/s
Twisted pair ethernet	10BaseT	2-pair Cat 3 TP	100m	10MB/s
Fibre ethernet	10BaseFX	Fibre optic pair	2km*	10MB/s
Fast ethernet	100BaseT	2-pair Cat 5 TP	100m	100MB/s
Fast fibre ethernet	100BaseFX	Fibre optic pair	412m	100MB/s

* depends on fibre type

The 100BaseT electrical characteristics are a peak differential output signal of 1 V into a 100 Ω characteristic impedance twisted pair; the 10BaseT level is 2.5 V. The rise and fall time and amplitude symmetries are also defined to achieve a high level of balance and hence common mode performance. It is normal to use a transformer and common mode choke to isolate the network connection from the driver electronics.

Peripheral Component Interconnect Express

An alternative to the traditional "wired" network is optical fiber, and one of the recent standards for use in Peripheral Component Interconnect (PCI) cards in PCs (hence the name PCI Express) has three basic standard definitions PCI Express (PCIe) 1, 2, and 3. There have been various subdivisions within these basic versions (such as 1.1). However, the goal is to achieve multiple optical channels driving data in the GB/s range per channel. For example, PCIe 2.0 has a specification of 5 GT/s (giga transfers per second), and PCIe 3.0 has been specified to carry 8 GT/s.

6.3 USING MICROCONTROLLERS

The subject area of microprocessors and microcontrollers is vast, and this book is not going to cover it all. What we can do, though, is look at some of the issues that arise in using these devices to fulfill functions that historically were the domain of the analogue circuit. As said at the start of this chapter, it is commonplace to implement analogue control functions with a microcontroller, because of the benefits in freedom from factors such as drift and temperature effects, and the flexibility that programming brings. But these advantages are gained at the cost of a number of other new limitations and these are the subject of this section.

There is no likelihood that analogue design will be completely replaced by digital, despite the ever increasing capability and speed of digital processing. In any design, you have to trade off what is done in analogue and what is done in digital. If you try and force unsuitable functions into the digital domain, the result will be suboptimum and probably lead to a redesign. For instance, filtering is a good example of something that can be done cheaply and accurately in the digital domain; but you cannot replace a low noise amplifier with digital processing, and you cannot A—D convert microvolt level signals, so there will always be a need for low noise amplifiers. Any transducer converts a physical parameter to an analogue voltage, which must be manipulated before passing it into the processor. And power management—necessary to keep digital circuits working properly—will always remain as an analogue function.

6.3.1 HOW A MICROCONTROLLER DOES YOUR JOB

The model of microcontroller operation can be generically described as in Fig. 6.23.

Input Processes

We covered some aspects of A-D conversion in Section 6.2.1. There are several different techniques for taking an analogue signal and creating a series of digital words from it, which you will select based on the desired speed and resolution, as well as cost. Essentially, the process is either externally clocked or controlled by the processor. In the first method, the ADC runs continuously and interrupts the processor whenever a conversion result is available; the processor then has a defined time to read the result and act on it, before the next result appears. In the second, the processor itself determines when to get an analogue value. It instructs the converter to perform a conversion, and then either waits for the result, if the conversion takes a definite, short time, or else is interrupted when the result is available.

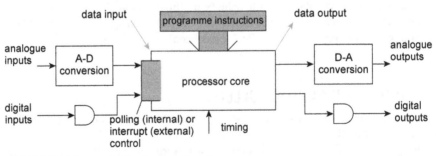

FIGURE 6.23 Microcontroller Process Model.

A-D, analogue to digital; *D-A*, digital to analogue.

Similar principles apply to digital inputs. The processor may poll some or all of its inputs in a periodic manner to check their states, and take action at its leisure depending on the values of these states. Or, if a particular input is time-critical, this input may be arranged to interrupt the processor so that action is taken immediately or at least within a defined short period.

Some of the more common A-D conversion methods are as follows:

Dual-slope: slow, but simple with good resolution; the input voltage drives an integrator for a fixed interval, after which its input is switched to a reference of opposite polarity, which drives it back to zero at a fixed rate. A counter determines the time to complete the process, and the count is the digital conversion value. Only needs an integrator, comparator, reference, and a few analogue switches, the rest of the work can be done by the processor. Variable execution time, suitable for instrumentation purposes.

Successive approximation: the reference voltage is driven closer and closer to the input voltage over a period of a few cycles under control of a comparator, which detects the polarity of the difference between the two. The number of cycles corresponds to the number of bits resolution required, and the speed at which it can be done is determined mainly by the settling time of the comparator. Fixed execution time and can be fast. Normally used in special-purpose ICs, which integrate all the functions together.

Flash: in which the conversion is performed almost instantaneously by a parallel group of comparators fed from a resistive ladder network voltage divider, which provides each comparator with a reference spaced one least significant bit from the next. The execution time is limited only by the settling time of the comparators, but the number of comparators rapidly becomes unwieldy if more than 8 bits of resolution (255 comparators) is needed. Typically used for video conversion and digital oscilloscopes.

Oversampling or sigma-delta: the output is only ever 1 bit, which signals "up" or "down." The comparator output is summed with the input, charging or discharging an integrator to keep its average output at zero. The resulting bitstream is digitally filtered to produce n-bit data at a rate less than half the sampling clock but much higher than twice the maximum signal bandwidth. This "oversampling" (a sampling clock of several MHz used for input bandwidths of a few kHz) allows for extremely high resolution and low distortion. Often used for audio applications for this reason and because of its good match with digital signal processing architectures.

Hybrids of these methods are also possible, for instance, a combination of flash and successive approximation, to achieve both faster speeds and higher resolution.

Instructions and Internal Processing

Once the input signals are in digital form they act as the data to the program operation. This can be pure signal processing—mathematical operations on the data—or it can be using these data as the source of output decisions, as in process control systems. These two broad applications tend to call for different architectures in the processor core: DSP (digital signal processor) on the one hand, microcontroller on the other. In either case, you will choose the core device according to its performance in terms of instruction processing and speed. Since microprocessor programming requires a significant investment in both software support and expertise, it is typical to stick with the same type of device through many projects. This approach is facilitated by manufacturers who offer a range of devices with a variety of memory sizes and peripherals, integrated with the same processor core and therefore using the same instruction set.

The main issue for this aspect of the design is the balance between hardware capabilities and the required software performance. That is, what is the maximum time available to perform time-critical software routines; how many instruction cycles does the most critical action take; and can the processor complete this many cycles within the required time at the intended clock frequency. If not, the principal choices are to increase the clock frequency (less time per instruction cycle), move to a more powerful processor that requires fewer instruction cycles, or segment the task among more than one processor.

Output Processes

There are some applications that do not require an analogue output: digital displays, for instance, or simple on/off controllers such as domestic heating or washing appliances or telecommunications devices whose function is limited to data crunching. Naturally these digital-only outputs can be accommodated easily from within the microcontroller. The software writes a value to a register, which sets the state of an output port or controls a data transfer in a serial transmitter, which drives a buffer to switch the required signal, and the job is done.

Providing an analogue output is not much more difficult. The basic issues are the same as for an analogue input: speed and resolution. To achieve a certain fidelity in the analogue output—that is, a lack of distortion in the required signal—you will need a minimum resolution of the converter, in number of bits. To achieve a desired bandwidth, your D-A process must be complete within a given time, determined by the Nyquist criterion, so that the next value can be set up. The processor must support these output requirements and its jobs of data input and internal processing.

6.3.2 TIMING AND QUANTIZATION CONSTRAINTS

Instruction Cycle Time

The most important constraint, which determines whether a digital solution using a particular architecture and speed is feasible, is the match between the tasks to be performed and the time allowed for them and the time per instruction of the processor. This is bound up with the efficiency of execution of the code (see Section 6.3.3 below). At an early stage of the design, you need to identify critical program subroutines, calculate their worst-case execution times, and compare these with the intended functional specification. For instance, an audio signal processor may need to carry out certain calculations on the incoming data stream, sampled at 44 kHz and output the results before the next output sample is due, i.e., 1/44 kHz = 23 µs later. If the instruction time is 30 ns then this allows 766 instructions absolute maximum.

In any complex software design this assessment is naturally approximate, and a healthy contingency is necessary to preempt later demands for increased speed.

Real-Time Interrupts and Latency

If all operations are carried out according to a strict schedule, determined by the processor's own timing, then calculating worst-case timing is in principle possible, if somewhat intensive. The real problems arise when the processor must respond to *external* events within a defined time frame. The occurrence of an interrupt in real time is signaled by a change of level at a specified pin on the processor package. Several such interrupt pins may be used, with different functions or meanings. When an interrupt occurs, the processor must perform a tight series of operations:

- stop execution of the current routine;
- save the status of its registers in an area of memory called the "stack";
- jump to the program segment mandated by the particular interrupt;
- carry out the operations demanded by the interrupt routine;
- recover the status data from the stack;
- resume execution of the interrupted operation.

This is illustrated diagrammatically in Fig. 6.24.
Two consequences flow from this:

(a) the routine that has been interrupted must be able to put up with a dead time equal to the total period of interruption
(b) the interrupt routine cannot produce any results, however important they are, for some period after the interrupt event, which is the sum of the execution time of the routine and the time to save the stack and jump to the interrupt code segment. This is known as "latency."

(a) above becomes particularly significant when you are trying to juggle more than one interrupt source, which can occur at any time, including at the same time.

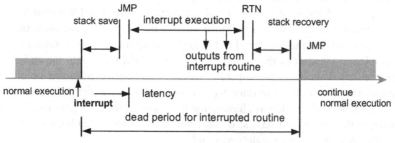

FIGURE 6.24 Interrupt Sequencing.

Firstly, all interrupts must be assigned a priority, so that if two or more do indeed occur together (within one cycle period), then the processor responds in an orderly and predictable fashion. Secondly, if a particular interrupt routine cannot tolerate delays, it will have to prohibit (mask) other interrupts for its critical period—and these other interrupts will have to accept that they may be delayed.

Analyzing interrupt timing requirements and assigning priorities and masking is a very important and challenging part of software design. Testing the resulting code is crucial. Unfortunately, it is quite possible that real-time interrupt-driven code cannot be shown to be deterministic—its outcomes cannot be predicted mathematically—and indeed testing can never be complete, since you cannot show that you have tested all combinations of interrupt timing and code execution. For this reason safety-critical software may forbid real-time interrupts.

Limits on A-D/D-A Conversion

There are a few issues that have to be addressed for any analogue input and output system.

Speed: if the digital data are to replicate the analogue signal accurately, it must convert a new word at least as fast as the change in the analogue signal; the Nyquist sampling theorem states that the sampling rate must be at least twice the highest frequency of the signal. At one extreme, slowly changing instrument transducer outputs may cope with a sample rate of 1 per second, but at the other, high quality video may need sampling in excess of 100 Msamples/s.

Resolution and scale: the full-scale range of the analogue signal must match that of the converter. The input stage gain must therefore be tailored to achieve this to make best use of the available converter range. For a given full-scale voltage the resolution of the converter—that is, the voltage step size between adjacent digital values—will be given by this voltage divided by 2^n where n is the number of bits offered by the converter. Table 6.1 shows some examples for a 10 V range. A steady DC signal should produce the same value each time but it may vary by at least one step, partly due to the inherent quantization uncertainty of $\pm\frac{1}{2}$LSB (least significant bit) and partly due to noise present at the analogue input. The higher the resolution, the stronger precautions must be taken in layout and filtering to noise-proof the input.

Sample and hold: many types of A-D conversion need a stable signal level for the duration of the conversion. Since there is no guarantee (and often no expectation) that the input will not change during this period, it must be sampled at the start of the period and the value held fixed until the end (Fig. 6.25). This requires a separate "sample and hold" conditioner before the converter, which may introduce its own sources of noise, drift, slewing, and offset errors and is often as challenging to design as the converter itself. Little wonder that integrated ADCs, which remove all of these challenges from the system designer, are so widely available!

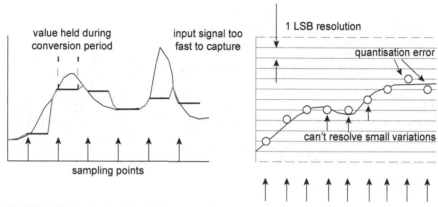

FIGURE 6.25 Resolution, Sample Rate, and Sample and Hold.

Aliasing: as stated above, the sample rate must be at least twice the highest signal frequency. Equally importantly, the signal *must not* contain frequencies higher than half the sample rate. If it does, these frequencies will be "aliased" into the base band of the digital data and create errors in the digital values. The aliased signals will have a spectrum equal to ($|f_{sig} - n \cdot f_{sample}|$) where n is any integer. So if, for instance, an audio feed sampled at 44 ksamples/s includes a frequency of either 40 or 48 kHz, the digitized output will contain an unwanted signal at 4 kHz. A signal exactly at the sampling frequency will cause a DC offset, whose magnitude depends on the phase shift between the two (Fig. 6.26). Input signals must therefore be band limited by input filtering before conversion if there is any threat of higher frequencies being present— including noise, since this will alias down to the base band as well.

Pulse Width Modulated—Style Outputs

Analogue outputs can be provided by integrated D-A converters. But a cheap and simple analogue output can also be delivered from a pulse width modulated (PWM) digital output. The digital level switches between logic 1 and logic 0 at a constant frequency and with a duty cycle depending on the desired analogue value. This is achieved with a programmed n-bit timer with a clock of 2^n times the output frequency; the output data value is repeatedly loaded into the timer on each output cycle. The value n sets the output resolution as with normal D-A converters, but it can easily be made 16-bit or greater at little or no cost within an integrated micro-controller. Naturally there is a trade-off between clock frequency, resolution, and output speed. A clock of 10 MHz with a 16-bit counter will give an output frequency of 152 Hz; if this is to be filtered to keep the output ripple to less than the available resolution it will be very slow indeed!

This method lends itself to isolated outputs since the PWM signal can be passed across an opto-isolated barrier with one optocoupler (see Section 6.2.4). Any offset

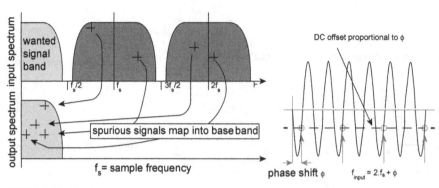

FIGURE 6.26 Aliasing.

induced by the asymmetrical switching delays of the optocoupler needs to be taken into account. The values of logic 1 and logic 0 must be defined to the requisite accuracy, the usual logic supplies are normally inadequate; the CMOS buffer B1 (Fig. 6.27) must be fed from a separate calibrated supply. Loading on the filtered output X1 must be kept to a minimum and this mandates a high-impedance input unity-gain analogue buffer B2 with acceptably low (or zeroed) offset.

Sleep and Wake-Up

It is common for low-power, low-speed applications to use a microcontroller for only a small part of its possible active period. The micro is "put to sleep" when quiescent, so that its power drain is minimal. An example would be a data logger that only needs to take a reading every few hours, say; the micro is woken up by an interrupt from a low-power clock, takes a reading and stores it, and goes back to sleep again waiting for the next interrupt.

Associated interface circuitry is usually powered from a switched output on the micro. Issues with this type of design usually revolve around the time needed to achieve full and accurate operation after wake-up, while keeping the awake period short to conserve power. For instance, the time constant of a 100 Ω output driving an analogue supply with a 100 nF decoupling capacitor is 10 μs. This will only reach

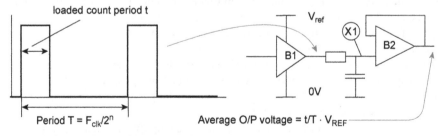

FIGURE 6.27 The Pulse Width Modulated Analogue Output.

99.6% of its steady-state value (8-bit accuracy) after 5.5 time constants, i.e., 55 μs. No accurate analogue A-D conversion should be performed before this, even if the digital circuit can do it.

6.3.3 PROGRAMMING CONSTRAINTS

High-Level Language or Assembler?

One of the issues that is created by the need to quantify the time taken by a given subroutine is that you have to know what, and how many, instructions are actually used. Having got this information, you may be faced with the need to shorten the overall time. This brings you up against a choice: do you use a high-level language (such as C) or do you use the microcontroller's own code (assembler language)? Programming in a high-level language is universally used because it is efficient in terms of software design resources—code can quickly be created and reused—and portable, in that the same code can be compiled for different machines. These advantages come at a price, which is that the code is not necessarily optimized for fast response in a real-time environment. The compiler will be able to give you the actual time for execution, but it would not necessarily code the execution in the fastest way. For those routines—particularly real-time interrupt routines—which must take the shortest possible execution time, it may be better to write the code by hand, taking shortcuts where necessary. This flies in the face of good software engineering practice, and the assembler code segments must be particularly well documented for maintainability.

6.4 MICROPROCESSOR WATCHDOGS AND SUPERVISION

Microprocessors and microcontrollers are exceptionally versatile devices. They can be applied to virtually any control, processing, or data acquisition task. But they are not entirely self-contained: like children, they need care and attention and occasional corrective action, if they are to function reliably and properly.

6.4.1 THE THREAT OF CORRUPTION

A microprocessor is a state machine. It steps predictably from state to state, its operation controlled entirely by the contents of the program counter and program memory. Provided that these are never misinterpreted, it will follow its program correctly and, assuming its software has been fully tested, will never deviate from its operational specification. If the digital circuit design rules outlined previously have been followed, there is no intrinsic reason why the data should be misinterpreted.

In the real world, though, there *is* a mechanism by which data get corrupted. This is when an external electromagnetic influence is coupled onto the signal, clock, or supply lines and overrides the transfer of stored program data (Fig. 6.28). The most usual threat is from a brief, submicrosecond transient due to a nearby

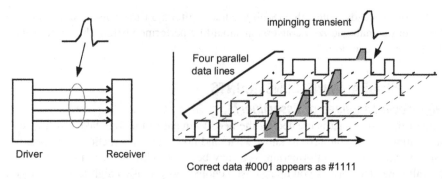

FIGURE 6.28 Transient Affecting Data Transfer.

electrostatic discharge or coupled in via mains or signal cables, but strong continuous or pulsed radio frequency fields can have the same effect. Nearly all microprocessor circuits operate with clock frequencies and data rates in excess of 1 MHz, some much higher. Possibly only one data bit need be corrupted to derail the program completely, and this can be achieved with a transient of only a fraction of a microsecond duration.

Chapter 8 deals with the circuit techniques to minimize the amplitude of such disruptive interference, and these go a long way toward "hardening" a microprocessor circuit against corruption. But they cannot *eliminate* the risk. The coincidence of a sufficiently high-amplitude transient with a vulnerable point in the data transfer, both in time and in the three dimensions of the pcb layout, is an entirely statistical affair. The most cost-effective way to ensure the reliability of a microprocessor-based product is to accept that the program *will* occasionally be corrupted and to provide a means whereby the program flow can be automatically recovered, preferably transparently to the user. This is the function of the microprocessor watchdog.

Power Rail Supervision

The other significant times when microprocessor operation may go astray is during the transitions of its power supply. All micros are characterized for a stable power rail of typically 3.0–3.6 or 4.75–5.25 V, though some allow wider tolerances. They have no guaranteed behavior below the specified minimum voltage, and so their operation while the power rail is ramping up or collapsing is an unknown quantity. Conventionally, the case of power-on has been treated by delaying the release of the micro's RESET input. It may seem that if the power has been switched off it does not much matter what the micro does, but the problem has been enormously compounded by the introduction of nonvolatile memory and real-time clocks. These should offer guaranteed security of data over power cycling if they are to be worth including, and so the micro must be effectively prevented from corrupting them under abnormal power conditions.

These considerations have led to the development of techniques for power rail supervision, including the functions of power-up and power-down reset, power fail detection, and write protection and battery backup switching for nonvolatile memory. They will be briefly touched on in the context of power supplies in Sections 7.3.4 and 7.3.5; here we shall discuss their application at the processor end.

6.4.2 WATCHDOG DESIGN

Many microcontrollers on the market include built-in watchdog devices, which may take the form of an illegal-opcode trap, or a timer that is repetitively reset by accessing a specific register address. These are more common in single-chip microcontrollers such as the Motorola 68HC11. This is an excellent example of IC designers listening to their customers, since early micros had no watchdog provision, and circuit designers had to learn the hard way about the need for one.

If your chosen micro has an onboard watchdog, use it without hesitation. It will be closely matched to the processor's peculiarities and requirements. If it has not, read on.

Basic Operation

The most serious result of a transient corruption is that the processor program counter or address register is upset, so that it starts interpreting data or empty memory as valid instructions. This causes the processor to enter an endless loop, either doing nothing or performing a few meaningless instructions. A similar effect can happen if the stack register or memory is corrupted. Either way, the processor will appear to be catatonic, in a state of "dynamic halt."

A watchdog guards against this eventuality by requiring the processor to execute a specific simple operation regularly, regardless of what else it is doing, on pain of consequent reset. The watchdog is actually a timer whose output is linked to the RESET input, and which itself is being constantly retriggered by the operation the processor performs, normally writing to a spare output port. This operation is shown schematically in Fig. 6.29.

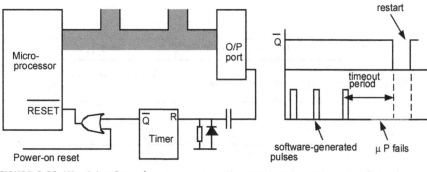

FIGURE 6.29 Watchdog Operation.

Timeout Period

If the timer does not receive a "kick" from the output port for more than its timeout period, its output goes low ("barks") and forces the microprocessor into reset. Clearly the timeout period is an important system parameter. It must be long enough so that servicing the port does not require the processor to interrupt time-critical tasks, and so that there is time for the processor to start the servicing routine when it comes out of reset (otherwise it would be continually barking and the system would never restart properly). On the other hand, it must not be so long that the operation of the equipment could be corrupted for a dangerous period. There is no one timeout period that is right for all applications, but usually it is somewhere between 10 ms and 1 s.

Timer Hardware

The watchdog circuit has to exceed the reliability of the rest of the circuit and so the simpler it is, the better. A standard timer IC such as the 555 or its derivatives is quite adequate. However, the 555 timeout period is set by an RC time constant and this may give an unacceptably wide variation in tolerance, besides needing extra discrete components. A digital divider such as the CMOS 4060B fed from a high-frequency clock and periodically reset by the report pulses may be a more attractive option, since no other components are needed. The clock has to have an assured reliability in the presence of transient interference, but such a clock may well already be present or could be derived from the unsmoothed AC input at 50/60 Hz.

An extra advantage of the digital divider approach is that its output in the absence of retriggering is a stream of pulses rather than a one-shot. Thus if the micro fails to be reset after the first pulse, or more probably is derailed by another burst of interference before it can retrigger the watchdog, the watchdog will continue to bark until it achieves success (Fig. 6.30). This is far more reliable than a monostable watchdog that only barks once and then shuts up.

On no account should you use a programmable timer to fulfill the watchdog function, however attractive it may be in terms of component count. It is quite possible that the transient corruption could result in the timer being programmed off, thereby completely silencing the watchdog.

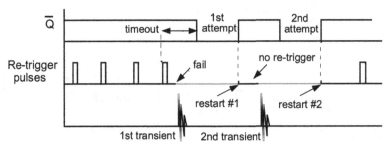

FIGURE 6.30 The Advantage of an Astable Watchdog.

Connection to the Microprocessor

Fig. 6.29 shows the watchdog's Q output being fed directly to the RESET input along with the power-on reset (POR) signal. In many cases it will be possible and preferable for you to trigger the timer's output from the POR signal, to assure a defined reset pulse width at the micro on power-up.

It is essential that you use the RESET input and not some other signal to the micro such as an interrupt, even a nonmaskable one. The processor may be in any conceivable state when the watchdog barks, and it must be returned to a fully characterized state. The only state that can guarantee a proper restart is RESET.

Source of the Retrigger Pulse

Equally important is that the micro should not be able to carry on kicking the watchdog when it is catatonic. At a minimum this demands AC coupling to the timer's retrigger input, as shown by the R-C-D network in Fig. 6.29. This ensures that only an edge will retrigger the watchdog and prevents an output that is stuck high or low from holding the timer off. The same effect is achieved with a timer whose retrigger input is edge-rather than level-sensitive.

Using a programmable port output in conjunction with AC coupling is attractive for two reasons. It needs two separate instructions to set and clear it, making it very much less likely to be toggled by the processor executing an endless loop; this is in contrast to designs that use an address decoder to produce a pulse whenever a given address is accessed, which practice is susceptible to the processor rampaging uncontrolled through the address space. Secondly, if the programmable port device is itself corrupted but processor operation otherwise continues properly, then the retrigger pulses may cease even though the processor is attempting to write to the port. The ensuing reset will ensure that the port is fully reinitialized. Conversely, you should make sure that the port output you select is not capable of being corrupted and reprogrammed to generate a square wave!

Generation of the Retrigger Pulses in Software

You should if possible generate the output pulse from two different software modules. The high-going edge should be generated in one module, perhaps labeled `kick_watchdog_high`, and the low-going edge in another, `kick_watchdog_low` (Fig. 6.31). With a port output as described above, both edges are necessary to keep the watchdog held off. This minimizes the chance of a rogue software loop generating a valid retrigger pulse. At least one edge should only be generated at one place in the code; if a real-time "tick" interrupt is used, this could be conveniently placed at the entry to the interrupt service routine, while the other is placed in the background service module. This has the added advantage of guarding against the interrupt being accidentally masked off.

Placing the watchdog retrigger pulse(s) in software is the most critical part of watchdog design and repays careful analysis. On the one hand, too many calls in different modules to the pulse generating routine will degrade the security and efficiency of the watchdog; but on the other hand, any nontrivial application software

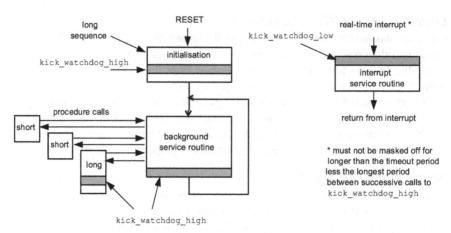

FIGURE 6.31 Generating the Watchdog Retrigger in Software.

will have execution times that vary and will use different modules at different times, so that pulses will have to be generated from several different places. Two frequent critical points are on initialization, and when writing to nonvolatile (EEPROM) memory. These processes may take several tens of milliseconds. Analyzing the optimum placement of retrigger pulses, and ensuring that under all correct operating conditions they are generated within the timeout period, is not a small task.

Testing the Watchdog

This is not at all simple, since the whole of the rest of the circuit design is bent toward making sure the watchdog never barks. Creating artificial conditions in the software is unsatisfactory. An adequate procedure for most purposes is to subject the equipment to repeated transient pulses, which are of a sufficient level to predictably corrupt the processor's operation, if necessary using specially "weakened" hardware. Be careful not to be overenthusiastic with the spikes, or you may end up destroying several good prototypes. Build up a large enough statistical base of events to have a good chance of covering all operational conditions, and check that after each one the processor has been correctly reset and has recovered to normal operation. (This is a good task for a junior technician.) An LED on the watchdog output is useful to detect its barks. Pay particular attention to what happens when you apply a burst of spikes, so that the processor is hit again just as it is recovering from the last one. This is a vulnerable condition but unhappily it is a common occurrence in practice.

As well as testing the reliability of the watchdog, remember to include a link to disable it so that you can test new versions of software.

6.4.3 SUPERVISOR DESIGN

The traditional method for POR is the simple R-C network across the power rail (Fig. 6.32A). This circuit delays the voltage rise at the RESET input for a given

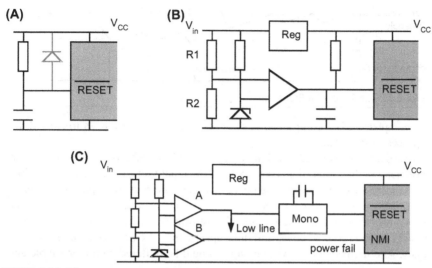

FIGURE 6.32 Microprocessor Reset Circuits.

(A) Simple Reset (B) Undervoltage Reset (C) Combined Undervoltage Reset and Power Fail Detect.

time, long enough for the micro's required start-up period. The diode, shown dotted here, is needed to discharge the capacitor rapidly in the event of a short V_{CC} interruption. Even so, the circuit is susceptible to interruptions or dips of up to a few milliseconds, when the capacitor cannot discharge fast enough to bring the RESET input below its required threshold. It also depends on a minimum rate of rise of V_{CC}, which may not be achieved in all circumstances, and it gives no early warning of an impending power failure. This simple approach suits consumer and gadget applications, where processor unreliability is no more than mildly inconvenient.

The undervoltage detector of Fig. 6.32B offers an improvement, whereby the RESET capacitor is held low until the input voltage to the regulator has reached a sufficiently high value for the regulator output to be stable. This point is set by R1 and R2. Any transient undervoltage will cause the comparator to rapidly discharge the capacitor and generate a reset, and the micro will also be forced to reset reliably as soon as the power fails. This still does not provide a power fail early warning, for which another comparator is needed as at Fig. 6.32C.

Undervoltage and Power Fail Monitor

Comparator A switches when the minimum regulator input voltage V_{in} for a stable output is reached and serves to trigger the monostable to provide a defined-length reset pulse. Comparator B switches at a higher level of V_{in}, but one that is still below the minimum operating level. Thus when the power fails, V_{in} ramps down and first comparator B switches then comparator A. Comparator B sets a nonmaskable interrupt at the micro, which triggers the power fail housekeeping functions. These have

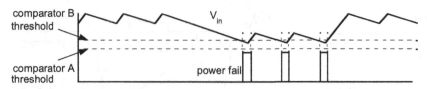

FIGURE 6.33 The Effect of a Brownout.

to be complete before comparator A switches, which triggers a reset to close down the processor completely.

A particular danger with this kind of circuit, which is effective under normal circumstances, is the threat of a "brownout." If the input voltage only droops but does not disappear altogether, the ripple on V_{in} will generate a string of power fail pulses without activating the RESET line (Fig. 6.33). Thus the software must not assume that a power fail signal is automatically followed by a reset and must recover smoothly from a series of power fail interrupts of unpredictable width at the line frequency. Alternatively, the power fail signal could be buffered by a monostable and the reset signal.

The circuit of Fig. 6.32C can easily be built from standard analogue ICs, although you need to check that the comparators work reliably at low voltage. The typical LM339-type is normally quite adequate, being specified down to 2 V, although their common mode input range vanishes at this point. However, they and their associated discrete components eat up board space, and you may well prefer to use one of the purpose-designed devices that are now on the market (cf. the comments in Section 7.3.4).

Protecting Nonvolatile Memory

Inadvertent write operations to nonvolatile components—EEPROM, battery-backed CMOS RAM, and real-time clocks—must be prevented in hardware when the supply rail is unstable. The behavior of the micro's write control line is not guaranteed under these conditions, and no precautions can be taken in software. The standard technique is to gate either the write-enable or the chip-enable line to the nonvolatile components, with the "low line" signal derived from comparator A in Fig. 6.32C. This ensures that no accesses are possible when low line is active. A complication with this method is that the (CMOS) gate package must be powered from the standby (battery backed) V_{CC} rail of the nonvolatile components. But the inputs of these gates are derived from the main circuit. If these inputs do not discharge fully to zero when the power is off, as may be the case with some power supply designs, then they may be unintentionally held in the CMOS gate threshold region, which can result in rapid power drain through the gate IC from the backup power rail. ("Rapid" is a relative term—it may discharge the backup battery in months rather than years, which you would not notice in prototype testing.) A similar effect may occur on the data/address lines to the RAM itself, though these are likely to drop to zero quickly. A pull-down (*not* pull-up) resistor on the vulnerable lines cures the problem (Fig. 6.34).

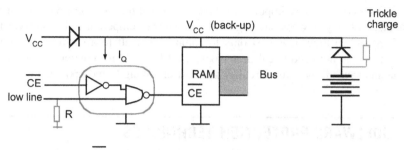

If either $\overline{\text{CE}}$ or low line are poorly defined with power-off, I_Q
may be high. Pull-down R is necessary

FIGURE 6.34 Nonvolatile Memory and the Backup Supply.

The minimum value of this pull-down is set by the output driver's current sourcing ability. The problem with using a pull-up at this interface, or on other data bus lines, is that it provides a sneak current path from the battery supply through the bus drivers' input/output protection diodes back to the main supply, which again will discharge the backup battery when the main supply is off (Fig. 6.35).

V_{CC} Differential

On the same subject of battery-backed RAM, another pitfall is that the RAM's V_{CC} in the circuit of Fig. 6.34 is one diode drop below the operating V_{CC}. (If you minimize this by using a Schottky diode, you pay the penalty of greater high-temperature reverse leakage current and hence reduced backup time.) CMOS RAMs, particularly the early types, are in danger of latch-up or malfunction when their input signals, in this case the microprocessor address/data bus, exceed their V_{CC} pin by 0.3 V. You may be able to source components, which do not suffer this problem, but you must ensure that the resulting current flow that passes through their input protection diodes to the battery supply is safely limited by the output impedance of the driver. Alternatively, rather than suffer this voltage difference, you can use an active switchover, provided, for instance, by a MOS transistor fed from the low line signal.

A battery-backup supply is essential if you need a real-time clock function and nonvolatile memory. Calculating the backup time is difficult because both battery

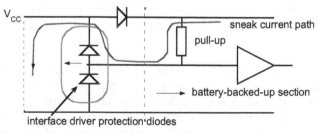

FIGURE 6.35 Sneak Current Path Through Pull-Up Resistor.

storage capacity and more importantly, leakage current through the CMOS components and isolating devices, are highly dependent on temperature; if you have to specify a backup time, it is best only to commit yourself to a room temperature figure. If you do not need a real-time clock, use EEPROM for nonvolatile memory storage and save all the extra circuitry and cost of having to incorporate a battery.

6.5 SOFTWARE PROTECTION TECHNIQUES

Some precautions against unwanted analogue effects can be taken in software, and it is relevant to look briefly at them here. The collection of techniques outlined below can be described as "defensive programming": recognizing the possibility of data corruption and guarding against its ill effects.

6.5.1 INPUT DATA VALIDATION AND AVERAGING

Frequently, you will know in advance the range of input data that is acceptable. For instance, a thermocouple will not be producing output voltages greater than a few tens of millivolts; a thermistor or platinum resistance thermometer will not show a near-zero or negative resistance. If you can set known limits on the figures that enter as digital input to the software, then you can reject data that are outside those limits.

When, as in most control or monitoring applications, each sensor inputs a continuous stream of data, this is simply a question of taking no action on false data. Since the most likely reason for false data is corruption by a noise burst or transient, subsequent data in the stream will probably be correct and nothing is lost by ignoring the bad item. Data logging applications might require you to flag or otherwise mark the bad data rather than merely ignore it.

This technique can be extended if you have a known limit to the maximum rate of change of the data. You can then ignore an input that exceeds this limit even though it may be still within the range limits. It is probably due to a noise burst. Alternatively, software averaging on a stream of data to smooth out process noise fluctuations can also help remove or mitigate the effect of invalid data.

It pays to be careful when using sophisticated software for error detection, which you do not lock out genuine errors that need flagging or corrective action, such as a sensor failure. The more complex the software algorithm is, the more it needs to be tested to ensure that these abnormal conditions are properly handled.

Digital Inputs

A similar checking process should be applied to digital inputs. In this case, you have only two states to check so range testing is inappropriate. Instead, given that the input ports are being polled at a sufficiently high rate, compare successive input values with each other and take no action until two or three consecutive values agree. This way, the processor will be "blind" to occasional noise glitches, which may coincide with the polling time slot. This does of course mean that the polling rate must be two or three times faster than the minimum required for the specified response time, which in turn may require a faster microprocessor than originally envisaged. It is not unknown for such unanticipated noise problems to force a complete system redesign late in the project. Including the solution at the beginning will avoid this.

The same technique can be applied directly to switch contact debouncing, as discussed in Section 6.2.2. Debouncing in software allows you to dispense with all the extra hardware and feed a switch input directly into a digital port, as in Fig. 6.15. This approach is understandably popular with keyboard users.

Interrupts

For similar reasons to those outlined above, it is preferable not to rely on edge-sensitive interrupt inputs. Such an interrupt can be set by a noise spike as readily as by its proper signal. Undoubtedly edge-sensitive interrupts are necessary in some applications, but in these cases you should treat them in the same way as clock inputs to latches or flip-flops and take extra precautions in layout and drive impedance to minimize their noise susceptibility. If you have a choice in the design implementation, then favor a level-sensitive interrupt input.

6.5.2 DATA AND MEMORY PROTECTION

Volatile memory (RAM, as distinct from ROM or EEPROM) is susceptible to various forms of data corruption. These vary from "soft" errors due to cosmic radiation particles, to unintentional write accesses in severe noise environments. You cannot prevent such corruption absolutely, but you can in some cases guard against its consequences.

If you place critical data in tables in RAM, each table can then be protected by a checksum, which is stored with the table. Checksum-checking diagnostics can be run by the background routine automatically at whatever interval is deemed necessary to catch RAM corruption, and an error can be flagged or a software reset can be generated as required. The absolute values of RAM data do not need to be known provided that the checksum is recalculated every time a table is modified. Beware that the diagnostic routine is not interrupted by a genuine table modification or vice versa, or errors will start appearing from nowhere! Of course, the actual partitioning of data into tables is a critical system design decision, as it will affect the overall robustness of the system.

Data Communication

The subject of data communications has a literature all to itself. The communication of digital data over long distances is prone to corruption in a statistically predictable way, and great effort has been expended to develop techniques to combat this corruption. These techniques range from simple parity checks on individual bytes through to sophisticated error detection and correction algorithms on large data blocks. This is not the place for a review of such methods. Many of the protocols mentioned in the standards of Section 6.2.6 include them. All that can be said here is that when your products use long-distance data communication, your software should incorporate some form of error detection on the received data to be at all reliable.

Unused Program Memory

One of the threats discussed in the section on watchdogs was the possibility of the microprocessor accessing unused memory space due to corruption of its program counter. If it does this, it will interpret whatever data it finds as a program instruction. In such circumstances it would be useful if this action had a predictable outcome.

Normally a bus access to a nonexistent address returns the data #FF$_H$, provided there is a passive pull-up on the bus, as is normal practice. Nothing can be done about this. However, unprogrammed ROM also returns #FF$_H$ and this can be changed. A good approach is to convert all unused #FF$_H$ locations to the processor's one-byte NOP (no operation) instruction (Fig. 6.36). The last few locations in ROM can be programmed with a JMP RESET instruction, normally 3 bytes, which will have the effect of resetting the processor. Then, if the processor is corrupted and accesses anywhere in unused memory, it finds a string of NOP instructions and executes these (safely) until it reaches the JMP RESET, at which point it restarts.

The effectiveness of this technique depends on how much of the total possible memory space is filled with NOPs, since the processor can be corrupted to a random address. If the processor accesses an empty bus, its action will depend on the meaning of the #FF$_H$ instruction. The relative cheapness of large ROMs and EPROMs means that you could consider using these, and filling the entire memory map with ROM, even if your program requirements are small.

6.5.3 REINITIALIZATION

As well as RAM data, you must remember to guard against corruption of the set-up conditions of programmable devices such as I/O ports or UARTs. Many programmers

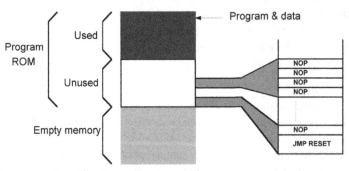

FIGURE 6.36 Protecting Unused Program Memory With No Operations (NOPs).

seem to assume that once an internal device control register has been set up (usually in the initialization routine) it will stay that way forever. This is a dangerous assumption. Experience shows that control registers can change their contents, even though they are not directly connected to an external bus, as a result of interference. This may have consequences that are not obvious to the processor: for instance, if an output port is reprogrammed as an input, the processor will happily continue writing data to it oblivious of its ineffectiveness.

The safest course is to periodically reinitialize all critical registers, perhaps in the main idling routine if one exists. Timers, of course, cannot be protected in this way. The period between successive reinitializations depends on how long the software can tolerate a corrupt register, versus the software overhead associated with the reinitialization.

6.6 CHOICE OF HARDWARE PLATFORM

There are numerous options for designers in selecting a hardware platform for custom electronics design, ranging from embedded processors, ASICs, programmable micro-processors (PICs), FPGAs, to programmable logic devices (PLDs). The decision to choose a specific technology such as an FPGA should depend primarily on the design requirements rather than a personal preference for one technique over another.

For example, if the design requires a programmable device with many design changes, and algorithms using complex operations such as multiplications and loop-ing, then it may make more sense to use a dedicated signal processor device such as a DSP that can be programmed and reprogrammed easily using C or some other high-level language. If the speed requirements are not particularly stringent, and a compact cheap platform is required, then a general-purpose microprocessor such as a PIC would be an ideal choice. Finally, if the hardware requirements require a higher level of performance, say up to several 100 MHz operation, then an FPGA offers a suitable level of performance, while still retaining the flexibility and reus-ability of programmable logic.

Other issues to consider are the level of optimization in the hardware design required. For example, a simple software program can be written in C, and then a PIC device programmed, but the performance may be limited by the inability of the processor to offer parallel operation of key functions. This can be implemented much more directly in an FPGA using parallelism and pipelining to achieve much greater throughput than would be possible using a PIC.

A general rule of thumb when choosing a hardware platform is to identify both the design requirements and the hardware options and then select a suitable platform based on those considerations. For example, consider the following:

- If the design requires a basic clock speed of up to 100 MHz, then an FPGA would be a suitable platform. If the clock speed could be 3–4 MHz, then the FPGA may be an expensive (overkill) option.

- If the design requires a flexible processor option, although the FPGAs available today support embedded processors, it probably makes sense to use a DSP or PIC.
- If the design requires dedicated hardware functionality, then an FPGA is the route to take.
- If the design requires specific hardware functions such as multiplication and addition, then a DSP may well be the best route; but if custom hardware design is required, then an FPGA would be the appropriate choice.
- If the design requires small simple hardware blocks, then a PLD or CPLD may be the best option (compact, simple programmable logic), however, if the design has multiple functions, or a combination of complex controller and specific hardware functions, then the FPGA is the route to take.

Examples of this kind of decision can be dependent on the complexity of the hardware involved. For example, a VGA controller will probably require an FPGA rather than a PLD device, simply due to the complexity of the hardware involved. Another related issue is that of flexibility and programmability. If an FPGA is used, and the resources are not used up on a specific device (say up to 60%), then if a communication protocol changes, or is updated, then the device may well have enough headroom to support several variants, or updates, in the future.

Using these simple guidelines, an intelligent choice can be made about the best platform to choose and also which hardware device to select based on these assumptions. The nice aspect of most synthesis software packages is that multiple design platforms can be tested for performance and utilization (PLD or FPGA, for example) prior to making a final decision on the hardware of choice.

6.7 PROGRAMMABLE LOGIC DEVICES

One of the most basic digital devices to be programmable from a hardware perspective was the Programmable Array Logic (PAL) device. This consists of an array of logic gates that could be connected using an array of connections. These devices could support a small number of flip-flops (usually <10) and were able to implement small state machines as seen in Fig. 6.37.

Complex Programmable Logic Devices (CPLD) were developed to address the limitations of simple PAL devices. These devices used the same basic principle as PALs, but had a series of macro blocks (each roughly equivalent to a PAL) and connected using routing blocks as shown in Fig. 6.38.

6.8 FIELD PROGRAMMABLE GATE ARRAYS

FPGAs were the next step from CPLD. Instead of a fixed array of gates, the FPGA uses the concept of a Complex Logic Block (CLB). This is configurable and allows not only routing on the device but also each logic block can be configured optimally. A typical CLB is shown in Fig. 6.39.

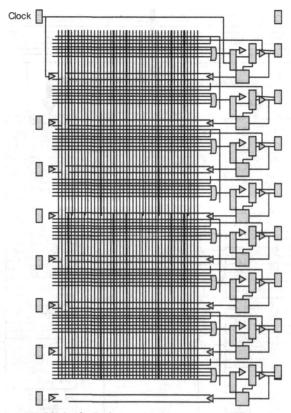

FIGURE 6.37 Programmable Logic Device.

The CLB has a lookup table (LUT) that can be configured to give a specific type of logic function when programmed. There is also a clocked d-type flip-flop that allows the CLB to be combinatorial (nonclocked) or synchronous (clocked), and there is also an enable signal. A Xilinx CLB is shown in Fig. 6.40 and this shows clearly the two 4 input LUTs and various multiplexers and flip-flops in a real device:

A typical FPGA will have hundreds or thousands of CLBs, of different types, on a single device allowing very complex devices to be implemented on a single chip and configured easily. Modern FPGAs have enough capacity to hold a number of 32-bit processors on a single device. The layout of a typical FPGA (in CLB terms) is shown in Fig. 6.41.

6.9 ANALOGUE TO DIGITAL CONVERSION

As we have seen briefly in this chapter, there are a number of key criteria that we require to understand to ensure we correctly specify the type and parameters of an

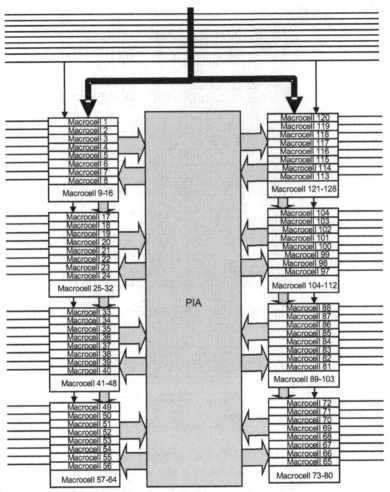

FIGURE 6.38 Complex Programmable Logic Device.

ADC. We can summarize these in the following list, and discuss them in more detail in this section.

- Number of bits (typically 8–20)
- Sampling rate (typically 50 Hz–100 MHz)
- Relative Accuracy: Deviation of the output from a straight line drawn through zero and full scale
  - Integral nonlinearity or linearity
  - Differential linearity: Measure of step size variation. Ideally each step is 1 bit, but in practice step sizes can vary significantly.
  - Usually converters are designed so that they have a linearity better than ½ bit (if this were not the case then the LSB is meaningless.)

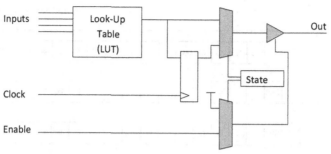

FIGURE 6.39 FPGA Complex Logic Block.

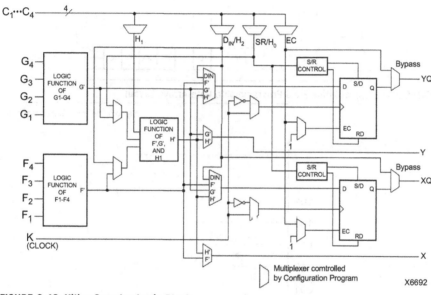

FIGURE 6.40 Xilinx Complex Logic Block.

Reproduced from Xilinx databook.

- Monotonicity: No missing codes (i.e., $1001 \rightarrow 1011$ is impossible).
- Signal-to-noise-ratio (SNR) (same as Dynamic Range)

Using these parameters we can make an informed decision as to the best choice of ADC and also how to specify its performance in detail.

6.9.1 DIGITIZATION

If we consider the most important stage of the conversion, the digitization phase, we can see how the number of bits available in the conversion limits the accuracy of the ADC, as shown in Fig. 6.42.

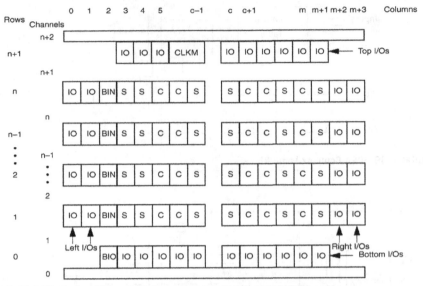

FIGURE 6.41 Typical Xilinx Field Programmable Gate Arrays Layout.

Reproduced from Xilinx databook.

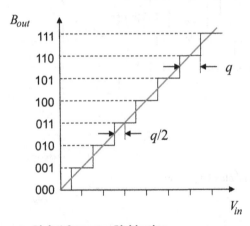

FIGURE 6.42 Analogue to Digital Converter Digitization.

The sampling did not result in any information loss (in an ideal world) in itself, but the digitizing will since only a limited number of bits is used to represent the analogue amplitude signal. This error manifests itself as noise and can be treated as white noise in many cases, and the maximum quantization error is defined as $\pm q/2$.

We can define some basic terminology as a result, most of which can be defined directly from this aspect of the ADC:

- The number of quantization levels for an N bit converter is 2^N
- The resolution is given by $V_{FS}/(2^N - 1)$; (V_{FS}: full-scale voltage). This is equivalent to the smallest increment level (or step size) q.

- MSB: Most significant bit: weighting of $2^{-1} V_{FS}$.
- LSB: weighting of $2^{-N} V_{FS}$.
- Oversampling ratio: OSR $= f_s/2f_m$
- Monotonicity: a monotonic converter is one in which the output never decreases as the input increases. For A/D converters this equivalent to saying that it does not have any *missing codes*.

Example

An analogue signal in the range 0 to +10 V is to be converted to an 8-bit digital signal.

(a) What is the resolution?

The resolution is given by $V_{FS}/(2^N - 1)$; (V_{FS}: full-scale voltage).
In this case, the full-scale voltage V_{FS} is 10 V and $N = 8$.
Resolution $= 10/(2^8 - 1) = 10/255 = 39.21$ mV

(b) What is the digital representation of an input signal of 6 V and of 6.2 V?

vfs	n	vb	Bit	Voltage
10	1	5	1	5
10	2	2.5	0	0
10	3	1.25	0	0
10	4	0.625	1	0.625
10	5	0.3125	1	0.3125
10	6	0.15625	0	0
10	7	0.078125	0	0
10	8	0.039063	1	0.039063
			Total	5.976563

(c) What is the error made for the quantization of 6.2 V in absolute terms and as a percent of the input?

Digital Value $= 6.210938$.
$=$ Absolute error of $+10.938$ mV
$=$ Percentage error of 6.2 V $= 0.176\%$

(d) What is the error made for the quantization of 6.2 V as a percent of full scale?

Digital Value $= 6.210938$.
$=$ Absolute error of $+10.938$ mV
$=$ Percentage error of 10 V $= 0.10938\%$

(e) What is the max. quantization error?

The maximum quantization error is the LSB change.
In this case the LSB corresponds to 39.063 mV.

If we assume that the quantization noise is uniformly distributed as shown in Fig. 6.43, the mean square value of the error can be calculated as shown below:

$$e_{qMS}^2 = \frac{1}{q} \int_{-q/2}^{q/2} e^2 de = \frac{q^2}{12} \tag{6.3}$$

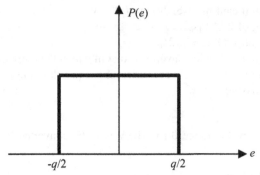

FIGURE 6.43 Quantization Noise.

therefore:

$$e_{qRMS} = \frac{q}{\sqrt{12}} \tag{6.4}$$

For a high number of bits the error is uncorrelated to the input signal ($N > 5$). In the frequency domain the error appears as white over the Nyquist range. This noise limits the S/N ratio of the digital system analogous to thermal noise in an analogue system.

The peak value of a full-scale sine wave (that is one whose peak to peak amplitude spans the whole range of the ADC) is given by

$$\frac{2^N q}{2} \tag{6.5}$$

The RMS of the sine wave is hence as follows:

$$V_{RMS} = \frac{2^N q}{2\sqrt{2}} \tag{6.6}$$

The signal-to-quantization-noise ratio (SQNR) is given by

$$SQNR = \frac{\left(2^N q/2\sqrt{2}\right)^2}{q^2/12} = \frac{3}{2} \times 2^{2N} \tag{6.7}$$

or in dB

$$SQNR = (6.02N + 1.76)\ dB \tag{6.8}$$

Thus each bit increases the SQNR by approximately 6 dB, so for an example 16 bit system the SQNR works out to be 98 dB. This equation is an extremely useful way of estimating rapidly the number of bits required to achieve a specified SQNR figure.

The previous calculation assumes that the input signal is sampled at the Nyquist rate, and the power spectral density of white quantization noise is given by

$$E^2(f) = \frac{2e^2 q_{\text{RMS}}}{f_s} \tag{6.9}$$

The OSR is given by

$$\text{OSR} = \frac{f_s}{2f_m} \tag{6.10}$$

The noise power is therefore given by

$$n_0^2 = \int_0^{f_m} E^2(f) df = e_{q\text{RMS}}^2 \left(\frac{2f_m}{f_s} \right) = \frac{e_{q\text{RMS}}^2}{\text{OSR}} \tag{6.11}$$

SQNR is then given by

$$\text{SQNR} = \frac{\left(2^N q/2\sqrt{2}\right)^2}{q^2/12\text{OSR}} = \frac{3}{2} \times 2^{2N} \times \text{OSR} \tag{6.12}$$

or in dB

$$\text{SQNR} = (6.02N + 1.76 + 10 \log_{10}(\text{OSR})) \text{ dB} \tag{6.13}$$

Thus doubling of the OSR increases the SQNR by approximately 3 dB or half a bit.

6.10 DIFFERENT TYPES OF ANALOGUE TO DIGITAL CONVERTER

As we have seen previously in this chapter, there are several different choices for ADC, and we will now take a quick look at some of these.

6.10.1 FLASH ANALOGUE TO DIGITAL CONVERTER

A "Flash" ADC is the quickest type of converter and a brief look at its architecture shows why. Essentially, it consists of multiple individual level comparators; and when the voltage input reaches each individual level, then a further bit is set by a comparator. Output logic then translates this code into a standard binary code as shown in Fig 6.44.

Clearly, the drawback with this type of approach is the resources required to achieve a relatively low number of bits—the flash architecture requires $2^N - 1$ comparators and 2^N resistors. It is, however, the fastest converter; and a conversion can be performed in one clock cycle. The sheer number of components leads to a high circuit complexity, and the converter accuracy depends on resistor matching and comparator performance (practical up to 8 bits).

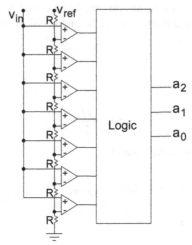

FIGURE 6.44 Flash Converter Architecture.

6.10.2 COUNTING ANALOGUE TO DIGITAL CONVERTER

A "counting" ADC is one of the simplest converters to implement and has the architecture shown in Fig. 6.45. It is easy to implement; however, a major drawback is that the conversion speed depends on the difference to previous sample. It is therefore very slow for fast varying signals and fast for slowly varying signals. It is also referred to as a "tracking or counting ADC."

6.10.3 SUCCESSIVE APPROXIMATION ANALOGUE TO DIGITAL CONVERTER

The Successive Approximation ADC is one of the most commonly used ADC types and has an excellent compromise between efficiency and accuracy. The architecture is shown in Fig. 6.46.

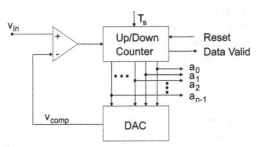

FIGURE 6.45 Counting Analogue to Digital Converter.

DAC, digital to analogue converter.

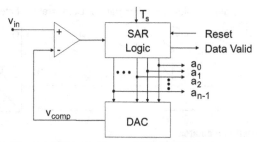

FIGURE 6.46 Successive Approximation Analogue to Digital Converter.

DAC, digital to analogue converter.

Unlike the "counting" ADC, it successively tries each bit in turn — so is much faster using a binary approach. It converts the MSB first and then progressively smaller bits — always taking N cycles, so has a uniform conversion time unlike the counting ADC.

6.10.4 DUAL-SLOPE ANALOGUE TO DIGITAL CONVERTER

An alternative approach is the "dual-slope" ADC, which uses an integrator to create the slope between input values as shown in Fig. 6.47. It has relatively high resolution (up to 14 bits), is independent of exact values of R and C, and is readily implemented in CMOS; however, it is relatively slow as it depends on the time constant of the integrator.

6.10.5 OVERSAMPLED OR SIGMA-DELTA CONVERTERS

Sigma-Delta Modulators are also called collectively "oversampled" converters. They turn an analogue signal into a bitstream corresponding to the input and consist of a sampler, quantizer, filter function, and feedback functions as shown in Fig 6.48.

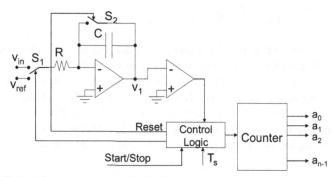

FIGURE 6.47 Dual Slope Analogue to Digital Converter.

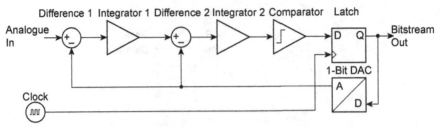

FIGURE 6.48 Sigma Delta Modulator.

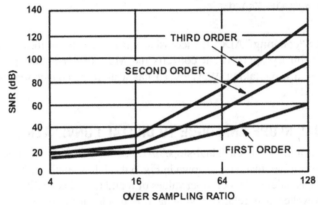

FIGURE 6.49 Relationship Between Signal-to-Noise Ratio (SNR), Order, and Oversampling Ratio.

Reproduced from AN9504: Intersil

If better noise shaping is required, the order can be increased from first order to any number; however, as the order increases, the potential for instability becomes much higher. If better noise shaping is required, a second order modulator can be used, and the RMS noise power can be calculated using the following expression:

$$n_0 = \int_0^{f_B} |N_q(f)|^2 df = e_{RMS}\frac{\pi^2}{\sqrt{5}}OSR^{-5/2} \qquad (6.14)$$

The implication is that both increasing the order and also the OSR leads to an improved SNR and this can be seen graphically in Fig. 6.49.

In summary, therefore, the advantages of a sigma-delta modulator are that you get a digital bitstream out automatically—making it an excellent choice for conversion and streaming. It is simple to implement in CMOS, scalable, and very easy to configure both the OSR and the order. One downside is that it can become complex to analyze for higher order converters—a particular problem for using in a sensor interface circuit.

Power Supplies

7

CHAPTER OUTLINE

The Circuit Designer's Companion. http://dx.doi.org/10.1016/B978-0-08-101764-7.00007-4

321

The power supply is a vital but often neglected part of any electronic product. It is the interface between the noisy, variable, and ill-defined power source from the outside world and the hopefully clear-cut requirements of the internal circuitry. For the purposes of this discussion, it is assumed that power is taken from the conventional ac mains supply. Other supply options are possible, for instance a low-voltage dc bus or a standard aircraft supply of 400 Hz 115 V. Batteries will be discussed separately at the end of this chapter.

7.1 GENERAL

A conceptual block diagram for the two common types of power supply—linear and switch-mode—is given in Fig. 7.1.

7.1.1 THE LINEAR SUPPLY

The component blocks of a linear supply are common to all variants and can be described as follows:

- input circuit: conditions the input power and protects the unit, typically voltage selector, fuse, on—off switching, filter, and transient suppressor

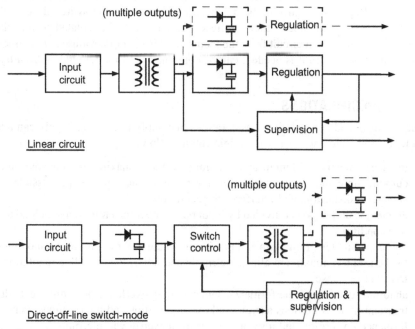

FIGURE 7.1 Power Supply Block Diagram.

- transformer: isolates the output circuitry from the ac input and steps down (or up) the voltage to the required operating level
- rectifier and reservoir: converts the ac transformer voltage to dc, reduces the ac ripple component of the dc, and determines the output hold-up time when the input is interrupted
- regulation: stabilizes the output voltage against input and load fluctuations
- supervision: protects against overvoltage and overcurrent on the output and signals the state of the power supply to other circuitry; often omitted on simpler circuits

7.1.2 THE SWITCH-MODE SUPPLY

The advantage of the direct-off-line switch-mode supply is that it eliminates the 50 Hz mains transformer and replaces it with one operating at a much higher frequency, typically 30–300 kHz. This greatly reduces its weight and volume. The component blocks are somewhat different from a linear supply. The input circuit performs a similar function but requires more stringent filtering. This is followed immediately by a rectifier and reservoir that must work at the full line voltage and feeds the switch element which chops the high-voltage dc at the chosen switching frequency.

The transformer performs the same function as in a linear supply but now operates with a high-frequency square wave instead of a low-frequency sine wave. The secondary output needs only a small-value reservoir capacitor because of the high frequency. Regulation can now be achieved by controlling the switch duty cycle against feedback from the output; the feedback path must be isolated so that the separation of the output circuit from the mains input is not compromised. The supervision function, where it is needed, can be combined with the regulation circuitry.

7.1.3 SPECIFICATIONS

The technical and commercial considerations that apply to a power supply can add up to a formidable list. Such a list might run as follows:

- input parameters: minimum and maximum voltage, maximum allowable input current, surge and continuous, frequency range, for ac supplies, permissible waveform distortion, and interference generation
- efficiency: output power divided by input power, over the entire range of load and line conditions
- output parameters: minimum and maximum voltage(s), minimum and maximum load current(s), maximum allowable ripple and noise, load and line regulation, transient response
- abnormal conditions: performance under output overload, performance under transient input conditions such as spikes, surges, dips and interruptions, performance on turn-on and turn-off: soft-start, power-down interrupts

- mechanical parameters: size and weight, thermal and environmental requirements, input and output connectors, screening
- safety approval requirements
- cost and availability requirements

7.1.4 OFF THE SHELF VERSUS ROLL YOUR OWN

The first rule of power supply design is: do not design one yourself if you can buy it off the shelf. There are many specialist power supply manufacturers who will be only too pleased to sell you one of their standard units or, if this does not fit the bill, to offer you a custom version.

The advantages of using a standard unit are that it saves a considerable amount of design and testing time, the resources for which may not be available in a small company with short timescales. This advantage extends into production—you are buying a completed and tested unit. Also your supplier should be able to offer a unit, which is already known to meet safety and electromagnetic compatibility (EMC) regulations, which can be a very substantial hidden bonus.

Costs

The major disadvantage will be unit cost, which will probably though not necessarily be more than the cost of an in-house designed and built power supply. The supplier must, after all, be able to make a profit. The exact economics depend very much on the eventual quantity of products that will be built; for lower volumes of a standard unit it will be cheaper to buy off the shelf, for high volumes or a custom-designed unit it may be cheaper to design your own. It may also be that a standard unit will not fulfill your requirements, though it is often worth bending the requirements by judicious circuit redesign until they match. For instance, the vast majority of standard units offer voltages of 3.3 or 5 V (for logic) and ±12 or 15 V (for analog and interface). Life is much easier if you can design your circuit around these voltages.

A graph of unit costs versus power rating for a selection of readily available single output standard units is shown in Fig. 7.2. Typically, you can budget for £1 per watt in the 50–200 W range. There is little cost difference between linear and switch-mode types. On the assumption that this has convinced you to roll your own, the next section will examine the specification parameters from the standpoint of design.

7.2 INPUT AND OUTPUT PARAMETERS

7.2.1 VOLTAGE

Typically you will be designing for 230 V ac in the United Kingdom and continental Europe and 115 V in the United States. Other countries have frustratingly minor differences. The usual supply voltage variability is ±10%, or sometimes +10%/−15%. In the United Kingdom the supply authorities are obliged to maintain their voltage at

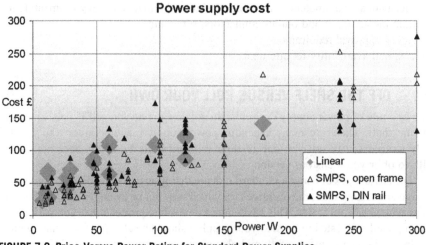

FIGURE 7.2 Price Versus Power Rating for Standard Power Supplies.

the point of connection to the customer's premises within ±6%, to which is added an allowance for local loading effects. If the voltage tolerance is applied to the UK/Europe nominal then the input voltage range becomes 207−253 V or 195−253 V. This range must be handled transparently by the power supply circuitry.

To cope simultaneously with both the American supply voltage, which may drop below 100 V, and the European voltages is difficult for a linear supply, although it is possible to design "universal" switch-mode circuits that can accept such a wide range (see the comment at the end of Section 7.2.5). Historically, this problem was handled by using a mains transformer with a split-primary (Fig. 7.3), which can be connected in series or parallel by means of a discreetly mounted voltage selector switch. This has the disadvantage that the switch may be so discreet that the user does not know about it, or else it may not be discreet enough and the user may be tempted to fiddle with it. This is not a real problem in the United States, but applying 230 V to a unit that is set for 115 V will at least annoy the user by blowing a fuse, and at worst cause real damage. Universal switch-mode supplies are therefore popular.

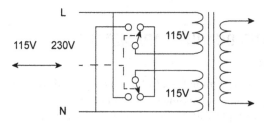

FIGURE 7.3 Split-Primary Transformer Winding.

7.2.2 **CURRENT**

The maximum continuous input current should be determined by the output load and the power conversion efficiency of the circuit. The main interest in this parameter is that it determines the rating of the input circuit components, especially the protective fuse. You have to decide whether an overload on the output will open the input circuit fuse or whether other protection measures, such as output current limiting, will operate. If the input fuse must blow, you need to characterize the input current very carefully over the entire range of input voltages. It is quite possible that the difference between maximum continuous current at full load, and minimum overload current at which the fuse should blow, is less than the fusing characteristics allow. Normally you need at least a 2:1 ratio between prospective fault current and maximum operating current. This may not be possible, in which case the input fuse protects the input circuit from faults only and some extra secondary circuit protection is necessary.

7.2.3 **FUSES**

A brief survey of fuse characteristics is useful here. The important characteristics that are specified by fuse manufacturers are the following:

- *Rated current I_N*: the value by which the fuse is characterized for its application and which is marked on the fuse. For fuses to IEC 60127, this is the maximum value which the fuse can carry continuously without opening and without reaching too high a temperature, and is typically 60% of its minimum fusing current. For fuses to the American UL-198-G standard the rated current is 85% −90% of its minimum fusing current, so that it runs hotter when carrying its rated current. The minimum fusing current is that at which the fusing element just reaches its melting temperature.
- *Time−current characteristic*: the prearcing time is the interval between the application of a current greater than the minimum fusing current and the instant at which an arc is initiated. This depends on the overcurrent to which the fuse is subjected, and manufacturers will normally provide curves of the time−current characteristic, in which the fuse current is normalized to its rated current as shown in Fig. 7.4. Several varieties of this characteristic are available:
  FF: very fast acting
  F: fast acting
  M: medium time lag
  T: time lag (or antisurge, slow blow)
  TT: long time lag
  Most applications can be satisfied with either type F or type T, and it is best to specify these if at all possible, since replacements are easily obtainable. Type FF is mainly used for protecting semiconductor circuits.
  The total operating time of the fuse is the sum of the prearcing time and the time for which the arc is maintained. Normally the latter must be taken into account only when interrupting high currents, typically more than 10 times the rated current.

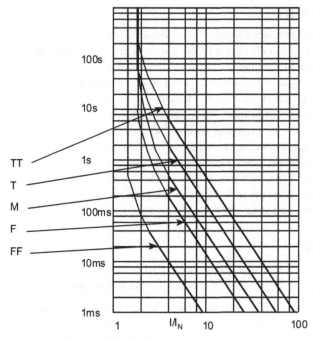

FIGURE 7.4 Typical Fuse Time–Current Curves.

The energy in a short-duration surge required to open the fuse depends on I^2t, and for pulse or surge applications you should consult the fuse's published I^2t rating. Current pulses that are not to open the fuse should have an I^2t value less than 50%–80% of the I^2t value of the fuse.

- *Breaking capacity*: breaking capacity is the maximum current the fuse can interrupt at its rated voltage. The rated voltage of the fuse should exceed the maximum system voltage. To select the proper breaking capacity you need to know the maximum prospective fault current in the circuit to be protected— which is usually determined in mains-powered electronic products by the characteristics of the next fuse upstream in the supply. Cartridge fuses fall into one of two categories: high breaking capacity, which are sand filled to quench the arc and have breaking capacities in the 1000 s of amps; and low breaking capacity, which are unquenched and have breaking capacities of a few tens of amps or less.

7.2.4 SWITCH-ON SURGE, OR INRUSH CURRENT

Continuous maximum input current is usually less than the input current experienced at switch-on. An unfortunate characteristic of mains power transformers is their low impedance when power is first applied. At the instant that voltage is applied

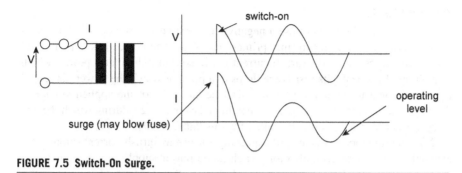

FIGURE 7.5 Switch-On Surge.

to the primary, the current through it is limited only by the source resistance, primary winding resistance, and the leakage inductance.

The effect is most noticeable on toroidal mains transformers when the mains voltage is applied at its peak halfway through the cycle, as in Fig. 7.5. The typical mains supply has an extremely low source impedance, so that the only current limiting is provided by the transformer primary resistance and by the fuse. Toroidals are particularly efficient and can be wound with relatively few turns, so that their series resistance and leakage inductance is low; the surge current can be more than 10 times the operating current of the transformer.[1] In these circumstances, the fuse usually loses out. The actual value of surge depends on where in the cycle the switch is closed, which is random; if it is near the zero crossing the surge is small or nonexistent, so it is possible for the problem to pass unnoticed if it is not thoroughly tested.

A separate component of this current is the abnormal secondary load due to the low impedance of the uncharged power supply reservoir capacitor. For the same reason, inrush current is also a problem in direct-off-line switch-mode supplies, where the reservoir capacitor is charged directly through the mains rectifier, and comparatively complex "soft-start" circuits may be needed to protect the input components.

Several simpler solutions are possible. One is to specify an antisurge or time-lag (type T or TT) fuse. This will rupture at around twice its rated current if sustained for tens or hundreds of seconds but will carry a short overload of 10 or 20 times rated current for a few milliseconds. Even so, it is not always easy to size the fuse so that it provides adequate protection without eventually failing in normal use, particularly with the high ratios of surge to operating current that can occur. A resettable thermal circuit breaker is sometimes more attractive than a fuse, especially as it is inherently insensitive to switch-on surges.

---

[1]The effect happens with all transformers but is more of a problem with toroidals.

Current Limiting

A more elegant solution is to use a negative-temperature-coefficient (NTC) thermistor in series with the transformer primary and fuse. The device has a high initial resistance, which limits the inrush current but in so doing dissipates power, which heats it up. As it heats, its resistance drops to a point at which the power dissipated is just sufficient to maintain the low resistance and most of the applied voltage is developed across the transformer. The heating takes one or 2 s during which the primary current increases gradually rather than instantaneously.

NTC thermistors characterized especially for use as inrush current limiters are available, and can be used also for switch-mode power supply inputs, motor soft-start, and filament-lamp applications. Although the concept of an automatic current-limiter is attractive, there are three major disadvantages:

- Because the devices operate on temperature rise, they are difficult to apply over a wide ambient temperature range.
- They run at a high temperature during normal operation, so require ventilation and must be kept away from other heat-sensitive components.
- They have a long cool-down period of several tens of seconds and so do not provide good protection against a short supply interruption.

Positive-Temperature-Coefficient Thermistor Limiting

Another solution to the inrush current problem is to use instead a *positive*-temperature-coefficient thermistor in place of the fuse. These are characterized such that provided the current remains below a given value self-heating is negligible, and the resistance of the device is low. When the current exceeds this value under fault conditions the thermistor starts to self-heat significantly, and its resistance increases until the current drops to a low value. Such a device does not protect against electric shock and so cannot replace a fuse in all applications, but because of its inherent insensitivity to surges it can be useful in local protection of a transformer winding.

A further more complex solution is to switch the ac input voltage only at the instant of zero crossing, using a triac. This results in a predictable switch-on characteristic and may be attractive if electronic switching is required for other reasons, such as standby control. Similarly, dc input supplies can use a power MOSFET to provide a controlled resistance at turn-on, as well as other circuitry such as reverse polarity protection and standby switching.

7.2.5 WAVEFORM DISTORTION AND INTERFERENCE

Interference

Electrical interference generated within equipment and conducted out through the mains supply port was already subject to regulation for some product sectors in some countries, and with the adoption of the European EMC Directive, it is mandatory for *all* electrical or electronic products to comply with interference limits. The usual method of reducing such interference is to use a radio frequency filter at the mains supply inlet, but good design practice also plays a substantial part.

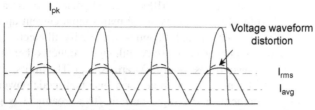

FIGURE 7.6 Peak Input Current in a Rectifier/Reservoir Power Supply.

Switch-mode power supplies are normally the worst offenders because they generate large interference currents at harmonics of the switching frequency well into the HF region. The size and weight advantages of switch-mode supplies are balanced by the need to fit larger filters to meet the interference limits.

Chapter 8 covers aspects of mains input filtering in greater detail.

Peak Current Summation

An increasing problem for electricity supply systems is the proportion of semiconductor-based equipment in the supply load. This is because the load current that such equipment takes is pulsed rather than sinusoidal. Current is only drawn at the peak of the input voltage, to charge the reservoir capacitors in the power supply. The normal RMS-to-average ratio of 1.11 for a sinusoidal current is considerably higher for this type of waveform (Fig. 7.6).

The ratio of the peak load current I_{pk} to I_{rms} is called the "crest factor," and here it depends on the input impedance of the reservoir circuit. The lower the impedance, the faster the reservoir capacitor(s) will charge, which results in lower output voltage ripple but higher peak current.

The significance of crest factor is that it affects the power handling capability of the supply network. A network of a given sinusoidal RMS current rating will show considerable extra losses when faced with loads of a high crest factor. The supply mains does not have zero impedance, and the result of the extra network voltage drop at each crest is a waveform distortion in which the sinusoidal peak is flattened. This is a form of harmonic distortion, and its seriousness depends on the susceptibility of other loads and components in the network.

Large systems installations, in which there are many electronic power supplies of fairly high rating fed from the same supply, are the main threat. In domestic premises, the switch-mode supplies of TV sets are the main offenders; in commercial buildings, the problem is worst with switch-mode supplies of PCs and their monitors, and fluorescent lamps with electronic ballasts. The current peaks always occur together and so reinforce each other. A network that is dominated by resistive loads, such as heating and filament-lamp lighting, can tolerate a proportion of high crest factor loads more easily.

Power Factor Correction

The "peakiness" of the input current waveform is best described in terms of its harmonic content and legislation now exists in Europe, under the EMC Directive, to control this. The European standard EN 61000-3-2:2000 places limits on the amplitude of each of the harmonic components of the mains input current up to the 40th (2 kHz at 50 Hz mains frequency), and it applies to virtually all electrical and electronic apparatus up to an input current of 16 A, although products other than lighting equipment with a rated power of less than 75 W are exempt. The limits, although not particularly stiff, are pretty much impossible for a switched-mode power supply to meet without some treatment of the input current. This treatment is generically known as "power factor correction (PFC)".

In this context, power factor (PF) is the ratio between the real power, as transferred through the power supply to its load with associated losses, and the apparent power drawn from the mains: RMS line voltage times RMS line current. A purely resistive load will have a PF of unity, but since peaks increase the RMS current, one drawing a peaky waveform will have a PF of 0.5—0.75. "Correcting" the PF toward unity requires that the input current waveform is made nearly sinusoidal, so that its harmonic content is much reduced. This is done by a second switching "pre-regulator" operating directly at the mains input. The usual topology is a boost regulator, as shown in Fig. 7.7.

The input rectifier supplies a full-wave-rectified half-sine voltage across C_{IN}. This capacitor is too low a value to affect the 50 Hz input current significantly, but high enough to act as an effective reservoir at the switching frequency (typically 50—100 kHz). One sense input of the switching controller comes from this input voltage, and the controller is designed to maintain an average input current through

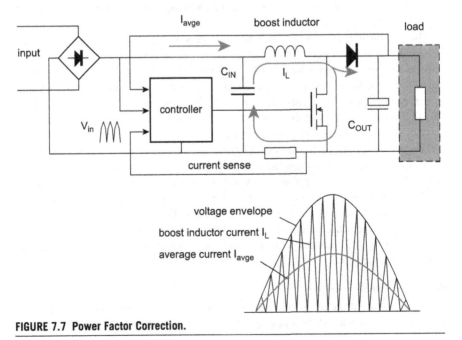

FIGURE 7.7 Power Factor Correction.

the inductor in phase with this voltage. It does this by varying the switching-pulse width or frequency as the input voltage changes. The rectified output is a dc voltage slightly higher than the highest peak supply voltage, which forms a reasonably well-regulated input to the main SMPS converter—which can of course be for any application, not just for an electronic power supply.

Naturally the addition of a second switching converter increases the cost of the total power supply and contributes to more interference which must be filtered out at the mains input. Neither of these disadvantages are excessive, and commercial PFC power supply modules are now widely available. If you are designing your own, several IC manufacturers offer controllers specifically for the purpose, such as the L4981 A/B, L6561, UC3853-5, and MC33626/33368. An extra advantage of the PFC preregulator is that almost by definition it will work over a wide input voltage range; so that a by-product of including it is that a single power supply will cover all worldwide markets (Section 7.2.1) and will also have a uniform and predictable response to dips and interruptions (Section 7.3.2).

7.2.6 FREQUENCY

The UK and European mains frequency is held to 50 Hz ± 1%. The American supply standard is 60 Hz. The difference in frequencies does not generally cause any problem for equipment that has to operate off either supply (provided that it is designed in Europe!), since mains transformers and reservoir circuits that perform correctly at 50 Hz will have no difficulty at 60 Hz. The sensitivity of the power supply circuits to supply voltage droops at 60 Hz should be less than 50 Hz since the ripple amplitude is only 83% of the 50 Hz figure, and the minimum voltage will thus be higher (Fig. 7.8).

The ±1% tolerance on the mains frequency is slightly misleading because the supply authorities maintain a long-term tolerance very much better than this. Diurnal variations are arranged to cancel out, and this allows the mains to be used as a timing source for clocks and other purposes. If you are planning to use the mains frequency for internal timing then you will need to incorporate some kind of switching arrangement if the equipment will be used on both US and European systems.

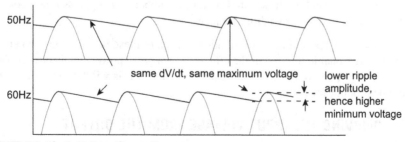

FIGURE 7.8 Ripple Voltage Versus Frequency.

7.2.7 EFFICIENCY

The efficiency of a power supply module is its output power divided by its input power. The difference between the two quantities is accounted for by power losses in the various components in the power supply as shown in Eq. (7.1).

$$\text{Efficiency } \eta = \frac{P_{out}}{P_{in}} = \frac{P_{out}}{(P_{out} - P_{loss})} \tag{7.1}$$

The efficiency normally worsens as the load is reduced because the various losses and quiescent operating currents assume a greater proportion of the input power. Therefore, if you are concerned about efficiency, do not use a power supply that is heavily overrated for its purpose. Linear supply efficiency also varies considerably with its input voltage, being worst at high voltages, because the excess must be lost across the regulator. Switch-mode supplies do not have this problem.

Normally efficiency is not of prime concern for mains power supplies, since it is not essential to make optimum use of the available power, although at higher powers the heat generated by an inefficient unit can be troublesome. It is far more important that a power converter for a portable instrument should be efficient because this directly affects useable battery life.

Linear power supplies are rarely more than 50% efficient unless they can be matched to a narrow input voltage range, whereas switch-mode supplies can easily exceed 70% and with careful design can reach 90%. This makes switch-mode supplies more popular, despite their greater complexity, at the higher power levels and for battery-powered units.

Sources of Power Loss

The components in a power supply, which make the major contribution to losses, are as follows:

- the transformer: core losses, determined by the operating level and core material, and copper losses, determined by I^2R where R is the winding resistance
- the rectifiers: diode forward voltage drop, V_F, multiplied by operating current; more significant at low output voltages
- linear regulator: the voltage dropped across the series-pass element multiplied by the operating current; greatest at high input voltages
- switching regulator: power dissipated in the switching element due to saturation voltage, plus switching losses in this and in snubber and suppressor components, proportional to switching frequency.

If you sum the approximate contribution of each of these factors, you can generally make a reasonable forecast of the efficiency of a given power supply design. The actual figure can be confirmed by measurement, and if it is wildly astray then you should be looking for the cause.

7.2.8 DERIVING THE INPUT VOLTAGE FROM THE OUTPUT

In a linear supply with a series-pass regulator element, the design must proceed from the minimum acceptable output voltage at maximum load current and minimum input voltage. These are the worst-case conditions and determine the input voltage step-down required.

The minimum dc input voltage is given by the minimum output voltage plus all the tolerances and voltage drops in series:

$$V_{in,dc} = V_{out(min)} + V_{tol,reg} + V_{series,reg} + V_{series,CS+...} \quad (7.2)$$

where $V_{out(min)}$ is the minimum acceptable output voltage; $V_{tol,reg}$ is the regulator voltage tolerance, assuming it is not adjustable; $V_{series,reg}$ is the voltage drop across the regulator series-pass element; $V_{series,\,CS}$ is the voltage drop across the current sense element if fitted.

All the above parameters are specified at full load current. This value for $V_{in,dc}$ is then the minimum input voltage allowed for a dc input supply, or it is the voltage at the minimum of the ripple trough for a rectified and smoothed ac input supply. This is related to the transformer secondary voltage as follows:

$$V_{tx} = \frac{(V_{in,dc} + V_{ripple} + V_D)}{0.92} \times \frac{(V_{ac(nom)})}{V_{ac(min)}} \times \frac{1}{\sqrt{2}} \quad (7.3)$$

where V_{ripple} is the peak ripple voltage across the reservoir capacitor; V_D is the voltage drop across the rectifier diode(s); V_{tx} is the RMS transformer secondary voltage; $V_{ac(nom)}$ is the specified transformer input voltage; $V_{ac(min)}$ is the minimum line input voltage; all parameters are at full load current.

The figure of 0.92 is an approximate allowance for full-wave rectifier efficiency with a single-capacitor reservoir. It can be more accurately derived using curves published by Schade.[2] Complications set in because the current drawn through the secondary is not sinusoidal, but occurs at the crest of the waveform (see Section 7.2.5). The extra peak current reduces the peak secondary voltage from its quoted value, if this value is specified for a resistive load. You can get around this either by knowing the transformer's losses in advance and allowing for the extra IR drop, or by specifying the transformer for a given circuit and letting the transformer supplier do the work for you, if you are buying a custom component. The transformer secondary RMS current rating is determined by the rectifier configuration (Fig. 7.9).

Take as an example a typical linear regulator circuit supplying 5 V ± 5% at 1 A (Fig. 7.10).

Here, $V_{out(min)}$ is allowed to be 5 V−5% = 4.75 V. The regulator we shall use is a standard 7805 type with ±4% tolerance and so $V_{tol,reg}$ is 5 V · 0.04 = 0.2 V. Its specified minimum series voltage drop (or dropout voltage) at 1 A and a junction temperature of 25°C (note the temperature restriction) is 2.5 V maximum. The required minimum input voltage is

$$V_{in,dc} = 4.75 + 0.2 + 2.5 = 7.45 \text{ V} \quad (7.4)$$

---

[2]O.H.Schade, *Analysis of Rectifier Operation.* Proc. IRE, vol 31 1943, pp. 341−361.

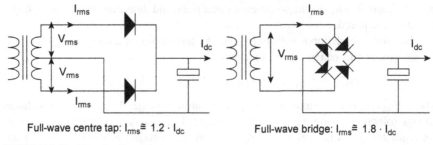

Full-wave centre tap: $I_{rms} \cong 1.2 \cdot I_{dc}$ Full-wave bridge: $I_{rms} \cong 1.8 \cdot I_{dc}$

FIGURE 7.9 Rectifier Configuration.

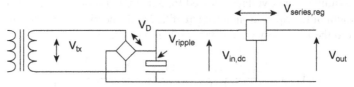

FIGURE 7.10 Linear Regulator Circuit Example.

If the peak ripple voltage is 2 V and each diode forward drop in the bridge is 1 V, then the transformer voltage with a 240 V nominal spec but a minimum line voltage of 195 V will need to be

$$V_{tx} = \frac{[7.45 + 2 + (2 \times 1)]}{0.92} \times \frac{240}{195} \times \frac{1}{\sqrt{2}} = 10.83 V_{rms} \qquad (7.5)$$

From this example you can see that the secondary-side input voltage needed to assure a given output voltage is very much higher than the actual output voltage. One of the major culprits is the dropout voltage of the regulator which in this example accounts for at least 50% of the output power, although it becomes proportionally less at higher output voltages. Low-dropout voltage regulators that use a PNP transistor as the series-pass element, such as National Semiconductor's LM2930 range, are popular for this reason and also where the minimum input voltage can be close to the output level, as in automotive applications.

Power Losses at High Input Voltage

You can also see more clearly in the above example where the power losses are which contribute to reduced efficiency. When the input voltage is increased to its maximum value the dissipation in the series-pass element is worst. In the above example with the mains input at 264 V, the average value of $V_{in,dc}$ rises to 12.5 V, and 7.45 V of this must be lost across the regulator, which because it is passing the full load current amounts to one-and-a-half times the load power! The advantage of the switch-mode supply is that it adjusts to varying input voltages by modifying its switching duty cycle, so that an increased input voltage automatically reduces the input current and the overall power taken by the unit remains roughly constant.

7.2.9 LOW-LOAD CONDITION

When the output load is removed or substantially reduced then the dissipation in the power supply will drop. This is good news for almost all parts of the circuit, except for the voltage rating of the components around $V_{in,dc}$. When there is a combination of low load and maximum supply input voltage, the peak value of $V_{in,dc}$ is highest. A crucial factor here is the transformer regulation. This is the ratio

$$\text{Regulation} = \frac{\left(V_{sec,unloaded} - V_{sec,loaded} \right)}{V_{sec,loaded}} \qquad (7.6)$$

and a small or poorly designed transformer can have a regulation exceeding 20%. If this figure is used for the transformer in the above example then the peak V_{tx} off-load at maximum input voltage will rise to 20.2 V. At the same time the diode forward voltage drops at low current will be much less, say 0.6 V each, so the possible maximum voltage at the reservoir capacitor could be around 19 V. Thus even the common 16 V rated electrolytic will not be adequate for this circuit. For higher voltage outputs, the maximum input voltage can even exceed the voltage rating of the regulator itself, and you have to invest in a preregulator to hold the maximum to a manageable level. Note that this condition is not the worst case for regulator power dissipation, because the regulator is not passing significant load current.

Maximum Regulator Dissipation

In fact maximum series-pass dissipation does not necessarily occur at full load current because as the current rises the voltage across the series-pass element falls. The maximum dissipation will occur at less than full output if the voltage dropped across the dc supply's equivalent series resistance is greater than half the difference between the no-load input voltage and the output voltage. Fig. 7.11 shows this graphically.

Minimum Load Requirement

A further problem, particularly with switch-mode supplies, is that the stability of the regulator cannot always be assured down to zero load. For this reason some rails have to be run with a minimum load, such as a bleed resistor, to remain within

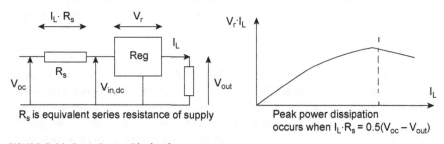

R_s is equivalent series resistance of supply

Peak power dissipation occurs when $I_L \cdot R_s = 0.5(V_{oc} - V_{out})$

FIGURE 7.11 Peak Power Dissipation.

specification, and this represents an unnecessary additional power drain. Many circuits, of course, always take a minimum current and so the minimum load is not then a problem.

7.2.10 RECTIFIER AND CAPACITOR SELECTION

The specification of the rectifiers and capacitors is dominated by surge and ripple current concerns.

Reservoir Capacitor
The minimum capacitor value is easily decided from the required ripple voltage:

$$C = \frac{I_L}{V_{\text{ripple}}} \times t \tag{7.7}$$

where I_L is the dc load current; V_{ripple} is the acceptable ripple voltage; For mains inputs, t is about 2 ms less than the ac input period, 8 ms for 50 Hz or 6 ms for 60 Hz full-wave.

A more accurate value can be derived from Schade's curves[2] which have been reprinted in numerous textbooks, but remember that the tolerance on reservoir capacitors is wide (typically ±20%) and accuracy is rarely needed.

For load currents exceeding 1 A, the ripple current rating tends to determine capacitor selection rather than ripple voltage (see Chapter 2 on passive components). As is made clear throughout this chapter, the peak current flow through the rectifier/capacitor circuit is many times higher than the dc current, due to the short time in each cycle for which the capacitor is charging. The RMS ripple current is two to three times higher than the dc load. Ripple current rating is directly related to temperature, and you may need to derate the component further if you need high ambient temperature and/or high reliability operation.

As an example, a load current of 2 A and a permissible ripple voltage of 3 V at 100 Hz suggests a 5300 µF capacitor. Typical capacitors of the next value up from this, 6800 µF, have 85°C ripple current ratings from 2 to 4 A. The higher ratings are larger and more expensive. But actual ripple current requirements will be 4−6 A. To meet this you will need to use either a much larger capacitor (typically 22,000 µF), or two smaller capacitors in parallel, or derate the operating temperature *and* use a slightly larger capacitor. If you do not do this, your design will become yet another statistic to prove that electrolytic capacitors are the prime cause of power supply failure.

Rectifiers
Although in the full-wave arrangements (Fig. 7.9) the diodes only conduct on alternate half cycles, because the RMS current is two to three times higher than the dc load current a rating of *at least* the full load current, and preferably twice it, is necessary. Surge current on turn-on may be much higher, especially in the higher power supplies where the ratio of reservoir capacitance to operating current is increased.

This is of even greater concern in direct-off-line switch-mode supplies where there is no transformer series resistance to limit the surge, and a diode rating of up to five times the average dc current is needed.

The maximum instantaneous surge current is V_{max}/R_s and the capacitor charges with a time constant of $\tau = C \cdot R_s$, where R_s is the circuit series resistance. As a conservative guide, the surge will not damage the diode if τ is less than a half cycle at mains frequency and V_{max}/R_s is less than the diode's rated I_{FSM}. All diode manufacturers publish I_{FSM} ratings for a given time constant; for example, the typical 1N5400 series with 3 A average rating have an I_{FSM} of 200 A. You may discover that you have to incorporate a small extra series resistance to limit the surge current, or use a larger diode, or apply the techniques discussed in Section 7.2.4.

The rectifier's peak-inverse-voltage (PIV) rating needs to be at least equivalent to the peak ac voltage for the full-wave bridge circuit or twice this for the full-wave center tap. But you should increase this considerably (by 50%—100%) to allow for linc transients. This is easy for low-voltage circuits, since 200 V diodes cost hardly any more than 50 V ones, and does not normally make much cost difference in mains circuits. For 240 V, a minimum of 600 V PIV and preferably 800 V PIV should be specified, even if you are using a transient suppressor at the input.

7.2.11 LOAD AND LINE REGULATION

Load regulation refers to the permissible shift in output voltage when the load is varied, usually from none to full. Line (or input) regulation similarly refers to the permissible shift in output voltage when the input is varied, usually from maximum to minimum. Provided that the design of the input circuit has been properly considered as described above, so that the input voltage never goes outside the regulator's operational range, these parameters should be wholly a function of the regulator circuit itself. The regulator is essentially a feedback circuit, which compares its output voltage against a reference voltage, so the regulation depends on two parameters: the stability of the reference and the gain of the feedback error amplifier. If you use a monolithic regulator IC, then these factors are taken into account by the manufacturer who will specify regulation as a data sheet parameter.

Thermal Regulation

A monolithic regulator IC includes the voltage reference on-chip, along with other circuitry and the series-pass element. This means that the reference is subject to a thermal shift when the power dissipation of the series-pass element changes. This gives rise to a separate longer term component of regulation, called thermal regulation, defined as the change in output voltage caused by a change in dissipated power for a specified time. Provided the chip has been well designed, thermal regulation is not a significant factor for most purposes, but it is rarely specified in data sheets and for some precision applications may render monolithic regulators unsuitable.

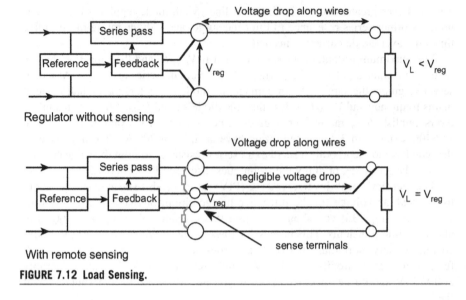

FIGURE 7.12 Load Sensing.

Load Sensing

No three-terminal regulator can maintain a constant voltage at anywhere other than its output terminals. It is common in larger systems for the load to be located at some distance from the power supply module, so that load-dependent voltage drops occur in the wiring connecting the load to the power supply output (cf. Section 1.1.5). This directly impacts the achievable load regulation.

The accepted way to overcome this problem is to split the regulator feedback path and incorporate two extra "sensing" terminals that are connected so as to sense the output voltage at the load itself (Fig. 7.12). The voltage drop across this extra pair of wires is negligible because they only carry the signal current. The voltage at the regulator output is adjusted so as to regulate the voltage at the sensing terminals.

The minimum voltage at the regulator input must be increased to allow for the extra output voltage drop. It is wise to connect coupling resistors (shown shaded in Fig. 7.12) from the output to sense terminals, so as to ensure correct operation when the sense terminals are accidentally or deliberately disconnected. Sensing can only offer remote load regulation at one point and so is not really suited when one power supply module feeds several loads at different points.

7.2.12 RIPPLE AND NOISE

Ripple is the component of the ac supply frequency (or more often its second harmonic), which is present on the output voltage; noise is all other ac contamination on the output. In a linear power supply, ripple is the predominant factor and is given by the ac across the reservoir capacitor reduced by the ripple rejection (typically 70–80 dB) of the regulator circuit. A figure of less than 1 mV RMS should be easy to obtain. HF noise is filtered by the reservoir and output capacitors, and there are no significant internal noise sources, provided that the regulator is not allowed to oscillate, so that apart from supply-frequency ripple linear power supplies are very "quiet" units.

Switching Noise

The same cannot be said for switch-mode power supplies. Here the noise is mainly due to output voltage spikes at the switching frequency, caused by fast-rise-time edges and HF ringing at these edges feeding through, or past, filtering components to the output. The ESR and ESL of typical output filter capacitors limits their ability to attenuate these spikes, while the self-inductance of ground wiring limits the high frequency effectiveness of ground decoupling anyway. Switch-mode output ripple and noise is typically 1% of the rail voltage, or 100–200 mV. In fact comparing ripple and noise specifications is the easiest way to distinguish a linear from a switch-mode unit, if there is no other obvious indication. The bandwidth over which the specification applies is important, since there is significant energy in the high-order harmonics of the switching noise, and at least 10 MHz is needed to get a true picture. Because of stray coupling over this extended bandwidth the noise frequently appears in common mode, on both supply and 0 V simultaneously, and is then very difficult to control. Differential mode noise spikes can be reduced dramatically by including a ferrite bead in series, and a small ceramic capacitor in parallel with the output capacitor.

The presence of switching noise is not a problem for digital circuits, but it creates difficulties for sensitive analog circuits if their bandwidth exceeds the switching frequency. It can cause interference on video signals, misclocking in pulse circuits and voltage shifts in dc amplifiers. These effects have to be treated as EMC phenomena (see Chapter 8) and can be cured by suitable layout, filtering and shielding, but if you have the option in the early stages to choose a linear supply instead, take it—you will save yourself a lot of trouble.

Layout to Avoid Ripple

Power supply output ripple is aggravated by incorrect layout of the wiring around the reservoir capacitor. This is a specific instance of the common-impedance interference coupling that was discussed in Chapter 1.

At first sight grounds A and B in Fig. 7.13 look equivalent. But there will be a potential between them of $I_R \cdot R_g$, where I_R is the capacitor ripple current and R_g is the track or wiring resistance common to the two grounds through which the ripple current flows. (The ripple current path is through the transformer, the two diodes, and the capacitor.) This current is only drawn on peaks of the ac input waveform to charge the reservoir capacitor, and its magnitude is only limited by the combined series resistance of the transformer winding, the diodes, capacitor, and track or wiring. If the steady-state dc current supplied is 1 A then the peak ripple current may be of the order of 5 A; thus 10 mΩ of R_g will give a peak difference of 50 mV between grounds A and B. If some parts of the circuit are grounded to A and some to B, then

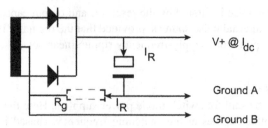

FIGURE 7.13 Incorrect Reservoir Connection.

tens of millivolts of hum injection are included in the design at no additional cost, and increasing the reservoir value to try and reduce it will actually make matters worse as the peak ripple current is increased. You can check the problem easily, by observing the output ripple on an oscilloscope; if it has a pulse shape then wiring is the problem, if it looks more like a sawtooth then you need more smoothing.

Correct Reservoir Connection

The solution to this problem, and the correct design approach, is to ground *all* parts of the supplied circuit on the supply side of the reservoir capacitor, so that the ripple current ground path is not common to any other part of the circuit (Fig. 7.14). The same applies to the V+ supply itself. The common impedance path is now reduced to the capacitor's own ESR, which is the best you can do.

7.2.13 TRANSIENT RESPONSE

The transient response of a power supply is a measure of how fast it reacts to a sudden change in load current. This is primarily a function of the bandwidth of the regulator's feedback loop. The regulator has to maintain a constant output in the face of load changes, and the speed at which it can do this is set by its frequency response as with any conventional operational amplifier. The trade-off that the power supply designer has to worry about is against the stability of the regulator under all load conditions; a regulator with a very fast response is likely to be unstable under some conditions of load, and so its bandwidth is "slugged" by a compensation

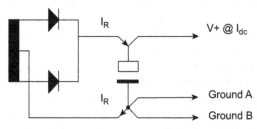

FIGURE 7.14 Correct Reservoir Connection.

capacitor within the regulator circuit. Too much of this and the transient response suffers. The same effect can be had by siting a large capacitor at the regulator output, but this is a brute-force and inefficient approach because its effect is heavily load-dependent. Note that the 78XX series of three-terminal regulators should have a small, typically 0.1 μF capacitor at the output for good transient response and HF noise decoupling. This is separate from the required 0.33−1 μF capacitor at the input to ensure stability.

Switch-Mode Versus Linear

The transient response of a switch-mode power supply is noticeably worse than that of a linear because the bandwidth of the feedback loop has to be considerably less than the switching frequency. Typically, switch-mode transient recovery time is measured in milliseconds while linear is in the tens of microseconds.

If your circuit only presents slowly varying loads then the power supply's transient response will not interest you. It becomes important when a large proportion of the load can be instantaneously switched—a relay coil or bank of LEDs for example—and the rest of the load is susceptible to short-duration over- or undervoltages.

Although load transient response is usually the most significant, a regulator also exhibits a delayed response to line transients, and this may become important when you are feeding it from a dc input that can change quickly. The line transient response is normally of the same order as the load response.

7.3 ABNORMAL CONDITIONS
7.3.1 OUTPUT OVERLOAD

At some point in its life, a power supply is almost sure to be faced with an overload on its output. This can take the form of a direct short circuit across its output due to the slip of a technician's screwdriver, or a reduced load resistance due to component failure in the load circuit, or incorrect connection of too many loads. It may also be mistaken connection to the output of another power supply. The overload can be transient or sustained, and at the very least any power supply should be designed to withstand a continuous short circuit at its output(s) without damage. This is almost universally achieved with one of two techniques: constant current limiting or fold-back current limiting.

Constant Current Limiting

Output overloads threaten mainly the series-pass element in a linear supply or the switching element in a switch-mode supply. In either case, an output overcurrent will subject the device to the maximum current that the input can supply while it is sustaining the full input−output differential voltage, and this will put its dissipation well outside its safe operating area (SOA) boundary (see Fig. 4.21). Swift destruction will ensue.

Constant current limiting operates by ensuring that the output current available from the power supply limits at a maximum that is only marginally above the full load rating of the unit. Fig. 7.15 shows this operation for a linear supply. This simple circuit works quite well, but the actual value of I_{SC} is very dependent on TR2's V_{BE}, and hence on temperature, so that either you must allow a large margin over full load current or use a more complex circuit.

Switch-mode current limiting is more complex yet because you need to limit on a cycle-by-cycle basis to protect the switching element properly; current sensing on the output line is insufficient. Several techniques have been evolved to achieve this; consult switching regulator design manuals for details.

Foldback Current Limiting

A disadvantage of constant current limiting is that to obtain sufficient SOA the pass element must have a much higher collector current capability than is needed for normal operation. "Foldback" current limiting reduces the short circuit current while still allowing full output current during normal regulator operation, thereby giving more efficient use of the pass element's SOA.

The development of the constant-current circuit to give foldback operation is shown in Fig. 7.16. Although foldback allows the use of a smaller series-pass element, it has its limitations. As the foldback ratio, I_K/I_{SC}, is increased, the required value of R_{SC} increases, and this calls for a greater input voltage at high foldback ratios. There is an absolute limit to the foldback ratio when R_{SC} is infinite of

$$\left[\frac{I_K}{I_{SC}}\right]_{max} = 1 + \left(V_{OUT}\Big/V_{BE(on)}\right) \tag{7.8}$$

and so foldback ratios of greater than two or three are impractical for low-voltage regulators (Fig. 7.17).

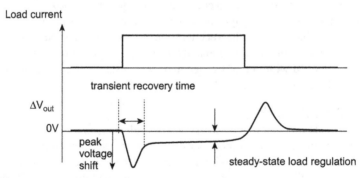

FIGURE 7.15 Load Transient Behavior.

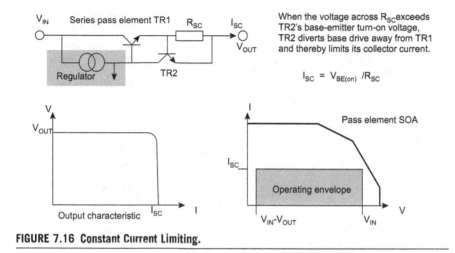

When the voltage across R_{SC} exceeds TR2's base-emitter turn-on voltage, TR2 diverts base drive away from TR1 and thereby limits its collector current.

$$I_{SC} = V_{BE(on)} / R_{SC}$$

FIGURE 7.16 Constant Current Limiting.

7.3.2 INPUT TRANSIENTS

Under this heading we need to consider spikes, surges, and interruptions on the input supply.

Interruptions

On the mains supply, dips ("brownouts") and outages of up to 500 ms are fairly common, due to surge currents and fault clearing in the supply network. Other sources of supply may also experience such interruptions. The occurrence of longer supply breaks depends very much on location. In the United Kingdom, the average consumer loses power for 90 min in the year, but a rural consumer on the end of a long overhead line may experience much longer interruptions, while an urban consumer with several redundant supply routes may see none at all.

A power supply should be able to cope with short interruptions and brownouts transparently, so that the load is unaffected by them. The "hold-up time" (Fig. 7.18) specifies for how long the output remains stable after loss of input, and it can be anywhere from a few to several hundred milliseconds. It is almost entirely determined by the size of the main reservoir capacitor, since this provides the only source of power when the input is removed.

A linear regulator can be considered as a constant-current sink discharging this capacitor and therefore it is easy to calculate the hold-up time for a given load and input voltage. A switch-mode regulator draws more current as its input voltage drops, so accurately determining hold-up time for this type requires the solution of a current–time integral. The higher the operating voltage, the easier it is to obtain a long hold-up time because energy storage in the reservoir is proportional to $0.5\,C\,V^2$. This gives another advantage to direct-off-line switching supplies, whose main reservoir operates at the full line voltage.

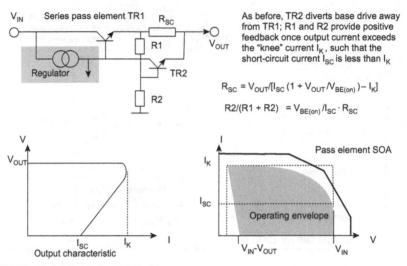

As before, TR2 diverts base drive away from TR1; R1 and R2 provide positive feedback once output current exceeds the "knee" current I_K, such that the short-circuit current I_{SC} is less than I_K

$$R_{SC} = V_{OUT}/[I_{SC}(1 + V_{OUT}/V_{BE(on)}) - I_K]$$

$$R2/(R1 + R2) = V_{BE(on)}/I_{SC} \cdot R_{SC}$$

FIGURE 7.17 Foldback Current Limiting.

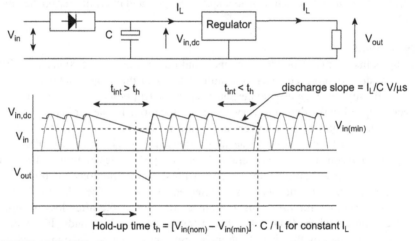

FIGURE 7.18 Hold-Up Time.

Taking the quoted parameters for the linear supply given in Section 7.2.8, what values does this give for its hold-up time at full load at 240 and 204 V?

The ripple on $V_{in,dc}$, of 2 V at 1 A, with a full-wave rectified supply so that its period is 10 ms, means that the reservoir capacitor is

$$C = \frac{It}{V} = 1 \times \frac{10^{-3}}{2} = 5000\ \mu F \tag{7.9}$$

At 240 V the minimum value for $V_{in,dc}$ at the ripple trough is

$$V_{in,dc(min)} = 14.05 - 2_{ripple} - 2_{diode} = 10.05 \text{ V} \qquad (7.10)$$

so the hold-up time at this voltage given a minimum requirement of 7.45 V at the regulator is

$$t_h = (10.05 - 7.45) \times 5000 \times \frac{10^{-6}}{1} = 13 \text{ ms} \qquad (7.11)$$

At 204 V (240 V−15%) the minimum value for $V_{in,dc}$ is 7.94 V, so the hold-up time is now

$$t_h = (7.94 - 7.45) \times 5000 \times \frac{10^{-6}}{1} = 2.5 \text{ ms} \qquad (7.12)$$

It is clear that hold-up time specified at nominal input voltage may be considerably less when the power supply is running at its minimum input voltage. In fact, the minimum input voltage as calculated in Section 7.2.8 is that for which the hold-up time is zero. All this is assuming the worst-case condition, that the supply is interrupted at the minimum of the ripple trough. If hold-up time is important for your circuit, you must decide at what input voltage it is to be specified.

Spikes and Surges

Chapter 8 discusses the occurrence of transient overvoltages on mains and automotive supplies. Some precautions need to be taken to prevent these as far as possible from propagating through the power supply and impacting the load circuit. Short, low-energy but fast rise-time transients can only be dealt with by good circuit layout, minimizing ground inductance and stray coupling, and by input filtering. Slower but higher-energy transients call for the use of transient suppressor devices at various points in the power supply, and for overvoltage protection.

7.3.3 TRANSIENT SUPPRESSORS

Fig. 7.19 shows three positions for transient suppressors in a linear supply. The advantages and disadvantages of each position can be summarized as follows:

- Z1: protects all components in the unit from differential-mode surges but is subject to the lowest source impedance. This means that it must have a high energy rating

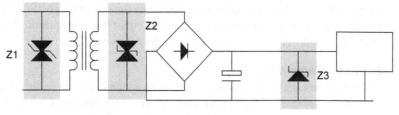

FIGURE 7.19 Transient Suppression in a Linear Supply.

to withstand the maximum expected surge without destruction, and it will have a fairly high ratio of clamped voltage to normal running voltage. In effect, voltage surges up to about twice the peak operating voltage will be let through.

- Z2: this is a more effective position as it still protects the vulnerable rectifiers, but is itself protected by the additional source impedance of the transformer. It can therefore be a smaller component but still have a good ratio of clamped to peak operating voltage. It has no effect on spikes, which may have been converted from differential to common mode by the interwinding capacitance of the transformer.
- Z3: this protects the regulator and subsequent circuitry but not the rectifiers. It is something of a "belt-and-braces" position, but it does suppress input common mode spikes that the previous positions would have let through. It should be sized so that its peak clamping voltage is just less than the absolute maximum input voltage of the regulator. Smaller surges then rely on the transient response of the regulator to contain them.

7.3.4 OVERVOLTAGE PROTECTION

If the circuit that your power supply is driving is very expensive and susceptible to overvoltages—for instance it may include a £100 microprocessor, which must not be subject to more than 7 V—then it is worth including extra circuitry at the power supply output for overvoltage protection. The first time that it operates, it will have saved the extra expense of designing it in.

This might be as simple as a 6.2- or 6.8-V zener diode across the output of a 5-V supply. See Section 4.1.8 for a discussion of how to size such a zener. This does not offer foolproof protection, because if the overvoltage is sustained and derives from a low source impedance—perhaps the series-pass element has failed—then the zener is likely to fail itself, and may fail open-circuit, in which case it has been wasted. Something more drastic is called for, and the conventional solution is a crowbar.

This gets its name from the time-honored method of ensuring that no voltage is present between two live terminals, by the simple expedient of putting a crowbar—which is assumed to be able to carry any likely short-circuit current—across them. In its more sophisticated version in electronic power supplies, the crowbar takes the form of a triggered thyristor. The thyristor is permanently in place across the output, or in some designs across the reservoir, but it is only triggered when a supervisory circuit detects the presence of an overvoltage. It then stays triggered, holding the output voltage to V_H, until the current through it is interrupted by external circumstances such as a power supply reset. Although this current may be high, the voltage across it is not, so its dissipation is fairly low. Obviously the power supply itself must be protected against a sustained output short circuit, either by current limiting or a fuse or preferably both. Fig. 7.20 shows the operating principle.

Crowbar Circuit Requirements

The thyristor must be capable of dumping, virtually instantaneously, both the continuous short-circuit current of the supply and the energy stored in the reservoir

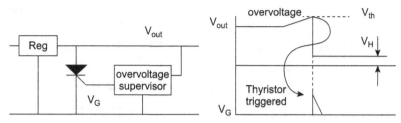

FIGURE 7.20 Overvoltage Protection.

capacitors. It must therefore have a high single-pulse I^2t and di/dt rating. Some manufacturers characterize devices especially for this purpose, and the di/dt performance is helped by making sure the trigger pulse has a fast edge and is well in excess of the minimum gate current requirement.

Both the supervisory circuit and the thyristor must be immune from false triggering due to short transients, as the nuisance value of an unnecessary shutdown may exceed that of a real overvoltage in some instances. Some degree of delay in the trigger pulse is essential, and characterizing the overall system (power supply plus crowbar protection plus load) for the acceptable and necessary delay and overvoltage threshold is the most critical part of overvoltage protection design.

7.3.5 TURN-ON AND TURN-OFF

Sometimes, the behavior of the power rails when the input power is applied or removed is important to the load circuit. The power rail never instantaneously reaches its operating level as soon as the input is applied. It will ramp up to the rail voltage as the reservoir and other capacitors charge, and it may overshoot its nominal voltage briefly if the regulator frequency compensation has not been optimized—this is a particular danger with some switch-mode circuits. It may suffer from noise or oscillation due to the switch-on process as it ramps up. Particularly if the load circuit includes a microprocessor, it will not be safe to start the circuit operation until the rail voltage has settled. You may require the power supply to have a flag output, which signals to the load circuit that all is well. This output is often connected to the micro's RESET input. (This is discussed from the micro's point of view in Section 6.4.3.)

Similarly, when the power is switched off, the microprocessor needs to be able to power down in an orderly fashion. This is best achieved by generating a power-fail interrupt as soon as a power failure is detected, followed by an undervoltage warning when the power rail starts to droop. The time delay between the two will be roughly equivalent to the hold-up time as discussed earlier, and this delay must be long enough to enable the micro to perform its power-down housekeeping functions. Required outputs are shown in Fig. 7.21.

Power Supply Unit Supervisor Circuits

All the functions of undervoltage and power-fail monitoring, and overvoltage protection, can be gathered up into a single power supply supervisory circuit, and

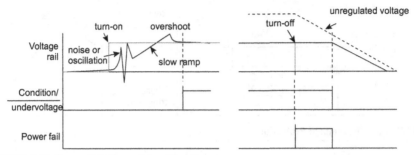

FIGURE 7.21 Power Rail Supervision Waveforms.

several ICs are on the market for this purpose. Examples are the MC3423, ICL7665 and 7673, TL7705, and MAX690 series. These chips are basically a collection of comparators and delay circuits, integrated into one package for ease of use. Unfortunately, there are a multiplicity of types and few second sources, and the parts cost may be greater than you would suffer when using standard comparators such as the LM339. In many cases you may still prefer to design your own supervisory circuit from such standard components.

A typical application will require the supervisory circuit to have inputs from the dc output rail for overvoltage protection, the reservoir capacitor for undervoltage warning, and the low-voltage ac input for power-fail detection. Outputs will go to the crowbar device and the load circuit. Bear in mind that the supervisor needs to operate reliably down to very low-supply voltages (Fig. 7.22).

7.4 MECHANICAL REQUIREMENTS
7.4.1 CASE SIZE AND CONSTRUCTION

If you are designing a supply as an integral part of the rest of the equipment then generally you will not need to consider its mechanical characteristics separately

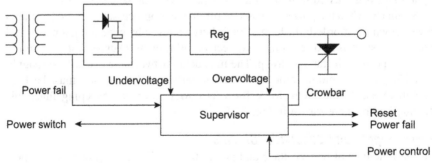

FIGURE 7.22 Configuration of Power Supply Supervisor.

from the equipment design. If you are buying in a standard unit, or designing yourself a modular unit, which will be used for different products, then case construction becomes important. Standard products tend to fall into one of four categories:

- open frame, chassis mounting
- enclosed, chassis mounting
- encapsulated, PCB or chassis mounting
- rack module

Both linear and switch-mode types are available in all these variants, but power rating, connections, and the need for screening play an important role in the final selection.

Open Frame

This is normally the cheapest option, since all that is supplied is a PCB mounted on a simple metal chassis that serves as a rudimentary heat sink. Connections are made by wiring to terminals or spade lugs mounted on the board. No environmental protection or screening is offered and the power unit must be enclosed completely within the equipment it is supplying. Open frame units are most popular in the 10–100 W range, with models available up to 250 W.

Enclosed

Cased power supplies are more common for power ratings above 100 W. They offer reasonably effective screening that is important for switch-mode supplies, and can incorporate a fan for efficient convective cooling, which is not possible with open or encapsulated types but is necessary at high powers. The greater electronics cost tends to mask the cost of the extra mechanical components. Connections are made to screw terminals on the outside of the case, and the internal circuitry is guarded from wandering fingers and other foreign bodies.

Encapsulated

Fully encapsulated units are available up to 40 W and can be either PCB mounted via pins or chassis mounted using screw terminals for the connections. Their great advantage is that they can be treated as just another component during equipment production, and do not need any further environmental protection for their internal circuit. Electromagnetic interference screening can be incorporated as part of the encapsulating box. Higher power ratings than 40 W require a heat sink outside the encapsulation. The encapsulation tends to provoke reliability problems when much heat has to be dissipated, and if you are going for a higher-power unit it would be wise to seek concrete reliability data. Encapsulation is particularly popular for low-power dc-to-dc converters, which can be incorporated within a system at board level, to generate different and/or regulated supplies from a common dc bus.

Rack Mounting Modules or Cassettes

With the increasing popularity of rack-mounted modular processor equipment, usually based on the Eurocard rack and DIN 41612 connector standard, there is a corresponding need for power supply modules that can share the same rack. These are available from 25 to 500 W. All but the smallest are switch-mode types, since space and thermal capacity are strictly limited, and applications are mainly digital. Connection is by mating plug and socket, mounted on the card frame, and it is vital to ensure that the connector used is capable of carrying the load current without loss, and is rated for mains voltages. The DIN 41612 H15 connector is widely used, with a leading earth pin to maintain safety when withdrawing or inserting the unit.

7.4.2 HEAT SINKING

A necessary requirement for the continued health of any semiconductor device, be it monolithic IC regulator, rectifier diode, or power transistor, is that its junction temperature should stay within safe limits. Junction temperature is directly related to power dissipation, thermal resistance, and ambient temperature, and the function of a heat sink is to provide the lowest possible thermal resistance between the junction and its environment—assuming the environment is always cooler.

Chapter 9 includes a discussion of thermal management, including methods of calculating the size requirements of heat sinks. Here it is enough to say that the power supply often represents the most concentrated source of heat in an item of equipment. As soon as its efficiency is roughly known, you should calculate the heat output and take steps to ensure that the mechanical arrangement will allow an efficient heat flow. At the minimum, this will involve ensuring that all components which will need heat sinking are positioned to allow this, and that the power unit's positioning within the overall equipment gives adequate thermal conductivity to the environment. Too many designs end up with a fan tacked onto the case as an afterthought!

7.4.3 SAFETY APPROVALS

Major safety risks for power supplies are the threat of electric shock due to contact with "live parts," and the threat of overheating and fire due to a fault. Safety is discussed in greater depth in Section 9.1. One of the important but forgotten functions of a power supply is to ensure a safe segregation of the low-voltage circuitry, which may be accessible to the user, from the high-voltage input, which must be inaccessible. Segregation is normally assured in a power supply by maintaining a minimum distance around all parts that are connected to the mains, including spacing between the primary and secondary of the transformer. This, of course, adds extra space to the design requirements. Insulation of at least a minimum thickness may be substituted for empty space.

There are many national and international authorities concerned with setting safety requirements. Foremost among these are UL in the United States, CSA in Canada, and the CENELEC safety standards, implementing the Low Voltage Directive in Europe. As designer, you can either choose to apply a particular set of requirements for your company's market, or if you plan to export worldwide, you can discover the most stringent requirements and apply these across the board. A common specification is EN 60950-1 (IEC 60950-1), which is the safety standard for information technology equipment and which is quoted by default by most off-the-shelf power supplies. If no safety specification is quoted, beware.

Most of the time it is legally necessary to have your product approved to safety regulations, often it is also commercially desirable. Using a bought-in supply that already has the right safety approval goes a long way to helping your own equipment achieve it. Note that there is a difference, on data sheets, between the words "designed to meet…" and "certified to…" The former means that, when you go for your own safety approval, the approvals agency will still want to satisfy themselves, at your expense, that the power supply does indeed meet their requirements. The latter means that this part of the approvals procedure can be bypassed. It therefore puts the unit cost of the power supply up, but saves you some part of your own approval expenses.

7.5 BATTERIES

Battery power is mainly used for portability or stand-by (float) purposes. All batteries operate on one or another variant of the principle of electrochemical reaction, in which anode (negative) and cathode (positive) terminals are separated by an electrolyte, which is the vehicle for the reaction. This basic arrangement forms a "cell," and a battery consists of one or more cells. The chemistry of the materials involved is such that a potential is developed between the electrodes, which is capable of sustaining a discharge current. The voltage output of a particular cell type is a complex function of time, temperature, discharge history, and state of charge.

The basic distinction is between primary (nonrechargeable) and secondary (rechargeable) cells. This section will survey the various types of each shortly, but first we shall make a few general observations on designing with batteries.

7.5.1 INITIAL CONSIDERATIONS

When you know you are going to use a battery, select the cell type as early as possible in the circuit and mechanical design. This allows you to take the battery's properties into account and increases the likelihood of a cost-effective result, as otherwise you will probably need a larger, or more expensive, battery or will suffer a reduced equipment specification. Having made the selection, you can then design the circuit so that it works over the widest possible part of the battery's available voltage range. Some of the cheaper types deliver useful power over quite a wide range, with an endpoint voltage of 60%–70% of nominal, and some of this energy will be lost if the design cannot cope with it. Also, check that the battery can deliver the circuit's load current requirements over the working temperature. This capability varies considerably for different chemical systems. Rechargeable batteries can often be recharged only over a narrower temperature range than they can be discharged.

Always aim to use standard types if your specification calls for the user to be able to replace the battery. Not only are they cheaper and better documented but also they are widely distributed and are likely to remain so for many years. You should only need to use special batteries if your environmental conditions or energy density requirements are extreme, in which case you have to make special provisions for replacement or else consider the equipment as a throwaway item.

Voltage and Capacity Ratings

Different types of battery have different nominal open-circuit voltages, and the actual terminal voltage falls as the stored energy is used. Manufacturers provide a discharge characteristic curve for each type, which indicates the behavior of voltage against time for given discharge conditions. Note that the open-circuit voltage can exceed the voltage under load by up to 15%, and the operating voltage may be significantly less than the nominal battery voltage for some of the duration.

The capacity of a battery is expressed in ampere-hours (Ah) or milliampere-hours (mAh). It may also be expressed in normalized form as the "C" figure, which is the nominal capacity at a given discharge rate. This is more frequently applied to rechargeable types. Capacity will be less than the C rating if the battery is discharged at a faster rate; for instance, a 15 Ah lead–acid type discharged at 15 A (1 C) will only last for about 20 min (Fig. 7.23).

Three typical modes under which a battery can be discharged are constant resistance, constant current, and constant power. For batteries with a sloping discharge characteristic, such as alkaline manganese, the constant power mode not only is the most efficient user of the battery's energy but also needs the most complex voltage regulating system to power the actual circuit.

Series and Parallel Connection

Cells can be connected in series to boost voltage output, but doing so decreases the reliability of the overall battery, and there is a risk of the weakest cell being driven into reverse voltage at the end of its life. This increases the likelihood of leakage or

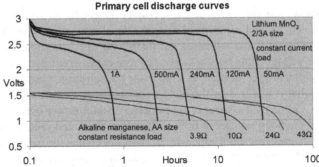

FIGURE 7.23 Load Discharge Characteristics for Lithium and Alkaline Manganese Primary Cells.

rupture and is the reason why manufacturers recommend that all cells should be replaced at the same time. Good design practice minimizes the number of series-connected cells. There are now several ICs which can be used to multiply the voltage output of even a single cell with high efficiency. It is not difficult to design a switching converter that simultaneously boosts and regulates the battery voltage.

Parallel connection can be used for some types to increase the capacity or discharge capability, or the reliability of the battery. Increased reliability requires a series diode in each parallel path to isolate failed cells. Recharging parallel cells is rarely recommended because of the uncertainty of charge distribution between the cells. It is therefore best to restrict parallel connection to specially assembled units.

On the same subject, reverse insertion of the whole battery will threaten your circuit, and if it is possible, the user will do it. Either incorporate assured polarity into the battery compartment or provide reverse polarity protection, such as a fuse, series diode or purpose-designed circuit, at the equipment power input.

Mechanical Design

Choose the battery contact material with care to avoid corrosion in the presence of moisture. The recommended materials for primary cells are nickel-plated steel, austenitic stainless steel or inconel, but definitely not copper or its alloys. The contacts should be springy to take up the dimensional tolerances between cells. Single-point contacts are adequate for low current loads, but you should consider multiple contacts for higher current loads. The simplest solution is to use ready-made battery compartments or holders, provided that they are properly matched to the types of cell you are using. PCB-mounting batteries have to be hand soldered in place after the rest of the board has been built, and you need to liaise well with the production department if you are going to specify these types.

Rechargeable batteries when under charge, and all types when under overload, have a tendency to out-gas. Always allow for safe venting of any gas, and since some gases will be flammable, do not position a battery near to any sparking or hot components. In any case, heat and batteries are incompatible: service life and efficiency will be greatly improved if the battery is kept cool. If severe vibration or shock is part of the environment, remember that batteries are heavy and will probably need extra anchorage and shock absorption material. Organic solvents and adhesives may affect the case material and should be kept away.

Dimensions of popular sizes of battery are shown in Table 7.1.

Storage, Shelf Life, and Disposal

Maximum shelf life is obtained if batteries are stored within a fairly restricted temperature and humidity range. Self-discharge rate invariably increases with temperature. Different chemical systems have varying requirements, but extreme temperature cycling should be avoided, and you should arrange for tight stock control with proper rotation of incoming and outgoing units, to ensure that an excessively aged battery is not used. Rechargeable types should be given a regular top-up charge.

Table 7.1 Sizes of Popular Primary batteries

Designation				Dimensions (mm)	
IEC	ANSI	Size	Voltage	Dia (or L × W)	Height
Alkaline Manganese Dioxide					
LR03	24A	AAA	1.5	10.5	44.5
LR6	15A	AA	1.5	14.5	50.5
LR14	14A	C	1.5	26.2	50
LR20	13A	D	1.5	34.2	61.5
6LR61	1604A	PP3	9	26.5 × 17.5	48.5
4LR25X	908A	Lamp	6	67 × 67	115
4LR25-2	918A	Lamp	6	136.5 × 73	127
Lithium Manganese Dioxide—Cylindrical Cell					
CR17345	5018LC	2/3A	3	17	34.5
CR11108	5008LC	1/3N	3	11.6	10.8
2CR11108	1406LC	2 × 1/3N	6	25.2	13
2CR5	5032LC	2 × 2/3A	6	17 × 34	45
CR-P2	5024LC	2 × 2/3A	6	19.5 × 35	36
Lithium Manganese Dioxide—Coin Cell					
CR2016	5000LC		3	20	1.6
CR2025	5003LC		3	20	2.5
CR2032	5004LC		3	20	3.2
CR2430	5011LC		3	24.5	3
CR2450	5029LC		3	24.5	5
Silver Oxide Button Cells (mAh)					
SR41	1135S0	42	1.55	7.87	3.6
SR43	1133S0	120	1.55	11.56	4.19
SR44	1131S0	165	1.55	11.56	5.58
SR48	1137S0	70	1.55	7.87	5.38
SR54	1138S0	70	1.55	11.56	3.05
SR55	1160S0	40	1.55	11.56	2.21
SR57	1165S0	55	1.55	9.5	2.69
SR59	1163S0	30	1.55	7.9	2.64
SR60	1175S0	18	1.55	6.8	2.15
SR66	1176S0	25	1.55	6.78	2.64

In the early 1990s, legislation appeared in many countries banning the use of some substances in batteries, particularly mercury, for environmental reasons. Thus mercuric oxide button cells were effectively outlawed and are not now obtainable. In Europe, this was achieved through the Batteries and Accumulators Directive (91/157/EEC).

This directive also encourages the collection of spent NiCad batteries with a view to recovery or disposal and their gradual reduction in household waste. In fact, what it has achieved is rather the development of alternative rechargeable technologies to NiCad, particularly NiMH and lithium. NiCads, though, are still widely used, despite the technical advantages of NiMH. The Batteries Directive is about to be updated, and it is likely to propose the following changes:

- EU member states to collect and recycle all batteries, with targets of 75% consumer (disposable or rechargeable) and 95% industrial batteries
- no less than 55% of all materials recovered from the collection of spent batteries to be recycled.

In the United Kingdom in 1999, 654 million consumer batteries were sold, but the rate for recycling consumer rechargeables is a mere 5%, and less than 1% of consumer batteries are collected for recycling. On the other hand, more than 90% of automotive batteries are recycled and 24% of other industrial batteries. Clearly, for consumer batteries at least, a sea change in disposal habits is expected.

7.5.2 PRIMARY CELLS

The most common chemical systems employed in primary, nonrechargeable cells are alkaline manganese dioxide, silver oxide, zinc air, and lithium manganese dioxide. Fig. 7.23 compares the typical discharge characteristics for lithium and alkaline types of roughly the same volume on various loads.

Alkaline Manganese Dioxide

The operating voltage range of this type, which uses a highly conductive aqueous solution of potassium hydroxide as its electrolyte, is 1.3 to 0.8 V per cell under normal load conditions, while its nominal voltage is 1.5 V. Recommended end voltage is 0.8 V per cell for up to six series cells at room temperature, increasing to 0.9 V when more cells are used. The alkaline battery is well suited to high-current discharge. It can operate between -30 and $+80°C$, but high relative humidity can cause external corrosion and should be avoided. Shelf life is good, typically 85% of stored energy being retained after 3 years at 20°C. Standard types are now widely and cheaply available in retail outlets, and it can therefore be confidently used in most general-purpose applications.

Silver Oxide

Zinc–silver oxide cells are used as button cells with similar dimensions and energy density to the older and now withdrawn mercury types. Their advantage is that they have a high capacity versus weight, offer a fairly high operating voltage, typically 1.5 V, which is stable for some time and then decays gradually, and can provide intermittent high pulse discharge rates and good low-temperature operation. They are popular for such applications as watches and photographic equipment. Typical shelf life is 2 years at room temperature.

Zinc Air

This type has the highest volumetric energy density, but is very specialized and not widely available. It is activated by atmospheric oxygen and can be stored in the sealed state for several years, but once the seal is broken it should be used within 2 months. It has a comparatively narrow environmental temperature and relative humidity range. Consequently its applications are somewhat limited. Its open-circuit voltage is typically 1.45 V, with the majority of its output delivered between 1.3 and 1.1 V. It cannot give sustained high output currents.

Lithium

Several battery systems are available based on the lithium anode with various electrolyte and cathode compounds. Lithium is the lightest known metal and the most electronegative element. Their common features are a high terminal voltage, very high energy density, wide operating temperature range, very low self-discharge and hence long shelf life, and relatively high cost. They have been used for military applications for some years. If abused, some types can be potentially very hazardous and may have restrictions on air transport. The lithium manganese dioxide $(LiMnO_2)$ couple has become established for a variety of applications, because of its high voltage and "fit-and-forget" lifetime characteristics. Operating voltages range from 2.5 to 3.5 V. Very high pulse discharge rates (up to 30 A) are possible. Widely available types are either coin cells, for memory backup, watches and calculators and other small, low-power devices; or cylindrical cells, which offer light weight combined with capacities up to 1.5 Ah and high pulse current capability, together with long shelf life and wide operating temperature range.

Other primary lithium chemistries are lithium thionyl chloride $(Li-SOCl_2)$ and lithium sulfur dioxide $(Li-SO_2)$. These give higher capacities and pulse capability and wider temperature range but are really only aimed at specialized applications.

7.5.3 SECONDARY CELLS

There have historically been two common rechargeable types: lead–acid and nickel–cadmium. These have quite different characteristics. Neither of them offer anywhere near the energy density of primary cells. At the same time, their heavy metal content and consequent exposure to environmental legislation (as discussed in Section 7.5.1) have spurred development of other technologies, of which NiMH and lithium ion are the front-runners.

Lead–Acid

The lead–acid battery is the type that is known and loved by millions all over the world, especially on cold mornings when it fails to start the car. As well as the conventional "wet" automotive version, it is widely available in a valve-regulated "dry" or "maintenance-free" variant in which the sulfuric acid electrolyte is retained in a glass mat and does not need topping-up. This version is of more interest to circuit designers as it is frequently used as the standby battery in mains-powered systems, which must survive a mains failure.

These types have a nominal voltage of 2 V, a typical open-circuit voltage of 2.15 V and an end-of-cycle voltage of 1.75 V per cell. They are commonly available in standard case sizes of 6 or 12 V nominal voltage, with capacities from 1 to 100 Ah. Typical discharge characteristics are as shown in Fig. 7.23. The value "C," as noted earlier, is the ampere-hour rating, conventionally quoted at the 20-h discharge rate (5-h discharge rate for nickel–cadmium and nickel–metal hydride). Ambient temperature range is typically from −30 to +50°C, though capacity is reduced to around 60%, and achievable discharge rate suffers, at the lower extreme.

Valve-regulated lead–acid types can be stored for a matter of months at temperatures up to 40°C, but will be damaged, perhaps irreversibly, if they are allowed to spend any length of time fully discharged. This is due to build-up of the sulfur in the electrolyte on the lead plates. Self-discharge is quite high—3% per month at 20°C is typical—and increases with temperature. You will therefore need to ensure that a recharging regime is followed for batteries in stock. For the same reason, equipment that uses these batteries should only have them fitted at the last moment, preferably when it is being dispatched to the customer or on installation.

Typical operational lifetime in standby float service is 4–5 years if proper float charging is followed, although extended lifetime types now claim up to 15 years. When the battery is frequently discharged a number of factors affect its service life, including temperature, discharge rates, and depth of discharge. A battery discharged repeatedly to 100% of its capacity will have only perhaps 15% of the cyclic service life of one that is discharged to 30% of its capacity. Overrating a battery for this type of duty has distinct advantages.

Nickel–Cadmium

NiCads, as they are universally known, are comparable in energy density and weight to their lead–acid competitors but address the lower end of the capacity range. Typically they are available from 0.15 to 7 Ah. Nominal cell voltage is 1.2 V, with an open-circuit voltage of 1.35–1.4 V and an end-of-cycle voltage of 1.0 V per cell. This makes them comparable to alkaline manganese types in voltage characteristics, and you can buy NiCads in the standard cell sizes from several sources, so that your equipment can work off primary or secondary battery power (Fig. 7.24).

NiCads offer an ambient temperature range from −40 to +50°C. They are widely used for memory backup purposes; batteries of two, three, or four cells are available with PCB-mounting terminals, which can be continuously trickle charged from the logic supply, and can instantly supply a lower backup voltage when this supply fails. Self-discharge rate is high, and a cell which is not trickle charged will only retain its charge for a few months at most. Unlike lead–acid types they are not damaged by long periods of full discharge, and because of their low internal resistance they can offer high discharge rates. On the other hand, they suffer from a "memory effect": a cell that is constantly being recharged before it has been completely discharged will lose voltage more quickly, and in fact it is better to recharge a NiCad from its fully discharged condition.

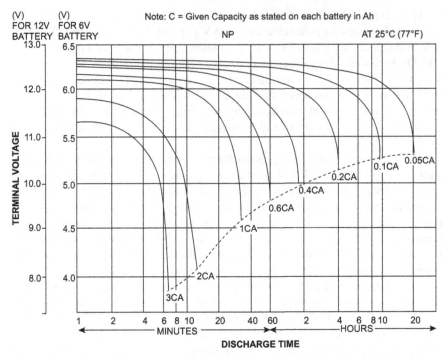

FIGURE 7.24 Discharge Characteristics for Sealed Lead—Acid Batteries (*Dotted Line* Indicates the Lowest Recommended Voltage Under Load).

From Yuasa.

However, NiCads are now frowned on because of their heavy metal content and hence the environmental consequences of their disposal to landfill. They are largely being superseded by nickel—metal hydride.

Nickel—Metal Hydride

The discharge characteristics of NiMH are very similar to those of NiCad. The charged open circuit, nominal and end-point voltages are the same. The voltage profile of both types throughout most of the discharge period is flat (Fig. 7.25). NiMH cells are generally specified from $-20°C$ to $+50°C$. They are around 20% heavier than their NiCad equivalents, but have about 40% more capacity. Also, they suffer less from the "memory effect" of NiCads (see above). On the other hand, they are less tolerant of trickle charging, and only very low trickle charge rates should be used if at all.

NiMH cells are available in a wide range of standard sizes, including button cells for memory backup, and are also frequently specified in multicell packs for common applications such as mobile phones, camcorders, and so on.

Lithium Ion

The Lithium-ion cell has considerable advantages over the types described above. Principally, it has a much higher gravimetric energy density (available energy for

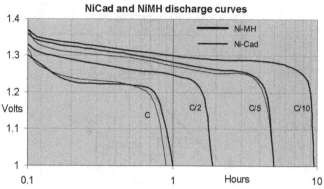

FIGURE 7.25 NiCad and NiMH Discharge Characteristics.

a given weight)—see Fig. 7.27, which compares approximate figures for the three types, drawn from various manufacturers' specifications. But also, its cell voltage is about three times that of nickel batteries, 3.6–3.7 V versus 1.2 V. Its discharge profile with time is reasonably flat with an endpoint of 3 V (Fig. 7.26), and it does not suffer from the NiCad memory effect.

These advantages come at a price, and Li-ion batteries are more expensive than the others. Also, they are much more susceptible to abuse in charging and discharging. The battery should be protected from overcharge, overdischarge, and overcurrent at all times, and this means that the best way to use it is as a battery pack, purpose designed for a given application, with charging and protection circuits built in to the pack. This prevents the user from replacing or accidentally degrading individual cells and gives the designer greater control over the expected performance of the battery. Since the high cost of a Li-ion battery pack makes it more suited to high-value applications such as laptops and mobile phones, the extra cost of the integrated control circuitry is marginal and acceptable.

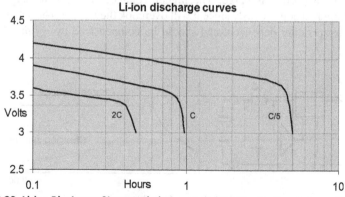

FIGURE 7.26 Li-ion Discharge Characteristics.

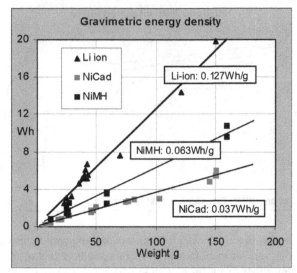

FIGURE 7.27 Comparison of Energy Density Versus Weight (Approximate Values).

7.5.4 CHARGING

The recharging procedures are quite different for the various types of secondary cell, and you will drastically shorten their lifetime if you follow the wrong one. The greatest danger is of overcharging. Briefly, NiCad and NiMH require constant current charging, while lead—acid needs constant voltage.

Lead—Acid

The recommended method for these types is to provide a current-limited constant voltage (Fig. 7.28). The initial charging current is limited to a set fraction of the C value, generally between 0.1 and 0.25. The constant voltage is set to 2.25—2.5 V per cell, depending on whether the intention is to trickle charge or recover from a cyclic discharge. The higher voltage for cyclic recharging must not be left applied continuously since it will overcook the battery. The actual voltage is mildly temperature dependent and should be compensated by −4 mV/°C when

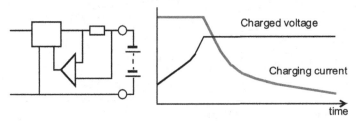

FIGURE 7.28 Current Limited Constant-Voltage Charging.

operating with extreme variations. This charging characteristic can easily be provided by a current-limited voltage regulator IC.

An elegant modification to this circuit is to arrange for the output voltage to drop from the cyclic charge level to the trickle charge level when the charging current has decreased sufficiently, typically to 0.05 C. This two-step charging gives the advantages of rapid recovery from deep discharge along with the benefit of trickle charging without threat to the battery's life. It does not work too well when the load circuit is connected, though.

The cheap-and-cheerful charging method is to charge the battery from a full-wave or half-wave rectified ac supply through a series resistance. This is known as "taper" charging because the applied voltage rises toward the constant voltage level as the charging current tapers away. Since it is cheap, it is very popular for automotive use, but manufacturers do not recommend it because of the risk of over-charging and the undesirable effects of the ac ripple current. Fluctuations in the mains supply can easily lead to overvoltage and the end current cannot be properly controlled. If you have to use it, include a timer to limit the overall charging time.

Lead–acid batteries *can* be charged from a constant current, typically between 0.05 and 0.2 C, subject to monitoring the cell voltage to detect full charge. The technique is not often used, but can be effective for charging several batteries in series at once.

NiCad and NiMH

Because the voltage characteristic during charging varies substantially and actually drops when the cell is overcharged, NiCad and NiMH batteries can only be charged from a constant current. Continuous charging at up to the 0.1 C rate is permissible without damage to the cell. A 0.1 C charge rate will recharge the cell in 16 h, not 10, due to the inefficiency of the charging process. An accelerated charge rate of up to 0.3 C is permissible for long periods without harm, but the cell temperature will rise when charging is complete. Higher rates for a rapid charge will work, but in this case it is essential to monitor the charging progress and terminate it before the battery overheats.

A series resistor from a voltage source at a significantly higher level than the battery terminal voltage, which can rise to 1.55 V per cell, is an adequate constant-current charger. For tighter control over the charging current, and especially for rapid charging, a voltage/current regulator with a battery temperature sensor is needed. Several dedicated battery charge controller ICs are available, such as the TEA1100, MC33340, and LT1510 series, and the MAX713.

Occasional overcharging, in which a partly discharged battery is put back on charge for longer than necessary, will not have much effect; but repeatedly recharging an already full battery will damage it and reduce its lifetime.

For NiMH, continuous long-term "trickle" charging at 0.1 or 0.05 C is not recommended. If trickle charging is necessary by design, it should be kept to C/250 or less, sufficient to replace losses due to self-discharge, but not enough to severely degrade the lifetime.

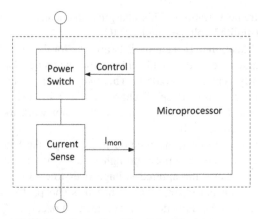

FIGURE 7.29 Typical Solid-State Protection Circuit Architecture.

Lithium Ion

Because of the high energy density of Li-ion cells, charging regimes must be very carefully controlled, both to get the best out of the battery and to prevent degradation and possible serious damage. In general it is best to integrate a constant voltage/constant current controller together with overcharge, overdischarge, and overcurrent protection, along with the battery pack. It is also necessary to ensure individual cells in a Li-ion (or Lithium-Polymer—LiPo) battery pack are balanced to avoid overheating due to excessive current flow between cells. Battery charges will therefore have an additional balancing connector for this purpose, and bespoke battery packs require a dedicated battery management system to be designed and implemented. Texas Instruments provides a range of useful ICs for this purpose that can monitor a number of cells (usually between 6 and 12) for larger battery packs.

7.6 ADVANCED CIRCUIT PROTECTION

With the increasing availability of very compact microprocessors and high-quality power devices, there has been the development of solid-state protection circuits ranging from simple fuses to advanced configurable protection devices. These solid-state devices can be considered under the generic term SSPC—solid-state protection circuit. The general form of the circuit is to have a main power switch (usually based on a power MOSFET device), a current sense circuit and an integrated microcontroller to calculate the equivalent fuse "blow" time. The architecture of a typical SSPC is given in Fig. 7.29.

Clearly, the major advantage of this approach is being able to program not only the absolute rating value of the "fuse," but also the i^2t characteristic of the device, and to be able to modify this when the device has been deployed in the field.

Electromagnetic Compatibility

CHAPTER OUTLINE

The Circuit Designer's Companion. http://dx.doi.org/10.1016/B978-0-08-101764-7.00008-6

8.1 THE NEED FOR ELECTROMAGNETIC COMPATIBILITY

All electrical and electronic devices generate electromagnetic (EM) interference and are susceptible to it. It is your job as product designer to reduce this generation and susceptibility to acceptable levels. With the increasing penetration of solid-state electronics into all areas of activity, acceptable levels of interference have become progressively tighter as physical separation between devices has reduced and reliance on their operation has increased. Solid-state, particularly integrated circuit technology, is more susceptible than the vacuum-tube devices of years ago, and the popularity of plastic cases with their lack of screening is a further factor. The ability of a device to operate within the limits of interference immunity and suppression is known as electromagnetic compatibility (EMC).

In some areas of electronics, EMC has been a product requirement for a long time. Military electronics has severe limitations imposed on it, often because of the proximity of high-power pulse equipment (radars) to sensitive signal processing equipment in the same aircraft, ship, or vehicle, and military EMC standards first appeared in the 1960s. The increasing use of walkie-talkies on process plant and elsewhere has prompted users of safety—critical instrumentation to specify a minimum immunity from radio frequency interference (RFI). Measuring and weighing equipment must be prevented from giving incorrect readings in the presence of interference, and domestic broadcast receivers should be able to work alongside home computers.

Radio frequencies (RF) are not the only source of interference. Transients can be generated by power switching circuits, lightning, electric motors, spark ignition devices, or electrostatic discharge (ESD) to name just a few. Microprocessor circuits are particularly susceptible to impulse interference and must be protected accordingly.

The Importance of Electromagnetic Compatibility

Many manufacturers have found out the hard way that poor EMC performance of a product can be extremely costly, both in terms of damaged reputation and in the measures needed to improve performance once a fault has been found. For this reason, many firms will test their products for EMC before releasing them even though there may be no applicable standard for that class of equipment.

However, in any case, virtually all electronic products are now subject to legislation from one source or another, demanding constraints on EMC performance. The technical constraints on equipment EMC are that the equipment should continue to function reliably in a hostile EM environment, and that it should not itself degrade that environment to the extent of causing unreliable operation in other equipment. EMC therefore splits neatly into two areas, labeled "immunity" and "emissions," and a list of some of the interference types and coupling paths is shown in Table 8.1.

8.1.1 IMMUNITY

The EM environment within which equipment will operate and which will determine its required immunity can vary widely, as can the permissible definition of reliable operation. For instance, the magnitude of radiated fields encountered depends critically on the distance from the source. Strong RF fields occur in the vicinity of radars, broadcast transmitters, and RF heating equipment (including microwave ovens at 2.45 GHz). The electric field strength falls off linearly with distance from the transmitting antenna, provided that your measuring point is in the far field,

Table 8.1 Electromagnetic Compatibility Phenomena

Immunity	Emissions
• mains voltage dropouts, dips, surges, and distortion • transients and radio frequency interference (RFI) conducted into the equipment via the mains supply • radiated transient or RFI picked up and conducted into the equipment via signal leads • RFI picked up directly by the equipment circuitry • electrostatic discharge	• mains distortion, transients, or RFI generated within the equipment and conducted out via the mains supply • transient or RFI generated within the equipment and conducted out via signal leads • RFI radiated directly from the equipment circuitry, enclosure, and cables

defined as greater than $\lambda/2\pi$ where λ is the wavelength. The value of the field strength in volts per meter can be calculated from

$$E = \sqrt{\frac{30 \times P}{d}} \qquad (8.1)$$

where P is the radiated power in watts or antenna feed power times the antenna gain; d is the distance from the antenna in meters.

In the near field closer than $\lambda/2\pi$ the field strength can be much greater, but it depends on the type of antenna and how it is driven.

Radio Transmitters

Amplitude modulation (AM) broadcast-band transmitter power levels tend to be around 100−500 kW, but the transmitters are usually located away from centers of population. Field strength levels in the 1−10 V/m range may be experienced occasionally, but at medium frequencies coupling to circuit components is usually inefficient, so these transmitters do not pose a great problem. TV and FM-band transmitters are more often found close to office or industrial environments and the greatest threat is to equipment on the upper floors of tall buildings, which may be in line of sight to a nearby transmitter of typically 10 kW. Field strengths greater than 10 V/m are possible, and although the building structure may give some attenuation, achieving immunity from even a 1 V/m field is not trivial at these frequencies, where cable and track lengths approach resonance and coupling is correspondingly more efficient.

Portable transmitters (walkie-talkies, cellphones) do not have a high radiated power, but they can be brought extremely close to susceptible equipment. Typical field strengths from a 1 W ultrahigh frequency (UHF) handheld transmitter are 5−7 V/m at half a meter distance.

Radars

Another serious threat is from radars in the 1−10 GHz range, particularly around airports. It is not difficult to find a pulsed 50 V/m field strength up to 3 km from these radars. Again, building attenuation may give some relief, but set against this is the problem that pulsed RFI is particularly upsetting to microprocessor circuits. As an aside, civil aircraft regulatory authorities have published a defined "RF environment"[1] in terms of field strength versus frequency that civil aircraft may be expected to meet and should therefore be protected against. The worst-case threat is from certain US ground-based radars in the 2−4 GHz range, where it is assumed that the aircraft can fly close to the antenna through the main beam, and the peak field

[1]Users's Guide, Protection of aircraft electrical and electronic systems against the effects of the external radio frequency environment, EUROCAE WG33 Subgroups 2 & 3.

Table 8.2 Average Rate of Occurrence of Mains Transients

Area Class	Average Rate of Occurrence (Transients/h)
Industrial	17.5
Business	2.8
Domestic	0.6
Laboratory	2.3

strength is defined to be 17 kV/m (no, that is not a misprint, seventeen thousand volts per meter).

From these considerations a minimum of 3 V/m from, say, 10 MHz to 1 GHz represents a reasonable design criterion for RFI immunity, with 10 V/m being preferred. Immunity from pulsed interference above 1 GHz is extremely hard to quantify.

Transients

Conducted transient immunity is important because microprocessor-based circuits are much more susceptible to transient upset than are analog circuits. Mains transients, from many sources, are far more common than is generally realized. A study by the German ZVEI[2] made a statistical survey of 28,000 live-to-earth transients exceeding 100 V, at 40 locations over a total measuring time of about 3400 h. Their results were analyzed for peak amplitude, rate of rise, and energy content. Table 8.2 shows the average rate of occurrence of transients for four classes of location, and Fig. 8.1 shows the relative number of transients as a function of maximum transient amplitude. This shows that the number of transients varies roughly in inverse proportion to the cube of peak voltage.

The rate of rise was found to increase roughly in proportion to the square root of peak voltage, being typically 3 V/ns for 200 V pulses and 10 V/ns for 2 kV pulses. Other field experience has shown that mechanical switching usually produces multiple transients (bursts) with rise times as short as a few nanoseconds and peak amplitudes of several hundred volts.

As a general guide, microprocessor equipment should be tested to withstand pulses at least up to 2 kV peak amplitude. Thresholds below 1 kV will give unacceptably frequent corruptions in nearly all environments, while between 1 and 2 kV occasional corruption will occur. For a belt-and-braces approach for high reliability equipment, a 4–6 kV threshold is not too much.

Other sources of conducted transients are telecommunication lines and the automotive 12 V supply. The automotive transient environment is particularly severe

---

[2]See **Transients in Low Voltage Supply Networks**, J.J. Goedbloed, IEEE Transactions on Electromagnetic Compatibility, Vol EMC-29 No 2, May 1987, pp. 104–115.

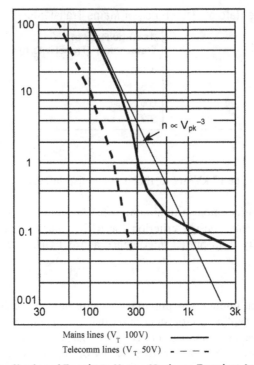

Mains lines (V_T 100V) ———

Telecomm lines (V_T 50V) - — — -

FIGURE 8.1 Relative Number of Transients Versus Maximum Transient Amplitude.

with respect to its nominal supply range. The most serious automotive transients (Fig. 8.2) are the load dump, which occurs when the alternator load is suddenly disconnected during heavy charging; switching of inductive loads, such as motors and solenoids; and alternator field decay that generates a negative voltage spike when the ignition switch is turned off.

Electrostatic Discharge

ESD is a further source of transient upset which most often occurs when a person who has been charged to a high potential by movement across an insulating surface then touches an earthed piece of equipment, thereby discharging themselves through the equipment. The achievable potential depends on relative humidity and the presence of synthetic materials (Fig. 8.3), and the human body is roughly equivalent to a 150 pF capacitance in series with 150 Ω resistance, so that currents of tens of amps can flow for a short period with a very fast (subnanosecond) rise time. Even though it may have low energy and be conducted to ground through the equipment case, such a current pulse couples very easily into the internal circuitry.

Determining and Specifying the Effects of Interference

Required reliability of operation depends very much on the application. Entertainment devices and gadgets come fairly close to the bottom of the scale, whereas

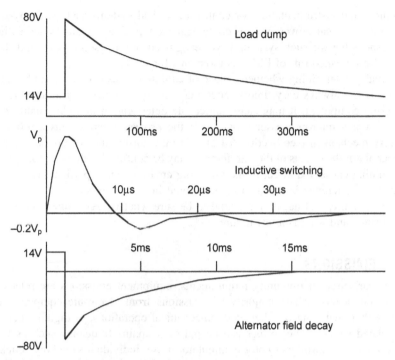

FIGURE 8.2 Automotive Transient Waveforms.

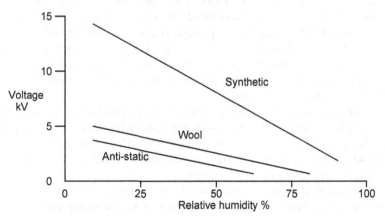

FIGURE 8.3 Electrostatic Charging Voltages.

Reproduced from IEC 61000-4-2.

computers and instrumentation for control of critical systems such as fly-by-wire aircraft and nuclear power systems come near the top. The purchasers and authorities responsible for such systems have recognized this for some years and there is a well-established raft of EMC requirements for them.

Actually determining whether a piece of equipment has been affected by interference is not always easy. Interference may cause a degradation in accuracy of measuring equipment, it may cause noticeable deterioration in audio quality or it may corrupt a microprocessor program. If the processor circuit has a program-recovery mechanism (see Section 6.4.2), this may correct the corruption before it is apparent, or the effects of the interference may be confused with a software glitch. Determining cause and effect in the operating environment is particularly difficult when the interference is transient or occasional in nature. When you are contemplating immunity testing, it is essential to be sure what will constitute acceptable performance and what is a test failure.

8.1.2 EMISSIONS

By comparison with immunity requirements, equipment emissions are relatively easy to characterize. The majority of emissions from electronic equipment are due to either switching or other electromechanical operations, or digital clock- or data-related signals. The former can be pulses at mains frequency such as from thyristor phase controllers, motor commutator noise, individual switching "clicks," or switch-mode power supply harmonics, and they are generally conducted out of the equipment via the mains lead. These emissions have been regulated for many years to minimize interference to AM broadcast and communications services.

Emissions From Digital Equipment

Digital equipment with high-frequency (hf) square wave clocks generates noise into the hundreds of MHz. The system clocks and their harmonics are the principal source because their energy is concentrated into a narrow band, but wideband noise from the data and address lines is also present. The noise amplitude and spectral distribution may vary depending on the operating mode of the circuit or its resident software. Emissions from some classes of digital equipment, such as personal computers, are particularly subject to regulation, again to control interference to broadcast, and communication services.

Equipment emissions can be either conducted or radiated. Commercial emission standards differentiate between the two on the basis of frequency; the breakpoint is universally accepted to be 30 MHz. This may appear to be arbitrary, but it has in fact been found empirically that the coupling mechanisms are predominantly by conduction below 30 MHz and by radiation above it.

The regulation of emissions is intended only to reduce the threat to innocent "victim" receivers. Some reasonable separation of the offending emitter and the victim is assumed. If you place a personal computer right next to a domestic radio set you can still expect there to be interference. Regulations have nothing to say on the subject of intrasystem interference, so that it is quite possible for two products both of which meet the appropriate standard to be incompatible when they are placed together in the same rack or cabinet.

8.2 ELECTROMAGNETIC COMPATIBILITY LEGISLATION AND STANDARDS

Legislative requirements can be discussed under the headings of the major trading blocs: the European Union, the United States, and Australasia.

The United States

In the United States, emissions from intentional and unintentional radiators are governed by Federal Communications Commission (FCC) rules part 15 subpart j. A subclass of these is a "digital device," which is any electronic device that generates or uses timing signals or pulses exceeding 9 kHz and uses digital techniques. There are some quite broad exemptions from the rules depending on application. Two classes are defined, depending on the intended market: class A for business, commercial, or industrial use, and class B for residential use. These classes are subject to different limits, class B being the stricter. Before being able to market his equipment in the United States, a manufacturer must either obtain certification approval from the FCC if it is a personal computer or must verify that the device complies with the applicable limits.

Australasia

Australia and New Zealand have an EMC compliance regime, which is known as the "C-tick" system. This requires a declaration of compliance against standards much like the EMC directive (see below) but with broader exemptions. Japan has a quasi-voluntary system for emissions limitation of information technology equipment under the aegis of the Voluntary Council for Control of Interference. Other Asian countries such as China, Taiwan, and South Korea have requirements based on international standards but generally mandate tests to be done in their own countries, at least for certain types of product.

8.2.1 THE ELECTROMAGNETIC COMPATIBILITY DIRECTIVE

The relaxed EMC regime that used to exist throughout Europe, with the exception of Germany, has now changed dramatically. In accord with the general objective of the single European market, the European Commission (EC) put forward an EMC directive whose purpose is to remove any barriers to trade on technical grounds relating to EMC. Thus, EC member states may not impede, for reasons relating to EMC, the free circulation on their territory of apparatus that satisfies the directive's requirements.

Scope and Coverage

The directive applies to all equipment placed on the market or taken into service, so that it includes systems as well as individual products. It operates as follows: it sets out the essential requirements; it requires a statement to the effect that the equipment complies with these requirements; and it provides alternative means of determining whether the essential requirements have been satisfied.

The essential requirements are that

The apparatus shall be so constructed that

(a) equipment shall not generate electromagnetic disturbances exceeding a level allowing radio and telecommunications equipment and other apparatus to operate as intended;

(b) equipment shall have an adequate level of intrinsic immunity from electromagnetic disturbances.

Thus protection is extended not only to radio and telecommunications but also to other equipment such as information technology and medical equipment—in fact any equipment that is susceptible to EM disturbances. The second requirement states that the equipment should not malfunction in whatever hostile EM environment it may reasonably be expected to operate.

The directive does not limit itself with respect to the range of EMC phenomena it covers, but those listed in Table 8.1 fit with the coverage of its harmonized standards. There are very few exceptions to the scope; virtually anything electrical or electronic is potentially included, unless it is unmistakably benign (unlikely to cause or suffer interference), although there are a number of other directives relating to EMC that take precedence for certain classes of product.

Routes to Compliance

Many manufacturers will not be able to assess whether their equipment is able to satisfy the two essential requirements, so the directive looks toward the development of European standards on EMC. Any equipment that complies with the relevant standards will be deemed to comply with the essential requirements. However, a manufacturer may choose to undertake his or her own technical assessment—indeed, and he or she may have to if there is no relevant EMC standard in existence. In this case, he or she is required to keep a technical construction file containing details of the test method used, the test results and a supporting statement by an independent competent body. This file must be at the disposal of the national administration.

However, the route that is followed by most manufacturers is self-certification to harmonized standards.

The potential advantage of certifying against standards from the manufacturer's point of view is that there is no mandatory requirement for testing by an independent test house. The only requirement is that the manufacturer makes a declaration of conformity, which references the standards against which compliance is claimed.

Of course the manufacturer will normally need to test the product to ensure that it actually does meet the requirements of the standards, but they can do this under their own responsibility. Companies that do not have sufficient expertise or facilities in-house to do this testing can take the product to an independent test house.

Harmonized Versus Nonharmonized Standards

Harmonized standards are those European standards whose references have been published in the Official Journal of the European Communities with respect to a particular directive. They can then be used to derive a *presumption of conformity* with the essential requirements of that directive. A European standard that has not been harmonized cannot be used for this purpose. The term "harmonized" is a legal qualification of the document to which a special meaning has been given by a New Approach Directive. IEC (international) standards themselves have no formal standing with regard to the EMC directive. It is not possible to self-certify compliance to an IEC document; only European standards can be harmonized.

The European standards body is the European Organization for Electrotechnical Standardization (CENELEC), which has UK representation from the British Standards Institution. Once CENELEC has produced a European EMC standard—which more often than not, is done by converting an IEC document into a European one—all the CENELEC countries are required to implement identical national standards, which will then be deemed to be "relevant standards" for the purpose of demonstrating compliance with the directive.

8.2.2 **EXISTING STANDARDS**

At an early stage in the implementation of the directive, CENELEC developed a set of "generic" standards for use where no other standards were appropriate, and these are still available. Their use is less necessary as the product-specific standards, which apply to many types of product, become more numerous. Such product standards take precedence over the generic.

Since emission standards have evolved over a number of years there is a good deal of agreement not only in methods of measurement but also in the limits themselves. Some of the principal standards are summarized in Table 8.3, and a comparison of the emission limits is shown in Fig. 8.4 for conducted and Fig. 8.5 for radiated emission. Measuring equipment is defined in CISPR publication 16-1, which specifies the bandwidth and detector characteristics of the measuring receiver (Table 8.4), the impedance of the artificial mains network and the construction of the coupling clamp.

Note that there are detailed differences in the methods of measurement allowed between the various emission standards. Measurements on conducted emissions below 30 MHz are made on the mains terminals and use an artificial mains network, or Line Impedance Stabilizing Network (LISN), to define the impedance of the mains supply. Radiated measurements above 30 MHz require the use of an open area test site to minimize reflections, with a calibrated attenuation characteristic. This can be literally an open area, or more commonly is an absorber-lined screened

Table 8.3 Summary of the Most Common Electromagnetic Compatibility Standards

RF Emission Standards			
Product Sector	**EN**	**CISPR**	**FCC (US)**
Industrial, scientific, and medical	EN55011	11	Part 18
Household appliances	EN55014-1	14–1	
Lighting equipment	EN55015	15	
Radio and TV receivers	EN55013	13	
Information technology equipment	EN55022	22	Part 15

Immunity Standards			
Product Sector	**EN**	**IEC/CISPR**	**Notes**
Information technology equipment	EN55024	CISPR 24	RFI, ESD, and transient
Household appliances	EN55014-2	CISPR 14-2	RFI, ESD, and transient
Lighting equipment	EN61547	IEC 61547	RFI, ESD, and transient
Radio and TV receivers	EN55020	CISPR 20	RFI, ESD, and transient, antenna terminals

ESD, *electrostatic discharge;* FCC, *Federal Communications Commission;* RFI, *radio frequency interference.*

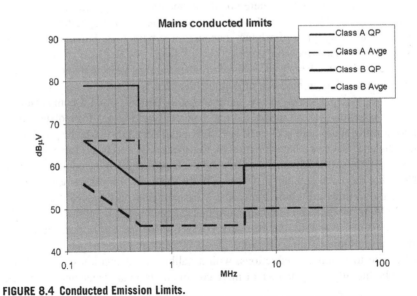

FIGURE 8.4 Conducted Emission Limits.

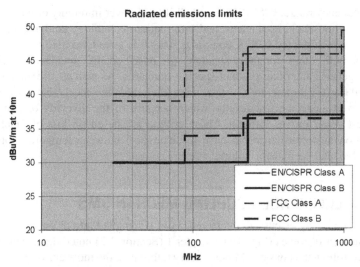

FIGURE 8.5 Radiated Emission Limits.

NB: Consult current specifications to confirm limit values.

Table 8.4 The CISPR 16-1 Quasi-peak Measuring Receiver

	Frequency Range		
Parameter	**9–150 kHz**	**0.15–30 MHz**	**30–1000 MHz**
Bandwidth	200 Hz	9 kHz	120 kHz
Charge time	45 ms	1 ms	1 ms
Discharge time	500 ms	160 ms	550 ms
Overload factor	24 dB	30 dB	43.5 dB

The CISPR 16-1 quasi-peak detector response allows for the subjective variability in annoyance of pulse-type interference. For continuous (narrowband) interference the response is essentially that of a peak detector. For pulse interference, the receiver is progressively desensitized as the pulse repetition frequency is reduced.

room. The limit levels shown in Fig. 8.5 have been normalized to a measurement distance of 10 m for ease of comparison; different standards call up distances of 3, 10 or 30 m. A direct translation of the levels (according to $1/d$) from 10 to 3 m is not strictly accurate because of the influence of near-field effects but is widely applied for practical test purposes.

Immunity

CISPR publication 20 lays down requirements for the immunity of radio and TV receivers and associated equipment to RFI and transients. Originally it was aimed primarily at immunity from citizens band transmitters, although it now covers a wider frequency range. CISPR 14-2 and CISPR 24 cover immunity of those classes of product, household appliances, and information technology equipment, respectively, for which related CISPR emissions standards already existed.

Virtually all other immunity standards are based on IEC documents, including both the generic and other product-specific standards. As well as referencing basic test standards for the methods of test, these have to define acceptable performance criteria to evaluate the test results, which because of the diversity of equipment they must cover proves difficult. These criteria are based on loss of function or degradation in performance but have to distinguish between temporary, operator recoverable, and permanent failure.

8.3 INTERFERENCE COUPLING MECHANISMS

As we have already discussed, interference can be coupled into or out of equipment over a number of routes (Fig. 8.6). Chapter 1 (Section 1.3) noted that electronic interactions follow the laws of EM field theory rather than the more convenient rules of circuit theory, and nowhere is this more evident than in EMC practice. At low frequencies the predominant modes of coupling are directly along the circuit wires or by magnetic induction, but at hfs each conductor, including the equipment housing if it is metallic, act as an aerial in its own right and will contribute to coupling.

8.3.1 CONDUCTED

A very frequent cause of conducted interference coupling is the existence of a common-impedance path between the interfering and victim circuits. This path is

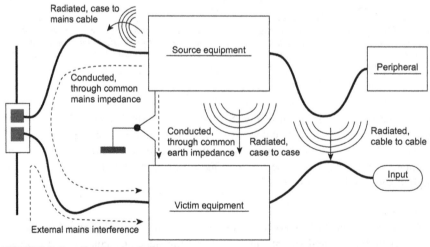

FIGURE 8.6 Coupling Mechanisms.

usually though not invariably in the ground return. This topic has been covered extensively in Chapter 1 and you are referred back to there for a full discussion. A typical case of common-impedance coupling might involve impulsive interference from a motor or switching circuit being fed into the 0 V rail of a microprocessor circuit because they share the same earth return. Because of resonances, the impedance of earthing and bonding conductors is high at frequencies for which their length is an odd multiple of quarter wavelengths.

Another coupling route is through the equipment power supply to or from the mains. The power supply forms the interface between the mains and the internal operation of the equipment, and as we have seen many emissions standards regulate the amount of interference that can be fed onto the mains via the power input leads. Power supply design was considered in more detail in Chapter 7. A great deal of work has been done to characterize the impedance of the mains supply, and perhaps surprisingly measurements in quite different environments show close agreement. This has allowed the development of the CISPR 16 artificial mains network (LISN) referred to in the last section. The impedance can be simulated by 50 Ω in parallel with 50 µH with respect to earth (Fig. 8.7).

Power cables tend to act as low-loss transmission lines up to 10 MHz so that interference can propagate quite readily around the power distribution network, mainly attenuated by the random connection of other loads rather than by the cable itself. Interference can appear either as differential (symmetric) currents or as common mode (asymmetric) currents, as in Fig. 8.8, and these need different treatment at the equipment, as we shall see when we discuss filters.

8.3.2 RADIATED

When source and victim are near one another, radiated coupling is predominantly due either to magnetic or electric induction. Magnetic induction occurs when the

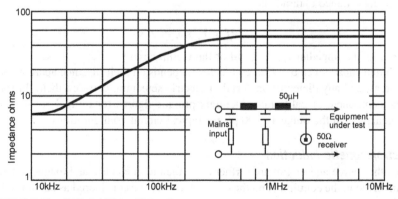

FIGURE 8.7 The Artificial Mains Network.

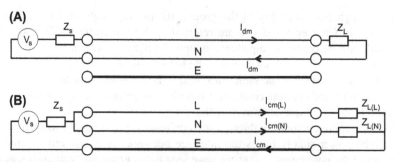

FIGURE 8.8 Interference Propagation Along Power Cables.

(A) Differential Mode and (B) Common Mode.

magnetic flux produced by a changing current in the source circuit links with the victim circuit (cf. Section 1.1.4). The voltage induced in the victim circuit by a sinusoidal current I_s at frequency f due to a mutual inductance L henries is

$$V = 2\pi f \times I_s \times L \text{ V} \tag{8.2}$$

Unfortunately calculating L accurately is difficult in most practical cases. It is proportional to the areas of the source and victim circuit loops, the distance between them, their relative orientation and the presence of any magnetic screening. As an example, short lengths of cable within the same wiring loom will have mutual inductances in the range of 0.1–3 µH.

You will note that magnetic induction is a current (low-impedance) phenomenon, and it increases with increasing current in the source circuit. Electric induction, or capacitive coupling, is a voltage phenomenon and occurs when changing electric fields from the source interact with the victim circuit. The induced voltage due to a sinusoidal voltage V_s of frequency f on the source conductor coupled through a mutual capacitance C farads is

$$V = 2\pi f \times V_s \times C \times Z \text{ V} \tag{8.3}$$

where Z is the impedance to ground of the victim circuit.

Mutual capacitance between conductors depends on their distance apart, respective areas, and any dielectric material or electric screening between them. In some cases, component or cable manufacturers provide figures for mutual capacitance. It is generally in the range 1–100 pF for typical circuit configurations.

Electromagnetic Induction
When the source and victim are further apart then both electric and magnetic fields are involved in the coupling, and the conductors must be considered as antennas. For conductors whose dimensions are much smaller than the wavelength then the

maximum electric field component in the far field at a distance d meters due to a current I A at frequency f Hz flowing in the conductor isfor a loop,

$$E = 131.6 \times 10^{-16}\frac{(f^2 Al)}{d} \text{ V/m} \qquad (8.4)$$

where A is the area of the loop in m^2 for a monopole against a ground plane,

$$E = 4\pi \times 10^{-7}\frac{(fLl)}{d} \text{ V/m} \qquad (8.5)$$

where L is the length of the conductor in m.

As we saw at the beginning of this chapter, in the *far field* the field strength falls off linearly with distance. The electric and magnetic field strengths are related by the impedance of free space, $E/H = 377\ \Omega$ (120π). By contrast, in the *near field*, a loop radiator will give a higher H (magnetic) field, and this will fall off proportional to $1/d^3$, while the E field falls off as $1/d^2$. Conversely, a short rod will give a high E field, which will fall off as $1/d^3$, while the H field falls off as $1/d^2$.

Once conductor lengths approach a quarter wavelength (1 m at 75 MHz) then they cannot be treated as "electrically small" and they couple much more efficiently with ambient fields.

8.4 CIRCUIT DESIGN AND LAYOUT

A common response of circuit designers when they discover EMC problems late in the day is to specify some extra shielding and filtering in the hope that this will provide a cure. Usually it does, but this brute force approach may not be necessary if you put in some extra thought at the early circuit design stage. Shielding and filtering costs money, circuit design does not.

Most design for EMC is just good circuit design practice anyway. The most fundamental point to consider is the circuit's grounding regime: authorities on EMC agree that the majority of postdesign interference problems can be traced to poor grounding. Printed circuit layout also has a significant impact. Chapter 1 considers ground design and Chapter 2 relates this to printed circuit boards (PCBs) and observations made there will not be repeated here, except to reiterate that short, direct tracks running close to their ground returns make very inefficient aerials and are therefore good for controlling both emissions and susceptibility.

8.4.1 CHOICE OF LOGIC

A careful choice of logic family will help to reduce hf emissions from digital equipment and may also improve RF and transient immunity. The harmonic spectrum of a trapezoidal wave, which approximates to a digital clock waveform, shows a roll-off of amplitude with frequency whose shape depends on the rise time (Fig. 8.9A). Using the slowest rise time compatible with reliable operation of your circuit will

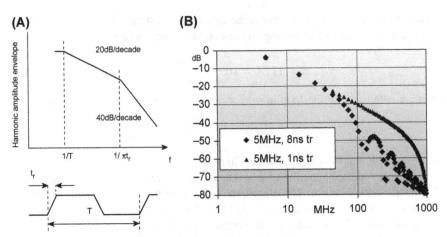

FIGURE 8.9 Harmonic Amplitudes for a Trapezoidal Waveshape.

minimize the amplitude of the higher-order harmonics where radiation is more efficient. Fig. 8.9B shows the calculated amplitude differences for a 5 MHz clock with rise times of 8 and 1 ns. An improvement approaching 20 dB is possible at frequencies around 200 MHz.

The advice based on this consideration is, use the slowest logic family that will do the job; do not use fast logic for no reason. Where parts of the circuit must operate at high speed, use fast logic only for those parts and keep the clocks local. A good way to control the speed of unnecessarily fast clock lines is to place a resistor of a few tens of ohms in series with the clock driver output. Acting together with the capacitance of all the inputs on that line, this reduces the hf energy that appears on the line.

Noise Margin, Clock Frequency, and Power Supply Noise

For good immunity, choose the logic family with the highest noise margin (see Section 6.1.1). Spurious signals coupled into a logic signal circuit will have no effect until they reach the logic threshold. At the same time, the amplitude of signals coupled into the circuit from a given field will depend on the impedance of the circuit, which is defined by the driver output impedance. In this context, faster drivers have lower output impedances so there is a compromise to be made here.

At the same time as you consider rise times, also think about the actual clock frequency. Lowest is best. You may be able to use low-frequency multiphase clocks rather than a single hf one. In some circumstances changing the spot frequency of the clock slightly may move its harmonics sufficiently far away from a particularly susceptible frequency, though this is more a case of EMC within a system than of meeting emission regulations.

The power supply rails also contribute significantly to RF disturbance problems with digital circuits. Section 6.1.3 discussed induced switching noise, which appears on both the ground and power rails. This must be prevented from propagating outside the immediate circuit area; the solution is a solid, unbroken ground plane in conjunction with local power plane segments (Section 2.2.4), and thorough decoupling of these segments (Section 6.1.4). As well as controlling emissions, this also prevents susceptibility to incoming RF and transients, so it should be applied even on low-speed circuits for best immunity.

8.4.2 ANALOG CIRCUITS

As we have seen elsewhere (Section 5.2.10), analog circuits are also capable of un-expected oscillation at RF. Gain stages should be properly decoupled, loaded, and laid out to avoid this. Check them at the prototype stage with an hf scope or spectrum analyzer, even if nothing appears to be wrong with the circuit function. Ringing on pulses transmitted along unterminated transmission lines will generate frequencies, which are related only to the length of the lines with perhaps sufficient amplitude to be troublesome. Terminate all long lines, particularly if they end at a complementary metal-oxide semiconductor input, which provides no inherent termination.

Good RF and transient immunity at interfaces calls for a consideration of signal bandwidth, balance, and level. Any cable connecting to a piece of equipment will conduct interference straight into the circuit and it is at the interface that protection is needed. This is achieved at the circuit design level by a number of possible strategies:

- minimize the signal bandwidth by passive resistor–capacitor (RC) or ferrite filtering, so that interfering signals outside the wanted frequency range are rejected—every analog amplifier should have some intentional bandwidth limitation;
- operate the interface at the highest possible power or voltage level consistent with other requirements, such as dynamic range, so that relatively more interference power is required to upset it;
- operate the interface where possible with the signal balanced, so that interference is injected in common mode and is therefore attenuated by the common mode rejection of the input circuit;
- in severe cases, galvanically isolate the input with an opto-isolator or transformer coupling, so that the only route for the interference is via the stray coupling capacitance of the isolation components.

Not so obvious is the overload performance of the circuit. If the interference drives the circuit into nonlinearity then it will distort the wanted signal, but if the circuit remains linear in the presence of interference then it may be filtered out in a later stage without ill effect. Thus, any circuit that has a good dynamic range and a high overload margin will also be relatively immune to interference.

8.4.3 SOFTWARE

If your circuit incorporates a microprocessor with resident embedded software then use all the available software tricks to overcome likely data corruption. This will not only be due primarily to transients but also to RFI. These are discussed in Chapter 6 but are reiterated briefly here as follows:

- incorporate a watchdog timer. Any microprocessor without some form of watchdog is inviting disaster when it is exposed to disruptive transients;
- type check and range check all input data to determine its reliability. If it is outside range, reject it;
- sample input data several times and either average it, for analogue data, or act only on two or three successive identical logic states, for digital data—this is similar to digital switch debouncing;
- incorporate parity checking and data checksums in all data transmission;
- protect data blocks in volatile memory with error detecting and correcting codes. How extensively you use this protection depends on allowable time and memory overheads;
- wherever possible rely on level- rather than edge-triggered interrupts;
- do not assume that programmable interface chips (PIAs, ACIAs, etc.) will maintain their initialized setup state forever. Periodically reinitialize them.

8.5 SHIELDING

If despite the best circuit design practices, your circuit still radiates unacceptable amounts of noise or is too susceptible to incoming radiated interference, the next step is to shield it. This involves placing a conductive surface around the critical parts of the circuit so that the EM field that couples to it is attenuated by a combination of reflection and absorption. The shield can be an all-metal enclosure if protection down to low frequencies is needed, but if only hf (>30 MHz) shielding will be enough then a thin conductive coating deposited on plastic is adequate.

Shielding Effectiveness

How well a shield attenuates an incident field is determined by its shielding effectiveness, which is the ratio of the field at a given point before and after the shield is in place. Shielding effectiveness of typical materials differs depending on whether the electric or magnetic component of the field is considered. Shielding effectiveness below 20 dB is considered minimal, between 20 and 80 dB is average, and 80 and 120 dB is above average. Above 120 dB is unachievable by cost-effective measures. The perfect electric shield consists of a seamless box with no apertures made from a zero-resistance material. This is known as a Faraday cage and does not exist. Michael Faraday described his attempt to build one as follows:

> I had a chamber built, being a cube of 12 feet, and copper wire passed along and across it in various directions, and supplied in every direction with bands of tin foil, that the whole might be brought into good metallic communication. I went

into the cube and lived in it, and using lighted candles, electrometers and all other tests of electrical states, I could not find the least influence on them, though all the time the outside of the cube was powerfully charged, and large sparks and brushes were darting off from every part of its outer surface.[3]

Any practical shield will depart from the ideal of infinite attenuation for two reasons:

- it is not made of perfectly conducting material
- it includes apertures and discontinuities.

Shielding effectiveness of a solid conductive barrier can be expressed as the sum of reflection, absorption, and rereflection losses:

$$SE_{(dB)} = R_{(dB)} + A_{(dB)} + B_{(dB)} \qquad (8.6)$$

The reflection loss depends on the ratio of wave impedance to barrier impedance, the barrier impedance being a function of its conductivity, and permeability, and of frequency. Reflection losses decrease with increasing frequency for the E field (electric) and increase for the H field (magnetic). In the near field, closer than $\lambda/2\pi$, the distance r between source and barrier also affects the reflection loss. Further away, the distance is not relevant because the wave impedance (for a "plane wave") is constant.

The rereflection loss B is insignificant in most cases where A is greater than 10 dB. A itself depends on the barrier thickness and its absorption coefficient. The inverse of the absorption coefficient is called the "skin depth" (δ). Skin depth is the measure of a magnetic phenomenon that tends to confine ac current to the surface of a conductor. The skin depth reduces as frequency, permeability, and conductivity increase, and fields are attenuated by 8.7 dB (1/e) for every skin depth of penetration, and is given in Eq. (8.7):

$$\delta = 6.61 \times (\mu_r \rho_r F) \times 0.5 \text{ cm} \qquad (8.7)$$

where μ_r is relative permeability (1 for air and copper); σ_r is relative conductivity (1 for copper); F is frequency in Hz.

This gives, for instance, a skin depth of 0.012 mm in copper at 30 MHz. Thus at hfs, absorption loss becomes the dominant term. Fig. 8.10 shows the combined reflection and absorption losses for 0.05 mm aluminum foil ($\sigma_r = 0.61$, $\mu_r = 1$) and 0.5 mm sheet steel ($\sigma_r = 0.1$, $\mu_r = 60$ at 10 kHz dropping to 1 above 1 MHz) versus frequency.

8.5.1 APERTURES

The curves in Fig. 8.11 suggest that upward of 200 dB attenuation is easily achievable using reasonable thicknesses of common materials. In fact, the practical shielding effectiveness is limited by necessary apertures and discontinuities in the shielding.

---

[3]**Experimental Researches in Electricity**, Michael Faraday, 1838, para 1173–1174.

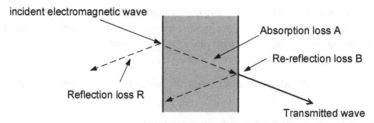

FIGURE 8.10 Absorption and Reflections of Electromagnetic Waves.

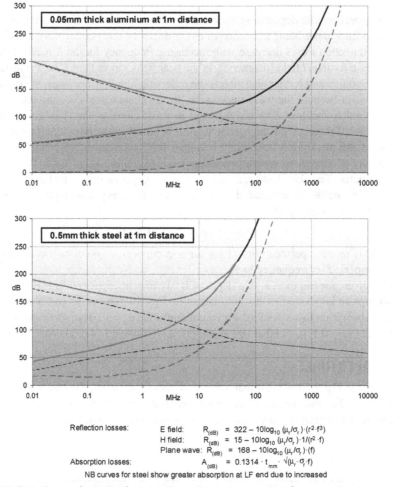

Reflection losses:

E field: $R_{(dB)} = 322 - 10\log_{10}(\mu_r/\sigma_r) \cdot (r^2 \cdot f^3)$

H field: $R_{(dB)} = 15 - 10\log_{10}(\mu_r/\sigma_r) \cdot 1/(r^2 \cdot f)$

Plane wave: $R_{(dB)} = 168 - 10\log_{10}(\mu_r/\sigma_r) \cdot (f)$

Absorption losses: $A_{(dB)} = 0.1314 \cdot t_{mm} \cdot \sqrt{(\mu_r \cdot \sigma_r \cdot f)}$

NB curves for steel show greater absorption at LF end due to increased

FIGURE 8.11 Composite Reflection and Absorption Losses for Aluminum and Steel.

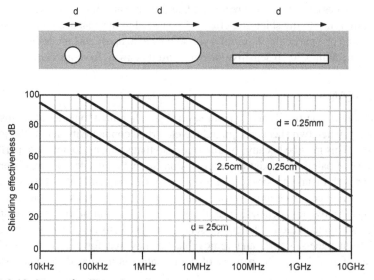

FIGURE 8.12 Attenuation Through an Aperture.

Apertures are needed for ventilation, for control access, and for viewing indicators. EM leakage through an aperture in a thin barrier depends on its longest dimension (d) and the minimum wavelength (λ) of the frequency band to be shielded against. For wavelengths less than or equal to twice the longest aperture dimension there is effectively no shielding. The frequency at which this occurs is the "cut-off frequency" of the aperture. For lower frequencies ($\lambda > 2d$) the shielding effectiveness increases linearly at a rate of 20 dB per decade (Fig. 8.12) up to the maximum possible for the barrier material. Comparing Figs. 8.11 and 8.12, you will see that for all practical purposes shielding effectiveness is determined by the apertures. For frequencies up to 1 GHz (the present upper limit for radiated emissions standards) and a minimum shielding of 20 dB the maximum hole size you can allow is 1.6 cm.

For viewing windows in particular this is unacceptable, and you then have to cover the window with a transparent conductive mesh, which must make good contact to the surrounding screen, or accept the penalty of shielding at lower frequencies only. You can cover ventilation holes with a perforated mesh screen without much trouble. If individual perforations are spaced close together (hole spacing $< \lambda/2$) then the reduction in shielding over a single hole is approximately proportional to the square root of the number of holes. Thus a mesh of 100 4 mm holes would have a shielding effectiveness 20 dB worse than a single 4 mm hole. Two similar apertures spaced greater than a half-wavelength apart do not suffer any significant extra shielding reduction.

The above shielding theory should not be taken too literally. For simplification, it assumes that the shielding barrier is a sheet of infinite extent, and also that the barrier is in the far field of both the impinging wave and the components to be shielded.

Neither is necessarily true for a real shield, and in fact proximity to apertures makes a huge difference to the shielding effect. If you need to have an accurate measure of the shielding of a particular assembly, you will need to apply EM modeling methods. As a rule of thumb, keep sensitive or noisy circuits away from apertures.

8.5.2 SEAMS

An EM shield is normally made from several panels joined together at seams. Unfortunately, when two sheets are joined the electrical conductivity across the joint is imperfect. This may be because of distortion, so that surfaces do not mate perfectly, or because of painting, anodizing, or corrosion, so that an insulating layer is present on one or both metal surfaces (cf. Section 1.1.3). Consequently, the shielding effectiveness is reduced by seams just as much as it is by apertures. The reduction of shielding effectiveness depending on the longest dimension of the aperture applies equally to a nonconductive length of seam. The problem is especially serious for hinged front panels, doors, and removable hatches that form part of a screened enclosure (Fig. 8.13A). The penalty of poor contact is mitigated to some extent if

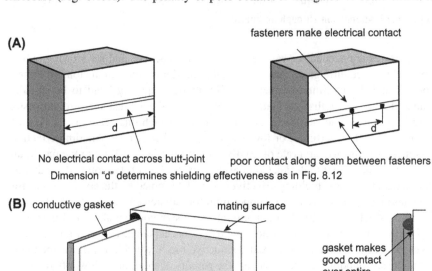

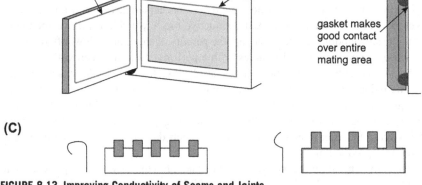

FIGURE 8.13 Improving Conductivity of Seams and Joints.

(B) Conductive Gasket and (C) Beryllium Copper Finger Stock.

the conductive sheets overlap since this forms a capacitor that provides a partial current path at hfs. There are a number of other design options you can take to improve the shielding effectiveness at the seams:

- Ensure that supposedly conductive surfaces remain conductive: do not paint or anodize them; alochrome is a suitable conductive finish for aluminum.
- Maximize the area of overlap of two joined sheets. This can be done by lapped or flanged joints.
- Where you are using screwed or riveted fastenings, space them as closely as possible. A good rule is no farther apart than $\lambda/20$, where λ is the wavelength at the highest frequency of interest.
- Where you need an environmental seal, or where you need good seam conductivity with only a few fasteners or without them altogether such as at a hinged panel, use conductive gaskets (Fig. 8.13B). These are available as knitted wire mesh, an elastomer loaded with a conductive material such as silver flakes or oriented wires, conductive fabric over foam, or other forms of construction including form-in-place conductive elastomer. Selection of the right gasket depends on mechanical and environmental factors and the required conductivity.
- Another way of improving shielding contact between two surfaces that frequently mate and unmate is to use beryllium copper "finger" stock (Fig. 8.13C) either as a continuous strip or spaced at intervals subject to the limitations on spacing given above for fasteners.

Above all, do not expect a supposedly shielded enclosure made of several ill-fitting painted panels with large holes and few fastenings to give you any serious shielding at all. To be fair, although as Fig 8.12 shows shielding effectiveness of such enclosures will be zero at UHF and above, in fact some equipment does not radiate sufficiently at these frequencies for shielding to be necessary to suppress radiated emissions. On the other hand, shielding against incoming RFI will also be minimal.

8.6 FILTERING

The purpose of filtering for EMC is almost invariably to attenuate hf components while passing low-frequency ones. It is also, almost invariably, to block interfering signals that are coupled onto cables that enter or leave the equipment enclosure. There is little point in applying good shielding and circuit design practice to guard against radiated coupling if you then allow interference to be coupled in or out via the external connections. Interference can be induced directly onto the cable or can be coupled through the external connection, and a proper filter can guard against either or both of these, but you need some knowledge of the characteristics of the circuit in which the filter will be embedded to design or select the best filtering device.

Filters fall broadly into three categories: those intended for mains terminals, for I/O connections, and for individual power or signal wires. The basic principle is the same for all of these.

8.6.1 THE LOW-PASS FILTER

This is not the place to go into the details of filter design, which are well covered in many books on circuit theory. The discussion here will confine itself to the practical aspects.

The simpler circuit arrangements that offer a low-pass response are shown in Fig. 8.14. The attenuation of any given filter is conventionally quoted in terms of its

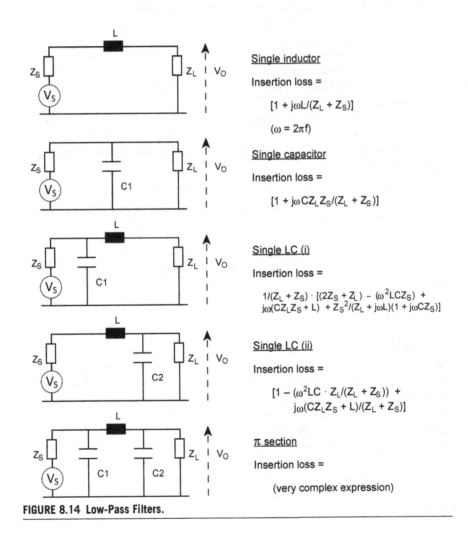

Single inductor

Insertion loss =

$$[1 + j\omega L/(Z_L + Z_S)]$$

$$(\omega = 2\pi f)$$

Single capacitor

Insertion loss =

$$[1 + j\omega C Z_L Z_S/(Z_L + Z_S)]$$

Single LC (i)

Insertion loss =

$$1/(Z_L + Z_S) \cdot [(2Z_S + Z_L) - (\omega^2 LCZ_S) + j\omega(CZ_LZ_S + L) + Z_S^2/(Z_L + j\omega L)(1 + j\omega CZ_S)]$$

Single LC (ii)

Insertion loss =

$$[1 - (\omega^2 LC \cdot Z_L/(Z_L + Z_S)) + j\omega(CZ_LZ_S + L)/(Z_L + Z_S)]$$

π section

Insertion loss =

(very complex expression)

FIGURE 8.14 Low-Pass Filters.

insertion loss that is the difference between the voltage across the load with the filter in and out of circuit. The first point to notice from the expressions to the right of the circuits in Fig. 8.14 is that the insertion loss depends not only on the filter components but also on the source and load impedances. The second point to notice is that, except for the simplest circuits, it's a lot easier either to work out the theoretical insertion loss by computer on a circuit simulator or to build the circuit and measure it!

Impedances

Knowledge of source and load impedances is fundamental. The simple inductor circuit, in which the inductor may be nothing more than a ferrite bead (see Chapter 3), will give good results—better than 40 dB attenuation—in a low-impedance circuit but will be quite useless at high impedances. Conversely, the simple capacitor will give good results at high impedances but will be useless at low ones. The multi-component filters will give better results provided that they are configured correctly; the capacitor should face a high impedance and the inductor a low one.

Frequently, Z_S and Z_L are complex and perhaps unknown at the frequencies of interest for suppression. We have seen earlier (Fig. 8.7) that the impedance of the mains supply is fairly predictable and can be quite easily modeled. You should be able to derive hf impedances for most signal circuits. It is not so easy for power supply inputs, as power components such as transformers, diodes, and reservoir capacitors are not characterized at hf, and you will either have to measure the impedances or make an intelligent guess at the likely values. When the source is taken to be a cable acting as an antenna any analytical method for deriving its impedance is likely to be wildly inaccurate when applied to the real situation, in which cable orientation and positioning is uncontrolled, and a nominal value has to be assumed. 50 Ω is usually specified as the test impedance for filter units, and this is as good a value as any to take for the external source impedance. You have to remember, though, that published insertion loss figures in a 50 Ω system are not likely to be obtained in the real application. This does not necessarily matter provided the filter-and-circuit combination is carefully characterized and tested as a whole.

Components and Layout

Filter components, like all others, are imperfect. Inductors have self-capacitance, capacitors have self-inductance. This complicates the equivalent circuit at hfs and means that a typical filter using discrete components will start to lose its performance above about 10 MHz. The larger the components are physically, the lower will be the break frequency. Chapter 3, Section 3.3.9 discusses the self-resonance effects of capacitors, and transposing the impedance curves of Fig. 3.19 to the low-pass filter circuits will show that as the frequency increases beyond capacitor self-resonance the impedance of the capacitors in the circuit actually rises, so that the insertion loss begins to fall. This can be countered by using special construction for the capacitors, as we shall see shortly when we look at feedthrough components.

The components are not the only cause of poor hf performance. Layout is another factor; lead inductance and stray capacitance can contribute as much degradation as

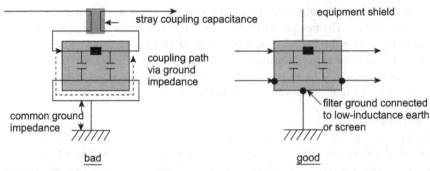

FIGURE 8.15 Filter Wiring and Layout.

component parasitics. Two common faults in filter applications are not to provide a decent low-inductance ground connection and to wire the input and output leads in the same loom or at least close to each other, as in Fig. 8.15.

A poor ground offers a common impedance that rises with frequency and couples hf interference straight through from one side to the other via the filter's local ground path. Common input—output wiring does the same thing through mutual capacitance and inductance. The cures are obvious: always mount the filter so that its ground node is directly coupled to the lowest inductance ground of the equipment, preferably the chassis. Keep the input/output (I/O) leads separate, preferably screened from each other. The best solution is to position the filter so that it straddles the equipment shielding.

8.6.2 MAINS FILTERS

RFI filters for mains supply inputs have developed as a separate species and are available in many physical and electrical forms from several specialist manufacturers. A typical "block" filter for European mains supplies with average insertion loss might cost around £5. The reasons for this separate development are

- mandatory RFI emission standards have concentrated on interference conducted out of equipment via the mains, and consequently the market for filters to block this interference is large and well established, with predictable performance needs;
- any components on the mains wiring side of equipment are exposed to an extra layer of safety regulatory requirements. Filter manufacturers are able to amortize the cost of designing and certifying their products to the required safety standards over a large number of units, thus relieving the equipment manufacturer to some extent of this particular burden;
- locating a filter directly at the mains inlet lends itself well to the provision of the whole input circuitry—connector, filter, fuse, on/off switch—as one block which the manufacturer can "fit and forget";
- many equipment designers are at a loss when it comes to RF filter design and prefer a bought-in solution.

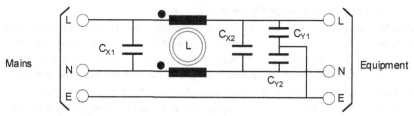

FIGURE 8.16 Typical Mains Filter Circuit.

On the other hand, mains filters *can* be designed in with the rest of the circuit, and this becomes a cost-effective approach for high-volume products and is almost always necessary if you need the optimum filter performance. A typical mains filter circuit (Fig. 8.16) includes components to block both common mode and differential mode interference currents (cf. Fig. 8.8).

The common mode choke L consists of two identical windings on a single high permeability, usually toroidal, core, configured in the circuit so that differential (line-to-neutral) currents cancel each other. This allows high inductance values, typically 1−10 mH, in a small volume without fear of choke saturation caused by the mains frequency supply current. The full inductance of each winding is available to attenuate common mode currents with respect to earth, but only the leakage inductance, which depends critically on the choke construction, will offer attenuation to differential mode interference.

Capacitors C_{X1} and C_{X2} attenuate differential mode only but can have fairly high values, 0.1−0.47 µF being typical. Either may be omitted depending on the detailed performance required, remembering that the source and load impedances may be too low for the capacitor to be useful. Capacitors C_{Y1} and C_{Y2} attenuate common mode interference and if C_{X2} is large have no significant effect on differential mode.

Safety Requirements

The values of $C_{Y1,2}$ are nearly always limited by the allowable earth return current, which is set by safety considerations. This current is due to the operating voltage at mains frequency developed across the capacitors. Several safety standards define maximum earth current levels and these depend on the safety class of the equipment (see Section 9.1.1) and on the actual application. Values range from 0.25 to 5 mA. BS613, which specifies requirements for mains RFI filters, gives the maximum value for Y-configured capacitors for class 1 appliances connected by a plug and socket as 0.005 µF, and this value is frequently found in general purpose filters. The quality of the components is also critical since they are continuously exposed to mains voltage; failure of either C_X or C_Y could result in a fire hazard, and failure of C_Y could also result in an electric shock hazard. You must therefore only use components that are rated for mains use in these positions (cf. Section 3.3.1).

Insertion Loss Versus Impedance and Current

Mains filter insertion loss is universally specified between 50 Ω terminations. As we noted earlier, the actual in-circuit performance will be different because your circuit is unlikely to look like a flat 50 Ω across the frequency range. It may also differ because of the working current. Increasing current through the inductor will eventually result in saturation, even of a dual wound common mode choke; once the core saturates, the inductance falls dramatically and attenuation is lost. All commercially available filters have a rms current rating and provided they are operated within this rating there should be no problem. Unfortunately, typical power supply input currents show a high crest factor (see Section 7.2.5), and the peak current may be three or more times the rms current. Although the filter may appear to be adequately rated on a rms basis, the peak current will overload it and it will be effectively useless.

8.6.3 INPUT/OUTPUT FILTERS

In contrast to the mains filter, a filter on an I/O line has to be more closely tailored to individual applications, and consequently ready-made filters are uncommon, except for a few well-defined high-volume applications. A major variable is the signal bandwidth. If this extends into the RF range, as for example 10 Mbit/s digital interfaces or video lines, then a simple low-pass filter using parallel capacitors cannot be used; except that common mode chokes, which are invisible to the signal but block common mode noise, are still suitable. Conversely, a slow signal such as from a transducer or switch can easily be filtered with a simple capacitor.

A low-pass filter may affect the signal waveshape even if its cut-off frequency is higher than the signal bandwidth. More complex filter components with very steep cut-off characteristics are becoming available to address this problem. I/O filters may also be required to clamp transients to a safe level determined by the over-voltage capability of the circuitry inboard of the filter. This is invariably achieved by a combination of low-pass filtering and transient suppression components, such as zener diodes (Section 4.1.7) or varistors. A discrete component approach may suffice for many applications, but where fast rising transients are expected the lead and wiring inductance can have a significant effect on the circuit's ability to clamp the edge of the pulse. In these cases a combined capacitor/varistor component, in which the dielectric has a predictable low-voltage breakdown characteristic, can offer a solution.

8.6.4 FEEDTHROUGH AND 3-TERMINAL CAPACITORS

Any low-pass filter configuration except for the simple inductor uses a capacitor in parallel with the signal path. A perfect capacitor would give an attenuation increasing at a constant 20 dB per decade as the frequency increased, but a practical wire-ended capacitor has some inherent lead inductance which in the conventional configuration puts a limit to its hf performance as a filter. The impedance characteristics given in Fig. 3.19 show a minimum at some frequency and rise with frequency

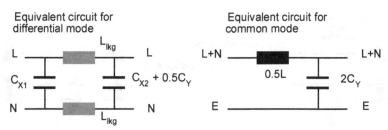

FIGURE 8.17 Filter Equivalent Circuits.

above this minimum. This lead inductance can be put to some use if the capacitor is given a three-terminal construction (Fig. 8.18).

The lead inductance now forms a T-filter with the capacitor, greatly improving its hf performance. Lead inductance can be enhanced by incorporating a ferrite bead on each of the upper leads. The ground connection, of course, must be as short as possible to maintain low inductance. The 3-terminal configuration can extend the effectiveness of a small ceramic capacitor from below 50 MHz to upward of 200 MHz, which is particularly useful for interference in the very high-frequency band. The construction can easily be implemented in surface mount form as also shown in Fig. 8.17, again with the essential proviso that the component's central ground terminal must be connected directly to a good ground plane.

Feedthroughs

Any leaded capacitor is still limited in effectiveness by the inductance of the connection to the ground point. For the ultimate performance, and especially where penetration of a screened enclosure must be protected at UHF and above (this is more often the case for military equipment) then a feedthrough construction is essential.

Here, the ground connection is made by the outer body of the capacitor being screwed or soldered directly to the metal screening or bulkhead (Fig. 8.19). Because

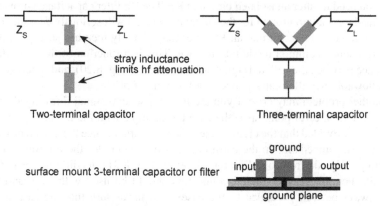

FIGURE 8.18 Two Versus Three Terminals.

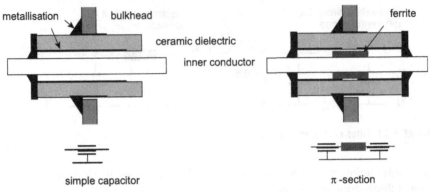

FIGURE 8.19 The Feedthrough Capacitor.

the current spreads out through 360 degrees around it, there is effectively no inductance associated with this terminal and the capacitor performance is maintained well into the GHz region. The inductance of the through lead can be increased, thereby creating a π-section filter, by separating the ceramic metallization into two parts and incorporating a ferrite bead within the construction.

Feedthrough capacitors are available in a wide range of voltage and capacitance ratings but their cost increases with size. Cheap solder-in types between 100 and 1000 pF may be had for a few tens of pence, but larger screw mounting components will cost £1 − £2. If you need good performance down to the low-MHz region then you will have to pay even more for it. A cheaper solution is to parallel a small feedthrough component with a larger, cheaper conventional unit that suppresses the lower frequencies at which physical construction is less critical.

Circuit Considerations

When using any form of capacitive filtering, you have to be sure that your circuit can handle the extra capacitance to ground. This factor can be particularly troublesome when you need to filter an isolated circuit at RF. The RF filter capacitance provides a ready-made ac path to ground for the signal circuit and will seriously degrade the ac isolation, to such an extent that an RF filter may actually *increase* susceptibility to lower frequency common mode interference. This is a function of the capacitance imbalance between the isolated signal and return lines (Fig. 8.20), and it may restrict your allowable RF filter capacitance to a few tens of pF.

Another problem may arise if you are filtering several signal lines together and using a common earth point, as is the case for example with the filtered-D range of connectors. Provided that the earth connection is low impedance there is no problem, but any series impedance in the earth path not only degrades the filtering but will also couple signals from one line into another (Fig. 8.21), leading to designed-in crosstalk. For this reason filtered connectors should be used with care, and they must always be well grounded to the case—and make sure that the case is also the signal ground!

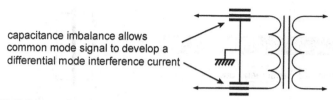

capacitance imbalance allows
common mode signal to develop a
differential mode interference current

FIGURE 8.20 Unbalance Between Feedthrough Capacitors.

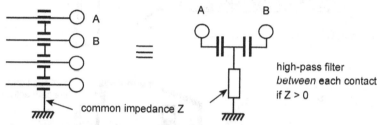

A

B

=

A   B

common impedance Z

high-pass filter
between each contact
if Z > 0

FIGURE 8.21 Common-Impedance Coupling Through Filter Capacitors.

8.7 CABLES AND CONNECTORS

The EMC of any given product is always affected by the configuration of the cables that are connected to it. In the presence of an EM field any cable acts as an antenna and energy from the field is coupled onto it. The current induced on the cable depends on its physical orientation with respect to the field and any nearby conductive objects and on its length. In a reciprocal manner, the field radiated by a cable carrying an RF current also depends on these parameters. Cable length is sometimes under your control, but orientation never is (unless you are a system designer). Therefore, you need to take steps to prevent interfering cable currents from affecting circuit operation or to prevent circuit operation from generating interfering cable currents.

Properly Terminating the Cable Shield

You can filter the signal lines at the point at which they enter or leave the cable, as we saw earlier. Where this is inadequate or impossible, the other approach is to surround the signal conductors with a conductive shield, which is grounded to the equipment screen, as is shown in Fig. 1.17. The function of this shield is to provide a return path for induced currents, which does not couple onto the signal conductors, or conversely to confine radiating currents that are present on the signal conductors and prevent them from creating external fields. Fig. 1.17 shows the ideal connection of a shield to screen in which there is no discontinuity between them. This can be best achieved when the cable is fixed to the equipment and led through a conductive

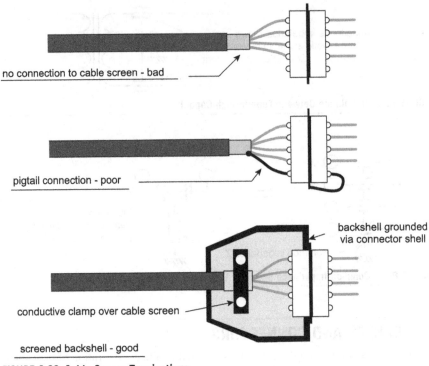

no connection to cable screen - bad

pigtail connection - poor

backshell grounded via connector shell

conductive clamp over cable screen

screened backshell - good

FIGURE 8.22 Cable Screen Terminations.

gland so that the cable screen makes contact all the way round its circumference, through 360 degrees. As soon as a connector is used, some compromise has to be made.

Military-style connectors are designed as above so that the cable screen makes 360 degrees contact, but they attract military-style prices and assembly costs. RF co-axial connectors, such as the common BNC type, also make 360 degrees contact, but they carry only one signal line at a time. Many multiway connectors do not have proper provision for terminating the shield, and this is where performance degradation creeps in. Only too often, the shield is brought down to a "pigtail" or drain wire and terminated to one of the connector pins—the EIA/RS-232F interface standard (Section 6.2.5) even has a pin allocated to this function, pin 1—or, worse still, it is not terminated at all (Fig. 8.22).

The effect of no shield connection at all is to nullify the shielding effectiveness at hf. This is perhaps to be expected, but what is less obvious is that the pigtail connection is almost as bad. The difference in effectiveness between a pigtail connection and a full 360 degrees connection is minimal below 3 MHz but can approach 40 dB at hfs. It is caused by resonant effects on the pigtail inductance and can show variations greater than 20 dB over small changes in frequency.

Screened Backshells

The best termination for a multiway connector is to use a screened conductive back-shell for the connector and to clamp the cable shield firmly to it. The backshell must make contact directly with the conductive shell of the connector itself, and this in turn must make good 360 degrees contact with the shell on its mating connector, which must be bolted firmly and conductively to the equipment case. Any departure from this practice—in terms of not bolting directly to the case, not having mating conductive shells on the connectors, and not using a screened backshell with the cable screen clamped to it—will compromise the hf shielding of the system.

Inexpensive subminiature multiway connectors are available whose construction adheres to these principles, and if they are used intelligently they can give good screening. It is quite possible to misuse them either in design or assembly and throw away all their advantages. Many other types of multiway connector, in particular the popular insulation displacement two-part units, have no potential for making shielded connections at all and should only be used on well-filtered interfaces or for interboard connections inside screened equipment.

8.8 ELECTROMAGNETIC COMPATIBILITY DESIGN CHECKLIST

- Design for EMC from the beginning; know what performance you require
- Select components and circuits with EMC in mind:
  - use slow and/or high-immunity logic
  - use good RF decoupling of power supplies
  - minimize signal bandwidths with RC filtering, maximize levels
  - used resistor buffering on long clock or data lines
  - incorporate a watchdog circuit on every microprocessor
- PCB layout:
  - keep interference paths segregated from sensitive circuits
  - minimize ground inductance with an unbroken ground plane or ground grid
  - minimize loop areas in high-current or sensitive circuits
  - minimize track and component leadout lengths
- Cables:
  - avoid parallel runs of signal and power cables
  - make sure that screens are 360 degrees bonded through properly designed connectors
  - use twisted pair for high-speed data or high-current switching
  - run internal cables away from apertures in shielded enclosures
  - use multiple ground wires or planes in ribbon or flexi cables
- Grounding:
  - ensure adequate bonding of screens, connectors, filters, cabinets, etc.
  - ensure that bonding methods will not deteriorate in adverse environments
  - mask paint from any intended conductive areas
  - keep earth straps short and wide: aim for a length/width ratio less than 3:1
  - route conductors to avoid common ground impedances

- Filters:
  - apply a mains filter for both emissions and immunity: check its required current rating
  - use correct components and filter configuration for I/O lines
  - ensure a good interface ground return for each filter group
  - ensure that filter input and output terminal wiring is kept separate
  - apply filtering to interference sources, such as switches or motors
- Shielding:
  - determine the type and extent of shielding required from the frequency range of interest
  - enclose particularly sensitive or noisy areas with extra internal shielding
  - avoid large or resonant apertures in the shield or take measures to mitigate them
  - use conductive gaskets where long (>l/20) gaps or seams are unavoidable
  - Test and evaluate for EMC continuously as the design progresses

General Product Design

CHAPTER OUTLINE

The Circuit Designer's Companion. http://dx.doi.org/10.1016/B978-0-08-101764-7.00009-8

9.1 SAFETY

Any electronic equipment must be designed for safe operation. Most countries have some form of product liability legislation that puts the onus on the manufacturer to ensure that his product is safe. The responsibility devolves onto the product design engineer, to take reasonable care over the safety of the design. This includes ensuring that the equipment is safe when used properly, that adequate information is provided to enable its safe use, and that adequate research has been carried out to discover, eliminate, or minimize risks due to the equipment.

There are various standards relating to safety requirements for different product sectors. In some cases, compliance with these standards is mandatory. In the European Community, the Low Voltage Directive (73/23/EEC) applies to all electrical equipment with a voltage rating between 50- and 1000-V ac or 75 and 1500-V dc, with a few exceptions and requires member states to take all appropriate measures

> to ensure that electrical equipment may be placed on the market only if, having been constructed in accordance with good engineering practice in safety matters in force in the Community, it does not endanger the safety of persons, domestic animals or property when properly installed and maintained and used in applications for which it was made.

If the equipment conforms to a harmonized CENELEC or internationally agreed standard then it is deemed to comply with the Directive. Examples of harmonized standards are EN 60065:1994, "Safety requirements for mains-operated electronic and related apparatus for household and similar general use," which is largely equivalent to IEC Publication 60065 of the same title; or EN 60950-1:2002, "Information technology equipment. Safety. General requirements," equivalent to IEC 60950-1. Proof of compliance can be by a Mark or Certificate of Compliance from a recognized laboratory, or by the manufacturer's own declaration of conformity. The Directive includes no requirement for compulsory approval for electrical safety.

9.2 THE HAZARDS OF ELECTRICITY

The chief dangers (but by no means the only ones, see Table 9.1) of electrical equipment are the risk of electric shock and the risk of a fire hazard. The threat to life from electric shock depends on the current that can flow in the body. For ac, currents less than 0.5 mA are harmless, while those greater than 50−500 mA (depending on duration) can be fatal.[1] Protection against shock can be achieved simply by limiting the current to a safe level, irrespective of the voltage. There is an old saying, "it's the volts that jolts, but the mils that kills." If the current is not limited, then the voltage level in conjunction with contact and body resistance determines the hazard. A voltage of less than 50-V ac rms, isolated from the supply mains or derived from an independent supply, is classified as a safety extra-low voltage (SELV) and equipment designed to operate from an SELV can have relaxed requirements against the user being able to contact live parts.

Aside from current and voltage limiting, other measures to protect against electric shock are as follows:

- Earthing and automatic supply disconnection in the event of a fault. See Section 1.1.12.
- Inaccessibility of live parts. A live part is any part, contact with which may cause electric shock, that is any conductor, which may be energized in normal use—not just the mains "live."

---

[1]IEC Publication 60479 gives further information.

Table 9.1 Some Safety Hazards Associated With Electronic Equipment

Hazard	Main Risk	Source
Electric shock	Electrocution, injury due to muscular contraction, burns	Accessible live parts
Heat or flammable gases	Fire, burns	Hot components, heat sinks, damaged or overloaded components and wiring
Toxic gases or fumes	Poisoning	Damaged or overloaded components and wiring
Moving parts, mechanical instability	Physical injury	Motors, parts with inadequate mechanical strength, heavy or sharp parts
Implosion/explosion	Physical injury due to flying glass or fragments	CRTs, vacuum tubes, overloaded capacitors, and batteries
Ionizing radiation	Radiation exposure	High-voltage CRTs, radioactive sources
Nonionizing radiation	RF burns, possible chronic effects	Power RF circuits, transmitters, antennas
Laser radiation	Damage to eyesight, burns	Lasers
Acoustic radiation	Hearing damage	Loudspeakers, ultrasonic transducers

9.2.1 SAFETY CLASSES

IEC publication 60536 classifies electrical equipment into four classes according to the method of connection to the electrical supply and gives guidance on forms of construction to use for each class. The classes are

Class 0: Protection relies on basic functional insulation only, without provision for an earth connection. This construction is unacceptable in the United Kingdom.

Class I: Equipment is designed to be earthed. Protection is afforded by basic insulation, but failure of this insulation is guarded against by bonding all accessible conductive parts to the protective earth conductor. It depends for its safety on a satisfactory earth conductive path being maintained for the life of the equipment.

Class II: The equipment has no provision for protective earthing, and protection is instead provided by additional insulation measures, such as double or reinforced insulation. Double insulation is functional insulation, plus a supplementary layer of insulation to provide protection if the functional insulation fails. Reinforced insulation is a single layer, which provides equivalent protection to double.

Class III: Protection relies on supply at SELV and voltages higher than SELV are not generated. Second-line defenses such as earthing or double insulation are not required.

9.2.2 INSULATION TYPES

As outlined above, the safety class structure places certain requirements on the insulation that protects against access to live parts. The basis of safety standards is that there should be at least two levels of protection between the casual user and the electrical hazard. The standards give details of the required strength for the different types of insulation, but the principles are straightforward.

Basic Insulation

Basic insulation provides one level of protection but is not considered fail-safe, and the other level is provided by safety earthing. A failure of the insulation is therefore protected against by the earthing system.

Double Insulation

Earthing is not required because the two levels of protection are provided by redundant insulation barriers, one layer of basic plus another supplementary; if one fails the other is still present, and so this system is regarded as fail-safe. The double-square symbol indicates the use of double insulation.

Reinforced Insulation

Two layers of insulation can be replaced by a single layer of greater strength to give an equivalent level of protection.

9.2.3 DESIGN CONSIDERATIONS FOR SAFETY PROTECTION

The requirement for inaccessibility has a number of implications. Any openings in the equipment case must be small enough that the standard test finger, whose dimensions are defined in those standards that call up its use, cannot contact a live part (Fig. 9.1). Worse, small suspended bodies (such as a necklace) that can be dropped through ventilation holes must not become live. This may force the use of internal baffles behind ventilation openings.

Protective covers, if they can be removed by hand, must not expose live parts. If they do, they must only be removable by use of a tool. Or, use extra internal covers over live portions of the circuit. It is anyway good practice to segregate high-voltage and mains sections from the rest of the circuit and provide them with separate covers. Most electronic equipment runs off voltages below 50 V and, provided the insulation offered by the mains isolating transformer is adequate, the signal circuitry can be regarded as being at SELV and therefore not live.

Any insulation must, in addition to providing the required insulation resistance and dielectric strength, be mechanically adequate. It will be dropped, impacted, scratched, and perhaps vibrated to prove this. It must also be adequate under humid

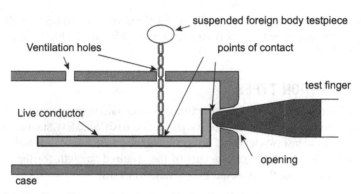

FIGURE 9.1 The Test Finger and the Suspended Foreign Body.

conditions: hygroscopic materials (those that absorb water readily, such as wood or paper) are out. Various standards define acceptable creepage and clearance distances versus the voltage proof required. As an example, EN 60065 allows 0.5 mm below 34 V rising to 3 mm at 354 V and extrapolated thereafter; distances between PCB conductors are slightly relaxed, being 0.5 mm up to 124 V, increasing to 3 mm at 1240 V. Creepage distance (Fig. 9.2) denotes the shortest distance between two conducting parts along the surface of an insulating material, while clearance distance denotes the shortest distance through air.

Easily discernible, legible, and indelible marking is required to identify the apparatus and its mains supply, and any protective earth or live terminals. Mains cables and terminations must be marked with a label to identify earth, neutral, and live conductors, and class I apparatus must have a label which states "WARNING: THIS APPARATUS MUST BE EARTHED". Fuse holders should also be marked with their ratings and mains switches should have their "off" position clearly shown. If user instructions are necessary for the safe operation of the equipment, they should preferably be marked permanently on the equipment.

Any connectors that incorporate live conductors must be arranged so that exposed pins are on the dead side of the connection when the connector is separated. When a connector includes a protective earth circuit, this should mate before the live terminals and unmate after them. (See the CEE-22 6A connector for an example.)

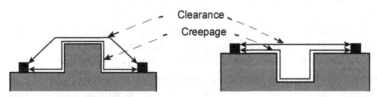

FIGURE 9.2 Creepage and Clearance Distance.

9.2.4 FIRE HAZARD

It is taken for granted that the equipment will not overheat during normal operation. But you must also take steps to ensure that it does not overheat or release flammable gases to the extent of creating a fire hazard under fault conditions. Any heat developed in the equipment must not impair its safety. Fault conditions are normally taken to mean short circuits across any component, set of terminals or insulation that could conceivably occur in practice (creepage and clearance distances are applied to define whether a short circuit would occur across insulation), stalled motors, failure of forced cooling, and so on.

The normal response of the equipment to these types of faults is a rise in operating current, leading to local heating in conductors. The normal protection method is by means of current limiting, fuses, thermal cutouts, or circuit breakers in the supply or at any other point in the circuit where overcurrent could be hazardous. As well as this, flame-retardant materials should be used wherever a threat of overheating exists, such as for PCB base laminates.

Fuses are cheap and simple but need careful selection in cases where the prospective fault current is not that much higher than the operating current. They must be easily replaceable, but this makes them subject to abuse from unqualified users (hands up anyone who has not heard of people replacing fuse links with bent nails or pieces of cigarette-packet foil). The manufacturer must protect his liability in these cases by clear labeling of fuse holders and instructions for fuse replacement. Fuse specification is covered in more detail in Section 7.2.3.

Thermal cutouts and circuit breakers are more expensive but offer the advantage of easy resetting once the fault has cleared. Thermal devices must obviously be mounted in close thermal contact with the component they are protecting, such as a motor or transformer.

9.3 DESIGN FOR PRODUCTION

Really, every chapter in this book has been about design for production. As was implied in the introduction, the ability that marks out a professional designer is the ability to design products or systems, which work under all relevant circumstances and which can be manufactured easily.

The sales and marketing engineer addresses the questions, "can I sell this product?" and "how much can I sell this product for?" This book has not touched on these issues, important though they are to designers; it has assumed that you have a good relationship with your marketing department and that your marketing colleagues are good at their job. But you as designer also have to address another set of questions, which includes the following:

- Can the purchasing department source the components quickly and cheaply?
- Can the production department make the product quickly and cheaply?
- Can the test department test it easily?
- Can the installation engineers or the customer install it successfully?

It is as well to bear all these questions in mind when you are designing a product, or even part of one. Your company's financial health, and consequently your and others' job security, ultimately depends on it. A good way to monitor these factors is to follow a checklist.

9.3.1 CHECKLIST

Sourcing

- Have you involved purchasing staff as the design progressed?
- Are the parts available from several vendors or manufacturers wherever possible? Have you made extensive use of industry standard devices?
- Where you have specified alternate sources, have you made sure that they are all compatible with the design?
- Have you made use of components that are already in use on other products?
- Have you specified close-tolerance components only where absolutely necessary?
- Where sole-sourced parts have to be used, do you have assurances from the vendor on price and lead time? How reliable are they? Have you checked that there is no warning, "not recommended for new designs" (implying limited availability), on each part?
- Does your company have a policy of vetting vendors for quality control? If so, have you added new vendors with this product, and will they need to be vetted?

Production

- Have you involved production staff as the design progressed?
- Are you sure that the mechanical and electrical design will work with all mechanical and electrical tolerances?
- Does the mechanical design allow the component parts to be fitted together easily?
- Are components, especially polarized ones, all oriented in the same direction on the PCB for ease of inspection and insertion?
- Are discrete components, notably resistors, capacitors, and transistors, specified to use identical pitch spacings and footprints as far as possible?
- Have you minimized wiring looms to front or rear panels and between PCBs, and used mass-termination connections (e.g., insulation displacement connector) wherever possible?
- Have you modularized the design as far as possible to make maximum use of multiple identical units?
- Is the soldering and assembly process (wave, infrared, autoinsert, pick and place, etc.) that you have specified compatible with the manufacturing capability? Will the placement machines cope with all the surface-mount components you have used?
- If the production calls for any special assembly procedures (e.g., potting or conformal coating), or if any components require special handling or assembly (MOSFETs, LEDs, batteries, relays, etc.) are the production and stores staff

fully conversant with these procedures and able to implement them? Have you minimized the need for such special procedures?

- Do all PCBs have adequate solder mask, track and hole dimensions, clearances, and silk screen legend for the soldering and assembly process? Are you sure that the test and assembly personnel are conversant with the legend symbols?
- Are your assembly drawings clear and easy to follow?

Testing and Calibration

- Have you involved test staff as the design progressed?
- Are all adjustment and test points clearly marked and easily accessible?
- Have you used easily set parts such as dual in line switches or linking connectors in preference to solder-in wire links?
- Does the circuit design allow for the selection of test signals, test subdivision, and stimulus/response testing (including boundary scan) where necessary?
- If you are specifying automatic testing with automatic test equipment (ATE), does the pc layout allow adequate access and tooling holes for bed-of-nails probing? Have you confirmed the validity of the ATE program and the functional test fixture?
- Have you written and validated a test software suite for microprocessor-based products?

Installation

- Is the product safe?
- Does the design have adequate electromagnetic compatibility?
- Are the installation instructions or user handbook clear, correct, and easy to follow?
- Do the installation requirements match the conditions, which will obtain on installation? For example, is the environmental range adequate, the power supply appropriate, the housing sufficient, etc.?

9.3.2 THE DANGERS OF ELECTROSTATIC DISCHARGE

There is one particular danger to electronic components and assemblies that is present in both the design lab and the production environment. This is damage from electrostatic discharge (ESD). This can cause complete component failure, as was discussed in Section 4.5.1, or worse, performance degradation that is difficult or impossible to detect. It can also cause transient malfunction in operating systems (cf. page 265).

Generation of Electrostatic Discharge

When two nonconductive materials are rubbed together, electrons from one material are transferred to the other. This results in the accumulation of *triboelectric* charge on the surface of the material. The amount of the charge caused by movement of the

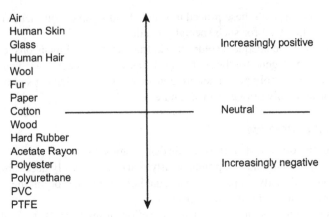

FIGURE 9.3 The Triboelectric Series.

materials is a function of the separation of the materials in the triboelectric series, an example of which is shown in Fig. 9.3. Additional factors are the closeness of contact, rate of separation, and humidity. Fig. 8.3 shows the electrostatic voltage related to materials and humidity, from which you can see that possible voltages can exceed 10 kV.

If this charge is built up on the human body, as a result of natural movements, it can then be discharged through a terminal of an electronic component. This will damage the component at quite low thresholds, easily less than 1 kV, depending on the device. Of the several contributory factors, low humidity is the most severe; if relative humidity is higher than 65% (which is frequent in maritime climates such as the UK's) then little damage is likely. Lower than 20%, as is common in continental climates such as the United States, is much more hazardous.

Gate-oxide breakdown of MOS or CMOS components is the most frequent, though not the only, failure mode. Static-damaged devices may show complete failure, intermittent failure, or degradation of performance. They may fail after one very high voltage discharge or because of the cumulative effect of several discharges of lower potential.

Static Protection

To protect against ESD damage, you need to prevent static buildup and to dissipate or neutralize existing charges. At the same time, operators (including yourself and your design colleagues) need to be aware of the potential hazard. The methods used to do this include the following:

- Package-sensitive devices or assemblies in conductive containers, keep them in these until use, and ensure they are clearly marked.
- Use conductive mats on the floor and workbench where sensitive devices are assembled, bonded to ground via a 1-MΩ resistor.

- Remove nonconductive items such as polystyrene cups, synthetic garments, wrapping film, etc. from the work area.
- Ground the assembly operator through a wrist strap, in series with a 1-MΩ resistor for electric shock protection.
- Ground soldering tool tips.
- Use ionized air to dissipate charge from nonconductors, or maintain a high relative humidity.
- Create and maintain a static-safe work area where these practices are adhered to.
- Ensure that all operators are familiar with the nature of the ESD problem.
- Mark areas of the circuit where a special ESD hazard exists; design circuits to minimize exposed high-impedance or unprotected nodes.

All production assembly areas should be divided into static-safe workstation positions. Your design and development prototyping lab should also follow these precautions, since it is quite possible to waste considerable time tracking down a fault in a prototype, which is due to a static-damaged device. A typical static-safe area layout is shown in Fig. 9.4. BS CECC 00015: Part 1:1991 gives a code of practice for the handling of electrostatic sensitive devices.

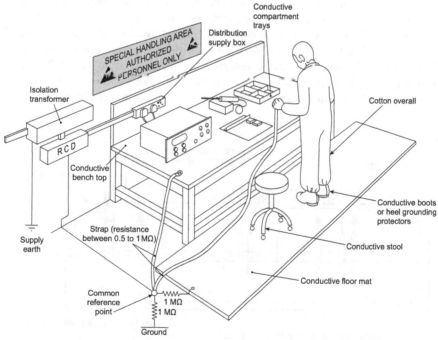

FIGURE 9.4 Static-Safe Workshop Layout.

Extract from BS5783:1987 reproduced with the permission of the BSI.

9.4　TESTABILITY

The previous section's checklist included some items that referred to the testability of the design. It is vital that you give sufficient thought throughout the design of the product as to how the assembled unit or units will be tested to prove their correct function. In the very early stages, you should already know whether your test department will be using in-circuit testing, manual functional testing, functional testing on ATE, boundary scan, or a combination of these methods. You should then be in a position to include test access points and circuits in the design as it progresses. This is a more effective way of incorporating testability than merely bolting it on at the end.

9.4.1　IN-CIRCUIT TESTING

The first test for an assembled PCB is to confirm that every component on it is correctly inserted, of the right type or value, and properly soldered in. It is quite possible for manual assembly personnel to insert the wrong component, or insert the right one incorrectly polarized, or even to omit a component or series of components. Automatic assembly is supposed to avoid such errors, but it is still possible to load the wrong component into the machine or for components to be marked incorrectly. Automatic soldering has a higher success rate than hand soldering but bad joints due to lead or pad contamination can still occur.

In-circuit testing lends itself to automatic test fixture and test program generation. Each node on the PCB has to be probed, which requires a bed-of-nails test fixture (Fig. 9.5), and this can be designed automatically from the PCB layout data. Similarly, the expected component characteristics between each node can be derived from the circuit schematic, using a component parameter library.

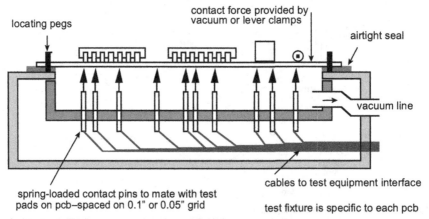

FIGURE 9.5 **Principle of the Bed-of-Nails Test Fixture.**

An in-circuit tester carries out an electrical test on each component, to verify its behavior, value, and orientation, by applying voltages to nodes that connect to each component and measuring the resulting current. Interactions with other components are prevented by guarding or back-driving. The technique is successful for discrete components but less so for integrated circuits, whose behavior cannot be described in terms of simple electrical characteristics. It is therefore most widely applied on boards, which contain predominantly discrete components, and which are produced in high volume, as the overhead involved in programming and building the test fixture is significant. It does not of itself guarantee a working PCB. For this, you need a functional test.

9.4.2 FUNCTIONAL TESTING

A functional test checks the behavior of the assembled board against its functional specification, with power applied and with simulated or special test signals connected to the input/output lines. It is often combined with calibration and setup adjustments. For low-volume products you will normally write a test procedure around individual test instruments, such as voltmeters, oscilloscopes, and signal generators. You may go so far as to build a special test jig to simulate some signals, interface others, provide monitored power, and make connections to the board under test. The test procedures will consist of a sequence of instructions to the test technician—apply voltage A, observe signal at B, adjust trimmer C for a minimum at D, and so on—along with limit values where measurements are made.

The disadvantage of this approach is that it is costly in terms of test time. This puts up the overhead cost of each board and affects the final cost price of the overall unit. It is cheap as far as instrumentation goes, since you only need a simple test jig, and you will normally expect the test department to have the appropriate lab equipment to hand. Hence it is best suited to low production volumes where you cannot amortize the cost of automatic test equipment.

A further, hidden, disadvantage may be that you do not have to define the testing absolutely rigorously but can rely on the experience of the test technician to make good any deficiencies in the test procedure or measurement limits. It is common for test personnel to develop a better "feel" for the quirks of a particular design's behavior under test than its designer ever could. Procedural errors and invalid test limits may be glossed over by a human tester, and if such information is not fed back to the designer then the opportunity to optimize that or subsequent designs is lost.

Automatic Test Equipment

Functional testing may more easily be carried out by ATE. In this case, the function of the human tester is reduced to that of loading and unloading the unit under test, pressing the "go" button and observing the pass/fail indicator. The testing is comprehensively deskilled; the total unit test time is reduced to a few minutes or less. This minimizes the test cost.

The costs occur instead at the beginning of the production phase, in programming the ATE and building a test fixture. The latter is similar to (in some cases may be identical to) the bed-of-nails fixture which would be used for in-circuit testing (Fig. 9.5). Or, if all required nodes are brought out to test connectors, the test fixture may consist of a jig, which automatically connects a suite of test instrumentation to the board under the command of a computer-based test program. The IEEE-488 standard bus allows interconnection of a desktop computer and remote-controlled meters, signal generators, and other instruments, for this purpose.

The skill required for a test technician now resides in the test program, which may have been written by you as designer or by a test engineer. In any case it needs careful validation before it is let loose on the product, since it does not have the skill or expertise to determine when it is making an invalid test. The cost involved in designing and building the test fixture, programming it, and validating the program and the capital cost of the ATE itself need to be carefully judged against the savings that will be made in test time per unit. It is normally only justified if high production volumes are expected.

9.4.3 BOUNDARY SCAN AND JOINT TEST ACTION GROUP

Many digital circuits are too complex to test by conventional in-circuit probing. Even if bed-of-nails contact could be made to the hundreds and sometimes thousands of test pads that would be needed, the IC functions and pin states would be so involved that no absolute conclusions could be drawn from the voltage and impedance states recorded. The increased use of small-sized PCBs, with surface mount, fine pitch components installed on both sides, presents the greatest problem. So a different method has been developed to address this problem, and it is known as boundary-scan testing.

History

In 1985, an ad hoc group created the Joint Test Action Group (JTAG). JTAG had over 200 members around the world, including major electronics and semiconductor manufacturers. This group met to establish a solution to the problems of board test and to promote a solution as an industry standard. They subsequently developed a standard for integrating hardware into compliant devices, which could be controlled by software. This was termed boundary-scan testing. The JTAG proposal was approved in 1990 by the IEEE and defined as IEEE Standard 1149.1-1990 *Test Access Port and Boundary Scan Architecture*. Since the 1990 approval, updates have been published in 1993 as supplement IEEE Std 1149.1a-1993 and in 1995 as IEEE Std 1149.1b-1994.

Description of the Boundary-Scan Method

Boundary scan is a special type of scan path with a register added at every I/O pin on a device. Although this requires special extra test latches on these pins, the technique offers several important benefits, the most obvious being that it allows fault isolation at the component level. A major problem driving the development of boundary scan has been the adverse effect on testability of surface-mount technology. The inclusion of a boundary-scan path in surface-mount components is sometimes the only way to perform continuity tests between devices. By placing a known value on an output buffer of one device and observing the input buffer of another interconnected device, it is easy to check the interconnection of the PCB net. Failure of this simple test indicates broken circuit traces, dry solder joints, solder bridges, or ESD-induced failures in an IC buffer—all common problems on PCBs.

Another advantage of the boundary-scan method is the ability to apply predeveloped functional pattern sets to the I/O pins of the IC by way of the scan path. IC manufacturers and ASIC developers create functional pattern sets for test purposes. Subsets of these patterns can be reused for in-circuit functional IC testing, which can show significant savings on development resources.

Each device to be included within the boundary scan has the normal application-logic section and related input and output, and in addition a boundary-scan path consisting of a series of boundary-scan cells (BSCs), typically one BSC per IC function pin (Fig. 9.6). The BSCs are interconnected to form a shift register scan path between the host IC's test data input (TDI) pin and test data output (TDO) pin. During normal IC operation, input and output signals pass freely through each BSC. However, when the boundary-test mode is entered, the IC's internal logic may be disconnected and its boundary controlled in such a way that test stimuli can be shifted in and applied from each BSC output, and test responses can be captured at each BSC input and shifted out for inspection. External testing of traces and neighboring ICs on a board assembly is achieved by applying test stimuli from the output BSCs and capturing responses at the input BSCs. If required, internal testing of the application logic can be achieved by

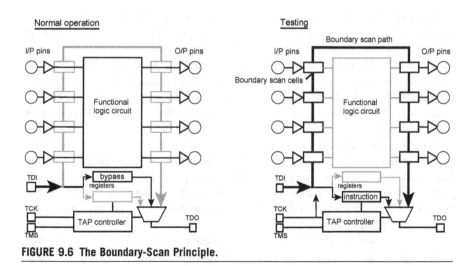

FIGURE 9.6 The Boundary-Scan Principle.

applying test stimuli from the input BSCs and capturing responses at the output BSCs. The implementation of a scan path at the boundary of IC designs provides an embedded testing capability that can overcome the physical access problems referred to above.

As well as performing boundary tests of each IC, ICs may also be instructed via the scan path to perform a built-in self-test operation, and the results inspected via the same path. Boundary scan is not limited to individual ICs; several ICs on a board will normally be linked together to offer an extended scan path (which could be partitioned or segmented to optimize testing speed). The whole board itself could be regarded as the system to be tested, with a scan path encompassing the connections at the board's boundary, and BSCs implemented at these connections using ICs designed for the purpose.

Devices

Every IEEE Std 1149.1-compatible device has four additional pins—two for control and one each for input and output serial test data. These are collectively referred to as the "test access port" (TAP). To be compatible, a component must have certain basic test features, but the standard allows designers to add test features to meet their own unique requirements. A JTAG-compliant device can be a microprocessor, microcontroller, programmable logic device, complex programmable logic device, field programmable gate array, ASIC, or any other discrete device that conforms to the 1149.1 specification. The TAP pins are:

- TCK—Test clock input. Shift register clock separate from the system clock.
- TDI—Data is shifted into the JTAG-compliant device via TDI.
- TDO—Data is shifted out of the device via TDO.
- TMS—Test mode select. TMS commands select test modes as defined in the JTAG specification.

The 1149.1 specification stipulates that at every digital pin of the IC, a single cell of a shift-register is designed into the IC logic. This single cell, known as the BSC, links the JTAG circuitry to the IC's internal core logic. All BSCs of a particular IC constitute the boundary-scan register (BSR), whose length is of course determined by the number of I/O pins that IC has. BSR logic becomes active when performing JTAG testing, otherwise it remains passive under normal IC operation. A 1-bit bypass register is also included to allow testing of other devices in the scan path.

You communicate with the JTAG-compliant device using a hardware controller that either inserts into a PC add-in card slot or by using a stand-alone programmer. The controller connects to the TAP on a JTAG-compliant PCB—which may be the port on a single device, or it may be the port created by linking a number of devices. You (or your test department) then must write the software to perform boundary-scan programming and testing operations.

As well as testing, the boundary-scan method can be used for various other purposes that require external access to a PCB, such as flash memory programming.

Deciding Whether or Not to Use Boundary Scan

Although the boundary-scan method has enormous advantages for designers faced with the testing of complex, tightly packed circuits, it is not without cost. There is a significant logic overhead in the ICs as well as a small overhead on the board in implementing the TAP, and there is the need for your test department to invest in the resources and become familiar with the method as well as programming each product. As a rough guide, you can use the following rule[2] (relating to ASIC design) to decide on whether or not the extra effort will be cost-effective:

- Designs with fewer than 10-K gates: not generally complex enough to require structured test approaches. The overhead impact is usually too high to justify them. Nonstructured, good design practices are usually sufficient.
- Designs with more than 10-K gates, but fewer than 20-K gates: Structured techniques should be considered for designs in this density. Nonstructured, good design practices are probably sufficient for highly combinatorial circuits without memory. Structured approaches should be considered as complexity is increased by the addition of sequential circuits, feedback, and memory. Consider boundary-scan testing for reduced cycle times and high fault grades.
- Designs with more than 20-K gates: The complexity of circuits this dense usually requires structured approaches to achieve high fault grades. At this density, it is often hard to control or observe deeply embedded circuits. The overhead associated with structured testability approaches is acceptable.

9.4.4 DESIGN TECHNIQUES

There are many ways in which you can design a PCB circuit to make it easy to test, or conversely hard to test. The first step is to decide how the board will be tested, which is determined by its complexity, expected production volume, and the capabilities of the test department.

Bed-of-Nails Probing

If you will be using a bed-of-nails fixture, then the PCB layout should allow this. Leave a large area around the outside of the board, and make sure there are no unfilled holes, to enable a good vacuum pressure to be developed to force the board onto the probes. Or if the board will be clamped to the fixture, make sure there is space on the top of the board for the clamps. Decide where your test nodes need to be electrically, and then lay out the board to include target pads on the underside for the probes. These pads should be spaced on a 0.1″ (2.5 mm) grid for accurate drilling of the test jig; down to 0.05″ or 1 mm is possible if the board layout is tight. It is not good practice to use component lead pads as targets, since pressure from the

---

[2]From "IEEE Std 1149.1 (JTAG) Testability Primer", Texas Instruments, 1997.

probe may cause a defective joint to appear good. Ensure that tooling holes are provided and are accurately aligned with the targets.

Remember that a bed-of-nails jig will connect several long, closely coupled wires to many nodes in the circuit. This will severely affect the circuit's stray reactances, and thereby modify its high frequency response. It is not really suitable for functionally testing high-frequency or high-speed digital circuits.

Test Connections

If your test department does not want to use bed-of-nails probing, then help them find the test points that are necessary by bringing them out to test connectors. These can be cheap-and-cheerful pin strips on the board since they will normally only be used once. The matching test jig can then take signals from these connections direct to the test instrumentation via a switch arrangement. Of course, preexisting connectors such as multiway or edge connectors can be used to bring out test signals on unused pins. Be careful, though, that you do not bring long test tracks from one side of the board to the other and thereby compromise the circuit's cross talk, noise susceptibility, and stability. Extra local test connectors are preferable.

Circuit Design

There are many design tricks to make testing easier. A simple one is to include a series resistor in circuits where you will want to back-drive against an output, or where you will want to measure a current (Fig. 9.7A). The cost of the resistor is minimal compared to the test time it might save. Of course, you must ensure that the resistor does not affect normal circuit operation. Also, unused digital gate inputs may be taken to a pull-up resistor rather than direct to supply or ground (cf. Section

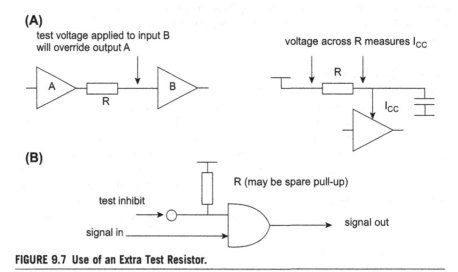

FIGURE 9.7 Use of an Extra Test Resistor.

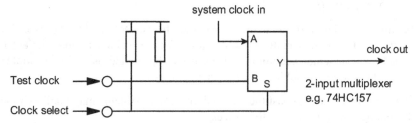

FIGURE 9.8 Test Clock Selection With Data Multiplexer.

6.1.5), and this point can then be used to inhibit or enable logic signals for testing purposes only (Fig. 9.7B).

The theme of Fig. 9.7B can be taken further to incorporate extra logic switching to allow data or timing signals to be derived either from the normal on-board source or from the external test equipment. This is particularly useful in situations where testing logic functions from the normal system clock would result either in too fast operation, or too slow. The clock source can be taken through a 2-input data multiplexer such as the 74HC157, one input of which is taken via a test connector to the external clock as shown in Fig. 9.8. In normal operation the clock select and test clock inputs are left unconnected, and the system clock is passed directly through the multiplexer.

When you are considering testing a microprocessor board, it is advantageous to have a small suite of test software resident on the main program PROM. This can be activated on start-up by reading a digital input, which is connected to a test link or test probe. If the test input is set, the program jumps to the test routines rather than to the main operating routines. These are arranged to exercise all inputs, outputs, and control signals continuously in a predictable manner, so that the test equipment can monitor them for the correct function. The test software operation depends of course on the core functions of the microprocessor, and its bus and control signal interconnections, being fault free.

More complex digital systems cannot easily be tested or, more to the point, debugged, with the techniques described so far. The boundary-scan methods described in Section 9.3.3 are aimed at these applications, and you need to design them in from the start, since they consume a serious amount of circuit overhead to function.

9.5 RELIABILITY

The reliability of electronic equipment can, to some extent, be quantified, and a separate discipline of reliability engineering has grown up to address it. This section will serve as an introduction to the subject for those designers who are not fortunate enough to have a reliability engineering department at their disposal.

9.5.1 DEFINITIONS

Reliability, itself, has a strictly defined meaning. This can be stated as "the probability that a system will operate without failure for a specified period, subject to specified environmental conditions." Thus it can be quoted as a single number, such as 90%, but this is subject to three qualifications:

- agreement as to what constitutes a "failure." Many systems may "fail" without becoming totally useless in the process.
- a specified operating lifetime. No equipment will operate forever; reliability must refer to the reasonably foreseen operating life of the equipment, or to some other agreed period. The age of the equipment, which may well affect failure rate, is not a factor in the reliability specification.
- agreement on environmental conditions. Temperature, moisture, corrosive atmospheres, dust, vibration, shock, supply, and electromagnetic disturbances all have an effect on equipment operation and reliability is meaningless if these are not quoted.

If you offer or purchase equipment whose reliability is quoted for one set of conditions and it is used under another set, you will not be able to extrapolate the reliability figure to the new conditions unless you know the behavior of those parameters that affect it.

Mean Time Between Failures

For most of the life of a piece of electronic equipment, its failure rate (denoted by λ) is constant. In the early stages of operation it could be high and decrease as weak components fail quickly and are replaced; late in its life components may begin to "wear out" or corrosion may take its toll, and the failure rate may start to rise again. The reciprocal of failure rate during the constant period is known as the mean time between failures (MTBFs). This is generally quoted in hours, while failure rate is quoted in faults per hour. For instance, an MTBF of 10,000 h is equivalent to a failure rate of 0.0001 faults per hour or 100 faults per 10^6 h. MTBF has the advantage that it does not depend on the operating period and is therefore more convenient to use than reliability.

Mean Time to Failure

MTBF measures equipment reliability on the assumption that it is repaired on each failure and put back into service. For components that are not repairable, their reliability is quoted as mean time to failure (MTTF). This can be calculated statistically by observing a sample from a batch of components and recording each one's working life, a procedure known as life testing. The MTTF for this batch is then given by the mean of the lifetimes.

Availability

System users need to know for what proportion of time their system will be available to them. This figure is given by the ratio of "uptime" (U), during which the system is switched on and working, to total operating time. The difference between the two is the "downtime" (D) during which the system is faulty and/or under repair. Therefore we can define the availability, A, using Eq. (9.1):

$$A = \frac{U}{U + D} \qquad (9.1)$$

The availability (A) can also be related to the MTBF figure and the mean time to repair (MTTR) figure by Eq. (9.2)

$$A = \frac{\text{MTBF}}{\text{MTBF} + \text{MTTR}} \qquad (9.2)$$

The availability of a particular system can be monitored by logging its operating data, and this can be used to validate calculated MTBF and MTTR figures. It can also be interpreted as a probability that at any given instant the system will be found to be working.

9.5.2 THE COST OF RELIABILITY

Reliability does not come for free. Design and development costs escalate as more effort is put into assuring it, and component costs increase if high performance is required of them. For instance, it would be quite possible to improve the reliability of, say, an audio power amplifier by using massively overrated output transistors, but these would add considerably to the selling cost of the amplifier. On the other hand, if the selling cost were reduced by specifying underrated transistors, the users would find their total operating costs mounting since the output transistors would have to be replaced more frequently. Thus there is a general trend of decreasing operating or "life-cycle" costs and increasing unit costs, as the designed-in reliability of a given system increases. This leads to the notion of an "optimum" reliability figure in terms of cost for a system. Fig. 9.9 illustrates this trend. The criterion of good design is then to approach this optimum as closely as possible.

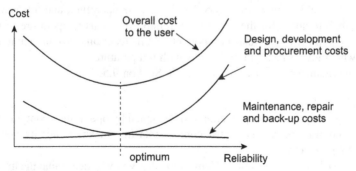

FIGURE 9.9 Reliability Versus Cost.

Of course, this argument only applies when the cost of unreliability is measured in strictly economic terms. Safety-critical systems, such as nuclear or chemical process plant controllers, railway signaling or flight-critical avionics, must instead meet a defined reliability standard, and the design criterion then becomes one of assuring this level of reliability, with cost being a secondary factor.

9.5.3 DESIGN FOR RELIABILITY

The goal of any circuit designer is to reduce the failure rate of his or her design to the minimum achievable within cost constraints. The factors that help in meeting this goal include the following:

- Use effective thermal management to minimize temperature rise.
- Derate susceptible components as far as possible.
- Specify high reliability or quality-assured components.
- Specify stress screening or burn-in tests.
- Keep circuits simple, and use the minimum number of components.
- Use redundancy techniques at the component level.

Temperature

High temperature is the biggest enemy of all electronic components and measures to keep it down are vital. Temperature rise accelerates component breakdown because chemical reactions occurring within the component, which govern bond fractures, growth of contamination or other processes, have an increased rate of reaction at higher temperature. The rate of reaction is determined by the Arrhenius equation,

$$\lambda = Ke^{\left(\frac{-E}{kT}\right)} \tag{9.3}$$

where λ gives a measure of failure rate; K is a constant depending on the component type; E is the reaction's activation energy; k is Boltzmann's constant, 1.38×10^{-23} J/K; T is absolute temperature.

Many reactions have activation energies around 0.5 eV, which results in an approximate doubling of λ with every 10°C rise in temperature, and this is a useful rule of thumb to apply for the decrease in reliability versus temperature of typical electronic equipment with many components. Some reactions have higher activation energy, which give a faster increase of λ with temperature.

Thermal management itself is covered in Section 9.5.

Derating

There is a very significant improvement to be gained by operating a component well within its nominal rating. For most components this means either its voltage or power rating, or both.

Take capacitors as an example. Conventionally, you will determine the maximum dc bias voltage a capacitor will have to withstand under worst-case conditions, and then select the next highest rating. Overspecifying the voltage rating may result in a larger and more costly component.

However, capacitor life tests show that as the maximum working voltage is approached, the failure rate increases as the fifth power of the voltage. Therefore, if you run the capacitor at half its rated voltage, you will observe a failure rate 32 times lower than if it is run at full-rated voltage. Given that a capacitor of double the required rating will not be as much as double the size, weight or cost, except at the extremes of range, the improvement in reliability is well worth having.

In many cases there is no difficulty in using a derated capacitor; small film capacitors for instance are rated at a minimum of 50 or 100 V and are frequently used in 5-V circuits. Electrolytics on the other hand are more likely to be run near their rating. These capacitors already have a much higher failure rate than other types because of their construction—the electrolyte has a tendency to "dry out," especially at high temperatures—and so you will achieve significant improvement, albeit at higher cost, if you heavily derate them.

Derating the power dissipation of resistors reduces their internal temperature and therefore their failure rate. In low-voltage circuits there is no need to check power rating for any except low-value parts; if for instance you use 0.4-W metal film resistors in a circuit with a maximum supply of 10 V, you can be sure that all resistors over 500 Ω will be derated by at least a factor of 2, which is normally enough.

Semiconductor devices are normally rated for power, current, and voltage, and derating on all of these will improve failure rate. The most important are power dissipation, which is closely linked to junction temperature rise and cooling provision, and operating voltage, especially in the presence of possible transient overvoltages.

High Reliability Components

Component manufacturers' reputations are seriously affected by the perceived reliability or otherwise of their product, so most will go to considerable effort not to ship defective parts. However, the cost of detecting and replacing a faulty part rises by an order of magnitude at each stage of the production process, starting at goods inwards inspection, proceeding through board assembly, test and final assembly, and ending up with field repair. You may therefore decide (even in the absence of mandatory procurement requirements on the part of your customer) that it is worth spending extra to specify and purchase parts with a guaranteed reliability specification at the "front end" of production.

CENELEC Electronic Components Committee

Initially it was military requirements, where reliability was more important than cost that drove forward schemes for assessed quality components. More recently many commercial customers have also found it necessary to specify such components. The need for a common standard of assessed quality is met in Europe by the CECC[3] scheme. This

---

[3]CENELEC Electronic Components Committee.

has superseded the earlier national BS9000 series of standards. CECC documents refer to a "harmonized system of quality assessment for electronic components."

Generic specifications are found in the CECC series for all types of component, which are covered by the scheme. These specify physical, mechanical, and electrical properties, and lay down test requirements. Individual component specifications are not found under the scheme.

Stress Screening and Burn-In

These specifications all include some degree of stress screening. This phrase refers to testing the components under some type of stress, typically at elevated temperature, under vibration or humidity and with maximum rated voltage applied, for a given period. This practice is also called "burning in." The principle is that weak components will fail early in their life and the failures can be accelerated by operating them under stress. These can then be weeded out before the parts are shipped from the manufacturer. A typical test might be 160 h at 125°C. Another common test is a repeated temperature cycle between the extremes of the permitted temperature range, which exposes failures due to poor bonding or other mechanical faults.

Such stress screening can be applied to any component, not just semiconductors, and also to entire assemblies. If you are unsure of the probable quality of early production output of a new design, specifying stress screening on the first few batches is a good way to discover any recurrent production faults before they are passed out to the customer. It is expensive in time, equipment, and inventory and should not be used as a crutch to compensate for poor production practices. It should only be employed as standard if the customer is willing to pay for it.

Simplicity

The failure rate of an electronic assembly is roughly equal to the sum of the failure rates of all its components. This assumes that a failure in any one component causes the failure of the whole assembly. This is not necessarily a valid assumption, but to assume otherwise you would have to work out the assembly's failure modes for each component failure and for combinations of failures, which is not practical unless your customer is prepared to pay for a great deal of development work.

If the assumption holds, then reducing the number of components will reduce the overall failure rate. This illustrates a very important principle in circuit design: *the highest reliability comes from the simplest circuits.* Apply Occam's razor ("entities should not be multiplied beyond necessity") and cut down the number of components to a minimum.

Redundancy

Redundancy is employed at the system level by connecting the outputs of two or more subsystems together such that if one fails, the others will continue to keep the system working. A typical example might be several power supplies, each connected to the same power distribution rail (via isolating diodes) and each capable

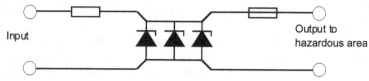

FIGURE 9.10 The Intrinsically Safe Zener Diode Barrier.

of supplying the full load. If the reliability of the interconnection is neglected, the probability of all supplies failing simultaneously is the product of the probabilities of failure of each supply on its own, assuming that a common-mode failure (such as the mains supply to all units going off) is ruled out.

The principle can also be applied at the component level. If the probability of a single component failing is too high then redundant components can be placed in parallel or series with it, depending on the required failure mode. This technique is mandatory in certain fields, such as intrinsically safe instrumentation. Fig. 9.10 illustrates redundant zener diode clamping. The zeners prevent the voltage across their terminals from rising to an unsafe value in the event of a fault voltage being applied at the input to the barrier. One zener alone would not offer the required level of reliability, so two further ones are placed in parallel, so that even with an open-circuit failure of two out of the three, the clamping action is maintained. The interconnections between the zeners must be solid enough not to materially affect the reliability of the combination.

Some provision must normally be made for detecting and indicating a failed component or subsystem so that it can be repaired or replaced. Otherwise, once a redundant part has failed, the overall reliability of the system is severely reduced.

9.5.4 THE VALUE OF MEAN-TIME-BETWEEN-FAILURE FIGURES

The mean-time-between-failure figure as defined in Section 9.4.1 can be calculated before the equipment is put into production by summing the failure rates of individual components to give an overall failure rate for the whole equipment. As discussed earlier, this assumes that a fault in any one component causes the failure of the whole assembly. This method presupposes adequate data on the expected failure rates of all components that will be used in the equipment.

Such sources of failure rate data for established component types are available. The most widely used is MIL-HDBK-217, now in its fifth revision, published by the United States Department of Defense. This handbook lists failure-rate models and tables for a wide variety of components, based on observed failure measurements. A failure rate for each component can be derived from its operating and environmental conditions, derating factor and method of construction or packaging. A further factor that is included for integrated circuits is their complexity and pinout. Another source of failure rate data, somewhat less comprehensive but widely used for telecommunications applications, is British Telecom's handbook HRD4.

The disadvantage with using such data is that it cannot be up-to-date. Proper failure rate data takes years to accumulate, and so data extracted from these tables for modern components will not be accurate. This is especially true for integrated circuits. Generally, figures based on obsolete failure-rate data will tend to be pessimistic, since the trend of component reliability is to improve.

Calculations of failure rates at component level are tedious, since operating conditions for each component, notably voltage and power dissipation, must be a part of the calculation to arrive at an accurate value. They do lend themselves to computer derivation, and software packages for reliability prediction are readily available. Since in many cases such operating conditions are highly variable, it is arguable that you will not obtain much more than an order-of-magnitude estimate of the true figure anyway.

A published MTBF figure does not tell you how long the unit will actually last, and it does not indicate how well the unit will perform in the field under different environmental and operating conditions. Such figures are mainly used by the marketing department to make the specification more attractive. But MTBF prediction *is* valuable for two purposes:

- for the designer, it gives an indication of where reliability improvements can most usefully be made. For instance, if as is often the case the electrolytic capacitors turn out to make the highest contribution to overall failure rate, you can easily evaluate the options available to you in terms of derating or adding redundant components. You need not waste effort on optimizing those components which have little effect overall.
- for the service engineer, it gives an idea of which components are likely to have failed if a breakdown occurs. This can be valuable in reducing servicing and repair time.

9.5.5 DESIGN FAULTS

Before leaving the subject of reliability design, we should briefly mention a very real problem, which is the fallibility of the designers themselves. There is no point in specifying highly reliable components or applying all manner of stress screening tests or redundancy techniques if the circuit is going to fail because it has been wrongly designed. Design faults can be due to inexperience, inattention, or incompetence on the part of the designer, or simply because the project timescale was too short to allow the necessary cross-checking. Computer-aided design techniques and simulators can reduce the risk, but they cannot eliminate the potential for human error completely.

The Design Review

An effective and relatively painless way of guarding against design faults is for your product development department to instigate a system of frequent design reviews. In these, a given designer's circuit is subjected to a peer critique to probe for flaws, which might not be apparent to the circuit's originator. The critique can check that the basic circuit concept is sound and cost effective, that all component tolerances have been accounted for, that parts will not be operated outside their ratings, and so on. The depth of the review is determined by the resources that are available within the group; the reviewers should preferably have no connection with the project being reviewed, so that they are able to question underlying and unstated assumptions. Naturally, the effectiveness of such a system depends on the resources a company is prepared to devote to it, and it also depends on the willingness of the designer to undergo a review. Personality clashes tend to surface on these occasions. Each designer develops pet techniques and idiosyncrasies during their career, and provided these are not actually wrong they should not attract criticism. Nevertheless, design reviews are valuable for testing the strength of a design before it gets to the stage where the cost of mistakes becomes significant.

9.6 THERMAL MANAGEMENT

It is in the nature of electronic components to dissipate power while they are operating. Any flow of current through a nonideal component will develop some power within that component, which in turn causes a rise in temperature. The rise may be no more than a small fraction of a degree Celsius when less than a milliwatt is dissipated, extending to several tens or even hundreds of degrees when the dissipation is measured in watts. Since excess temperature kills components, some way must be found to maintain the component-operating temperature at a reasonable level. This is known as thermal management.

9.6.1 USING THERMAL RESISTANCE

Heat transfer through the thermal interface is accomplished by one or more of three mechanisms: conduction, convection, and radiation. The attractiveness of thermal analysis to electronics designers is that it can easily be understood by means of an electrical analogue. Visualize the flow of heat as emanating from the component, which is dissipating power, passing through some form of thermal interface and out to the environment, which is assumed to have a constant ambient temperature T_A and infinite ability to sink heat. Then the heat source can be represented electrically as a current source; the thermal impedances as resistances; the temperature at any given point is the voltage with respect to 0 V; and thermal inertia can be represented by capacitance with respect to 0 V. The 0-V reference itself does not have an exact thermal analogue, but it is convenient to represent it as 0°C, so that temperature in degree Celsius is given exactly by a potential in volts. All these correspondences are summarized in Table 9.2.

Table 9.2 Thermal and Electrical Equivalences

Thermal Parameter	Units	Electrical Analogue	Units
Temperature difference	°C	Potential difference	Volts
Thermal resistance	°C/W	Resistance	Ohms
Heat flow	J/s (W)	Current	Amps
Heat capacity	J/°C	Capacitance	Farads

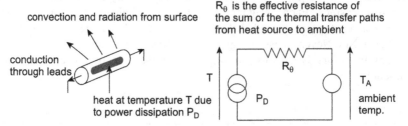

FIGURE 9.11 Heat Transfer From Hot Component to Ambient.

Fig. 9.11 shows the simplest general model and its electrical analogue. The model can be analyzed using conventional circuit theory and yields the following Eq. (9.4) for the temperature at the heat source:

$$T = P_D \times R_\theta + T_A \tag{9.4}$$

This temperature is the critical factor for electronic design purposes, since it determines the reliability of the component. Reducing any of P_D, R_θ, or T_A will minimize T. Ambient temperature is not normally under your control but is instead a specification parameter (but see Section 9.5.4). The usual assumption is that the ambient air (or other cooling medium) has an infinite heat capacity, and therefore its temperature stays constant no matter how much heat your product puts into it. The intended operating environment will determine the ambient temperature range, and for heat calculations only the extreme of this range is of interest; the closer this gets to the maximum allowable value of T the harder is your task. Since you are normally attempting to manage a given power dissipation, the only parameter that you are free to modify is the thermal resistance R_θ. This is achieved by heat sinking.

There are more general ways of analyzing heat flow and temperature rise, using thermal conductivity and the area involved in the heat transfer. However, component manufacturers normally offer data in terms of thermal resistance and maximum permitted temperature, so it is easiest to perform the calculations in these terms.

Partitioning the Heat Path

When you have data on the component's thermal resistance directly to ambient, and your mounting method is simple, then the basic model of Fig. 9.11 is adequate. For components that require more sophisticated mounting and whose heat transfer paths are more complicated, you can extend the model easily. The most common application is the power semiconductor mounted via an insulating washer to a heat sink (Fig. 9.12A).

The equivalent electrical model is shown in Fig. 9.12B. Here, T_j is the junction temperature and $R_{\theta j-c}$ represents the thermal resistance from junction to case of the device. All manufacturers of power devices will include $R_{\theta j-c}$ in their data sheets, and it can often be found in low-power data as well. Sometimes it is disguised as a power derating figure, expressed in W/°C. The maximum allowable value of T_j is published in the maximum ratings section of each data sheet, and this is the parameter that your thermal calculations must ensure is not exceeded.

$R_{\theta c-h}$ and $R_{\theta h-a}$ are the thermal resistances of the interface between the case and the heat sink, and of the heat sink to ambient, respectively. $R_{\theta c-a}$ represents the thermal resistance due to convection directly from case to ambient and can be neglected if you are using a large heat sink.

An example should help to make the calculation clear.

An IRF640 power MOSFET dissipates a maximum of 35 W steady state. It is mounted on a heat sink with a specified thermal resistance of 0.5°C/W, via an insulating pad with a thermal resistance of 0.8°C/W. The maximum ambient temperature is 70°C. What will be the maximum junction temperature?

From the above conditions, $R_{\theta c-h} + R_{\theta h-a} = 1.3°C/W$. The IRF640 data quotes a junction-to-case thermal resistance ($R_{\theta j-c}$) of 1.0°C/W.

So the junction temperature is defined by Eq. (9.5):

$$T_j = 35 \times [1.3 + 1.0] + 70 = 150.5 \,°C \tag{9.5}$$

This is just over the maximum permitted junction temperature of 150°C so reliability is marginal, and you need a bigger heat sink. However, we have neglected the junction-to-ambient thermal resistance, quoted at 80°C/W. This is in parallel with the other thermal resistances. If it is included, the calculation becomes

$$T_j = 35 \times \frac{[2.3 \times 80]}{2.3 + 80} + 70 = 148.25 \,°C \tag{9.6}$$

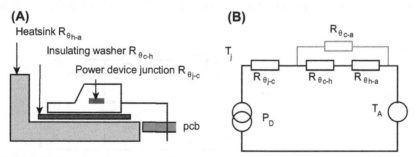

(A)
Heatsink $R_{\theta h-a}$
Insulating washer $R_{\theta c-h}$
Power device junction $R_{\theta j-c}$
pcb

(B)
$R_{\theta c-a}$
T_j
$R_{\theta j-c}$ $R_{\theta c-h}$ $R_{\theta h-a}$
P_D
T_A

FIGURE 9.12 Heat Transfer for a Power Device on a Heat Sink.

which is a very minor improvement and not enough to rely on!

This example illustrates a common misconception about power ratings. The IRF640 is rated at 125 W dissipation, yet even with a fairly massive heat sink (0.5°C/W will require a heat sink area of around 80 square inches) it cannot safely dissipate more than 35 W at an ambient of 70°C. The fact is that the rating is specified *at* 25°C *case temperature*; higher case temperatures require derating because of the thermal resistance from junction to case. You will not be able to maintain 25°C at the case under any practical application conditions, except possibly outside in the Arctic. Power device manufacturers publish derating curves in their data sheets: rely on these rather than the absolute maximum power rating on the front of the specification.

Incidentally, if having followed these thermal design steps you find that the needed heat sink is too large or bulky, it will be far cheaper to reduce the thermal resistance of the total system. You do this by using two (or more) transistors in parallel in place of a single device. Although the thermal resistances for each of the transistors stay the same, the resultant heat flow for each is effectively halved because each transistor is only dissipating half the total power, and therefore the junction temperature rise is also half.

Thermal Capacity

The previous analysis assumed a steady-state heat flow, in other words constant power dissipation. If this is not a good description of your application, you may need to take account of the thermal capacity of the heat sink. The electrical analogue circuit of Fig. 9.12 can be modified according to Fig. 9.13:

From this you can see that a step increase in dissipated power will actually cause a gradual rise in heat sink temperature T_h. This will be reflected at T_j, modified by $R_{\theta j-c}$ and $R_{\theta c-h}$, which will take typically several minutes, possibly hours, to reach its maximum temperature. The value of C_h depends on the mass of heat sink metal, and its heat storage capacity. Values of this parameter for common metals are given in Table 9.3. As an example, a 1°C/W aluminum heat sink might have a volume of 120 cm^3, which has a heat capacity of 296 J/°C. Multiplying the thermal resistance by the heat capacity gives an idea of the time constant, of 296 s.

The thermal capacity will not affect the end-point steady-state temperature, only the time taken to reach it. But if the heat input is transient, with a low duty cycle to

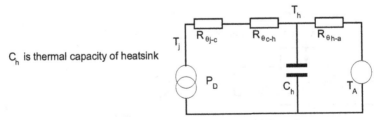

FIGURE 9.13 Electrical Analogue Modified to Include Thermal Capacity.

Table 9.3 Thermal Properties of Common Metals

Metal	Finish	Heat Capacity (J/cm^3/°C)	Bulk Thermal Conductivity (W/°C/m)	Surface Emissivity ε (Black Body = 1)
Aluminum	Polished	2.47	210	0.04
	Unfinished			0.06
	Painted			0.9
	Matt anodized			0.8
Copper	Polished	3.5	380	0.03
	Machined			0.07
	Black oxidized			0.78
Steel	Plain	3.8	40–60	0.5
	Painted			0.8
Zinc	Gray oxidized	2.78	113	0.23 0.28

allow plenty of cooling time, then a larger thermal capacity will reduce the maximum temperatures T_h and T_j reached during a heat pulse. You can analyze this if necessary with the equivalent circuit of Fig. 9.13. Strictly, the other heat transfer components also have an associated thermal capacity, which could be included in the analysis if necessary.

Transient Thermal Characteristics of the Power Device

In applications where the power dissipated in the device consists of continuous low duty cycle periodic pulses, faster than the heat sink thermal time constant, the instantaneous or peak junction temperature may be the limiting condition rather than the average temperature. In this case you need to consult curves for transient thermal resistance. These curves are normally provided by power semiconductor manufacturers in the form of a correction factor that multiplies $R_{\theta j-c}$ to allow for the duty cycle of the power dissipation. Fig. 9.13 shows a family of such curves for the IRF640. Because the period for most pulsed applications is much shorter than the heat sink's thermal time constant, the values of $R_{\theta h-a}$ and $R_{\theta c-h}$ can be multiplied directly by the duty cycle. Then the junction temperature can now be calculated from Eq. (9.7)

$$T_j = P_{D(\mathrm{max})} \times \left[KR_{\theta j-c} + \delta \times (R_{\theta c-h} + R_{\theta h-a}) \right] + T_A \qquad (9.7)$$

where δ is the duty cycle, and K is derived from curves as in Fig. 9.14 for a particular value of δ. $P_{D\mathrm{max}}$ is still the maximum power dissipated during the conduction period, not the power averaged over the whole cycle. At frequencies greater than a few kHz, and duty cycles more than 20%, cycle-by-cycle temperature fluctuations are small enough that the peak junction temperature is determined by the average power dissipation, so that K tends toward δ.

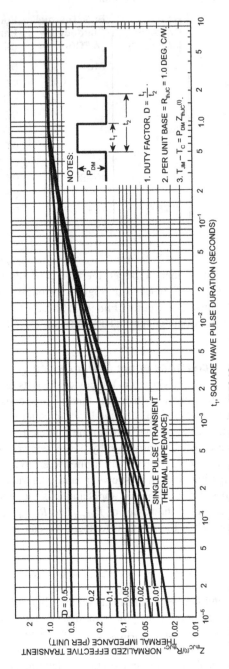

FIGURE 9.14 Transient Thermal Impedance Curves for the IRF640.

From International Rectifier.

Some applications, notably RF amplifiers or switches driving highly inductive loads, may create severe current crowding conditions on the semiconductor die, which invalidate methods based on thermal resistance or transient thermal impedance. Safe operating areas and *di/dt* limits must be observed in these cases.

9.6.2 HEAT SINKS

As the previous section implied, the purpose of a heat sink is to provide a low thermal resistance path between the heat source and the ambient. Strictly speaking, it is the ambient environment that is the heat *sink*; what we conventionally refer to as a heat sink is actually only a heat *exchanger*. It does not itself sink the heat, except temporarily. In most cases the ambient sink will be air, though not invariably: this author recalls one somewhat tongue-in-cheek design for a 1-kW-rated audio amplifier, which suggested bolting the power transistors to a central heating radiator with continuous water cooling! Some designs with a very high power density need to adopt such measures to ensure adequate heat removal.

A wide range of proprietary heat sinks is available from many manufacturers. Several types are predrilled to accept common power device packages. All are characterized to give a specification figure for thermal resistance, usually quoted in free air with fins vertical. Unless your requirements are either very specialized or very high volume, it is unlikely to be worth designing your own heat sink, especially as you will have to go through the effort of testing its thermal characteristics yourself. Custom heat sink design is covered in the application notes of several power-device manufacturers.

A heat sink transfers heat to ambient air primarily by convection and to a lesser degree by radiation. Its efficiency at doing so is directly related to the surface area in contact with the convective medium. Thus heat sink construction seeks to maximize surface area for a given volume and weight; hence the preponderance of finned designs. Orientation of the fins is important because convection requires air to move past the surface and become heated as it does so. As air is heated it rises. Therefore the best convective efficiency is obtained by orienting the fins vertically to obtain maximum air flow across them; horizontal mounting reduces the efficiency by up to 30%.

Convection cooling efficiency falls at higher altitudes. Atmospheric pressure decreases at a rate of 1 mb per 30 ft height gain, from a sea level standard pressure of 1013 mb. Since the heat transfer properties are proportional to the air density, this translates to a cooling efficiency reduction as shown in Table 9.4.

Table 9.4 Free Air Cooling Efficiency Versus Altitude

Sea Level	2,000 ft	5,000 ft	10,000 ft	20,000 ft
100%	97%	90%	80%	63%

The most common material for heat sinks is black anodized aluminum. Aluminum offers a good balance between cost, weight, and thermal conductivity. Black anodizing provides an attractive and durable surface finish and also improves radiative efficiency by 10–15 times over polished aluminum. Copper can be used as a heat sink material when the optimum thermal conductivity is required, but it is heavier and more expensive.

The cooling efficiency does not increase linearly with size, for two principal reasons:

- longer heat sinks (in the direction of the fins) will suffer reduced efficiency at the end where the air leaves the heat sink, since the air has been heated as it flows along the surface;
- the thermal resistance through the bulk of the metal creates a falling temperature gradient away from the heat source, which also reduces the efficiency at the extremities; this resistance is not included in the simple model of Fig. 9.12.

The first of the above reasons means that it is better to reduce the thermal resistance by making a shorter, wider heat sink than by a longer one. The average performance of a typical heat sink is linearly proportional to its width in the direction perpendicular to the air flow and approximately proportional to the square root of the fin length in the direction parallel to the flow.

Also, the thermal resistance of any given heat sink is affected by the temperature differential between it and the surrounding air. This is due to both increased radiation (see below) and increased convection turbulence as the temperature difference increases. This can lead to a drop in $R_{\theta h-a}$ at 20°C difference to 80% of the value at 10°C difference. Or put the other way around, the $R_{\theta h-a}$ at 10°C difference may be 25% higher than that quoted at 20°C difference.

Forced Air Cooling

Convective heat loss from a heat sink can be enhanced by forcing the convective medium across its surface. Detailed design of forced air-cooled heat sinks is best done empirically. Simulation software is available to map heat flow and the resulting thermal transfer in complex assemblies; most heat sink applications will be too involved for simple analytical methods to give better than ballpark results. It is not too difficult to use a thermocouple to measure the temperature rise of a prototype design with a given dissipation, most easily generated by a power resistor attached to a dc supply.

Fig. 9.15 shows the improvement in thermal resistance that can be gained by passing air over a square flat plate, and of course at least a similar magnitude can be expected for any finned design. Optimizing the placement of the fins requires either experimentation or simulation, although staggering the fins will improve the heat transfer. When you use forced air cooling, radiative cooling becomes negligible, and it is not necessary to treat the surface of the heat sink to improve radiation; unfinished aluminum will be as effective as black anodized.

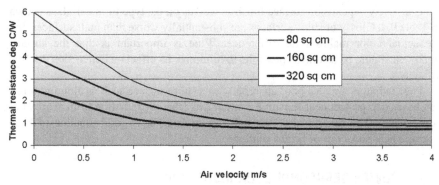

FIGURE 9.15 Thermal Resistance Versus Air Velocity for Various Flat Plate Sizes.

Another common use of forced air cooling is ventilation of a closed equipment cabinet by a fan. The capacity of the fan is quoted as the volumetric flow rate in cubic feet per minute (CFM) or cubic meters per hour (1 CFM = 1.7 m³/h). The volumetric flow rate required to limit the internal temperature rise of an enclosure in which P_D watts of heat is dissipated to $\theta°$C above ambient is defined by Eq. (9.8):

$$\text{Flow rate} = 3600 \times \frac{P_D}{(\rho \times c \times \theta)} \; \text{m}^3/\text{h} \qquad (9.8)$$

where ρ is the density of the medium (air at 30°C and atmospheric pressure is 1.3 kg/m³); c is the specific heat capacity of the medium (air at 30°C is around 1000 J/kg°C).

Fan performance is shown as volumetric flow rate versus pressure drop across the fan. The pressure differential is a function of the total resistance to air flow through the enclosure, presented by obstacles such as air filters, louvers, and PCBs. You generally need to derive pressure differential empirically for any design with a nontrivial air-flow path.

Radiation

Radiative cooling is something of a mixed blessing. Radiant heat travels in line of sight and is therefore as likely to raise the temperature of other components in an assembly as to be dissipated to ambient. For the same reason, radiation is rarely a significant contribution to cooling by a finned heat sink, since the finned areas that make up most of the surface merely heat each other. However, radiation can be used to good effect when a clear radiant path to ambient can be established, particularly for high-temperature components in a restricted air flow. The thermal loss through radiation is defined in Eq. (9.9)

$$Q = 5.7 \times 10^{-12} \times \Delta T^4 \times \varepsilon \qquad (9.9)$$

where ε is the emissivity of the surface, compared to a black body; ΔT is the temperature difference between the component and environment; Q is given in watts per second per square centimeter.

Emissivity depends on surface finish as well as on the type of material, as shown in Table 9.4. Glossy or shiny surfaces are substantially worse than matt surfaces, but the actual color makes little difference. What is important is that the surface treatment should be as thin as possible, to minimize its effect on convection cooling efficiency.

Poor radiators are also poor absorbers, so a shiny surface such as aluminum foil can be used to protect heat-sensitive components from the radiation from nearby hot components. The reverse also holds, so for instance it is good practice to keep external heat sinks out of bright sunlight.

9.6.3 POWER SEMICONDUCTOR MOUNTING

The way in which a power device package is mounted to its heat sink affects both actual heat transfer efficiency and long-term reliability. Faulty mounting of metal-packaged devices mainly causes unnecessarily high junction temperature, shortening device lifetime. Plastic packages (such as the common TO-220 outline) are much more susceptible to mechanical damage, which allows moisture into the case and can even crack the semiconductor die.

The factors that you should consider when deciding on a mounting method are summarized in Fig. 9.16 for a typical plastic-packaged device.

Heat Sink Surface Preparation

The heat sink should have a flatness and finish comparable to that of the device package. The higher the power dissipation, the more attention needs to be paid to surface

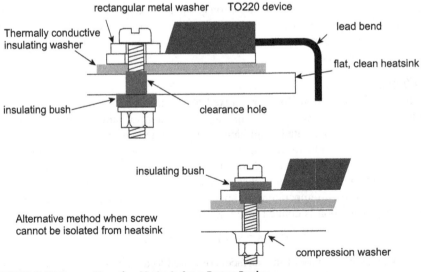

FIGURE 9.16 Screw-Mounting Methods for a Power Device.

finish. A finish of 50–60 microinches is adequate for most purposes. Surface flatness, which is the deviation in surface height across the device-mounting area, should be less than 4 mils (0.004″) per inch.

The mounting hole(s) should only be large enough to allow clearance of the fastener, plus insulating bush if one is fitted. Too large a hole, if the screw is torqued too tightly, will cause the mounting tab to deform into the hole. This runs the risk of cracking the die, as well as lifting the major part of the package, which is directly under the die off the heat sink in cantilever fashion, seriously affecting thermal transfer impedance. Chamfers on the hole must be avoided for the same reason, but deburring is essential to avoid puncturing insulation material and to maintain good thermal contact. The surface should be cleaned of dust, grease, and swarf immediately before assembly.

Lead Bend

Bending the leads of any semiconductor package stresses the lead interface and can result in cracking and consequent unreliability. If possible, mount your devices upright on the PCB so that lead bending is unnecessary. Plastic-packaged devices (TO220, TO126 etc.) can have their leads bent, provided that

- the minimum distance between the plastic body and the bend is 4 mm
- the minimum bend radius is 2 mm
- maximum bend angle is no greater than 90 degrees
- leads are not repeatedly bent at the same point
- no axial strain is applied to the leads, relative to each other or the package

Use round-nosed pliers or a proper lead-forming jig to ensure that these conditions are met. Metal-cased devices must not have their leads bent, as this is almost certain to damage the glass seal.

When the device is inserted into the board, the leads should always be soldered after the mechanical fastening has been made and tightened. Some manufacturing departments prefer not to run cadmium-plated screws through a solder bath because it contaminates the solder, and they may decide to put the screws in after the mass soldering stage. Do not allow this: insist on hand soldering or use different screws.

The Insulating Washer

In most devices, the heat transfer tab or case is connected directly to one of the device terminals, and this raises the problem of isolating the case. The best solution from the point of view of thermal resistance is to isolate the entire heat sink rather than use any insulating device between the package and the heat sink. This is often not possible, for electromagnetic interference or safety reasons, because the chassis serves as the heat sink, or because several devices share the same heat sink. Some devices are now available in fully isolated packages, but if you are not using one of these you will have to incorporate an insulating washer under the package.

Insulating washers for all standard packages are available in many different materials: polyimide film, mica, hard anodized aluminum, and reinforced silicone rubber are the most popular. The first three of these require the use of a thermally conductive grease (heat sink compound) between the mating surfaces, to fill the minor voids that exist and that would otherwise increase the thermal resistance across the interface. This is messy and increases the variability and cost of the production stage. If excess grease is left around the device, it may accumulate dust and swarf and lead to insulation breakdown across the interface. Silicone rubber, being somewhat conformal under pressure, can be used dry and some types will outperform mica and grease.

Table 9.5 shows the approximate range of interface thermal resistances ($R_{\theta c-h}$) that may be expected. Note that the actual values will vary quite widely depending on contact pressure; a minimum force of 20 N should be maintained by the mounting method, but higher values will give better results provided they do not lead to damage. When thermally conductive grease is not used, wide variations in thermal resistance will be encountered because of differences in surface finish and the micro-air gaps that result. Thermal grease fills these gaps and reduces the resistance across the interface. Despite its name, it is not any more thermally conductive than the washer it is coating; it should only be applied very thinly, sufficient to fill the air gaps but no more, so that the total thickness between the case and the heat sink is hardly increased. In this context, more is definitely not better. The mounting hole(s) in the washer should be no larger than the device's holes, otherwise flashover to the exposed metal (which should be carefully deburred) is likely.

Mounting Hardware

A combination of machine screws, compression washers, flat washers, and nuts is satisfactory for any type of package that has mounting holes. Check the specified

Table 9.5 Interface Thermal Resistances for Various Mounting Methods

Package Type	Metal-to-Metal		2-mil Mica		2-mil Polyimide		6-mil Silicone Rubber
	Dry	Greased	Dry	Greased	Dry	Greased	
Metal, Flanged							
TO204AA (TO3)	0.5	0.1	1.2	0.5	1.5	0.55	0.4–0.6
TO213AA (TO66)	1.5	0.5	2.3	0.9			
Plastic							
TO126	2.0	1.3	4.3	3.3			4.8
TO220AB	1.2	0.6	3.4	1.6	4.5	2.2	1.8

mounting hole tolerances carefully; there is a surprisingly wide variation in hole dimensions for the same nominal package type across different manufacturers. A flat, preferably rectangular (in the case of plastic packages) washer under the screw head is vital to give a properly distributed pressure, otherwise cracking of the package is likely. A conical compression washer is a very useful device for ensuring that the correct torque is applied. This applies a constant pressure over a wide range of physical deflection and allows proper assembly by semiskilled operators without using a torque wrench or driver. Tightening the fasteners to the correct torque is very important; too little torque results in a high thermal impedance and long-term unreliability due to overtemperature, while too much can overstress the package and result in long-term unreliability due to package failure.

When screw-mounting a device that has to be isolated from the heat sink, you need to use an insulating bush either in the device tab or the heat sink. The preferred method is to put the bush in the heat sink, and use large flat washers to distribute the mounting force over the package. You can also use larger screws this way. The bush material should be of a type that will not flow or creep under compression; glass-filled nylon or polycarbonate is acceptable, but unfilled nylon should be avoided. The bush should be long enough to overlap between the transistor and the heat sink, to prevent flashover between the two exposed metal surfaces.

A fast, economical, and effective alternative is a mounting clip. When only a few watts are being dissipated, you can use board mounting or free-standing dissipators with an integral clip. A separate clip can be used for larger heat sinks and higher powers. The clip must be matched to the package and heat sink thickness to obtain the proper pressure. It can actually offer a lower thermal resistance than other methods for plastic packages, because it can be designed to bear directly down on the top of the plastic over the die, rather than concentrating the mounting pressure at the hole in the tab. It also removes the threat of flashover around the mounting hole, since no hole is needed in the insulating washer.

When you have to mount several identical flat (e.g., TO-220) packages to a single heat sink, a natural development of the clip is a single clamping bar that is placed across all the packages together (Fig. 9.17). The bar must be rigid enough and fixed at enough places with the correct torque to provide a constant and predictable clamping pressure for each package. With suitably ingenious mechanical design, it may also contribute to the total thermal performance of the whole assembly.

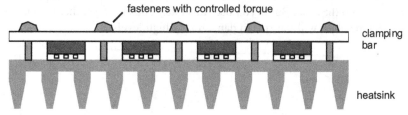

FIGURE 9.17 Package Clamping Bar.

9.6.4 PLACEMENT AND LAYOUT

If you are only concerned with designing circuits that run at slow speeds with CMOS logic and draw no more than a few milliamps, then thermal layout considerations will not interest you. As soon as dissipation raises the temperature of your components more than a few tens of degrees above ambient, it pays to look at your equipment and PCB layout in terms of heat transfer. As was shown in Section 9.4.3, this will ultimately reflect in the reliability of the equipment.

Some practices that will improve thermal performance are as follows:

- Mount PCBs vertically rather than horizontally. This is standard in card cages and similar equipment practice, and it allows a much freer convective air flow over the components. If you are going to do this, do not then block off the air flow by putting solid metal screens above or below the boards; use punched, louvered, or mesh screens.
- Put hot components near the edge of the board, to encourage a good air flow around them and their heat sinks. If the board will be vertically mounted, put them at the top of the board.
- Keep hot components as far away as possible from sensitive devices such as precision op-amps or high failure rate parts such as electrolytic capacitors. Put them above such components if the board is vertical.
- Heat sinks perform best in low ambient temperatures. If you are using a heat sink within an enclosure without forced air cooling, remember to allow for the steady-state temperature rise inside the enclosure. However, do not position a heat sink near to the air inlet, as it will heat the air that is circulating through the rest of the enclosure; put it near the outlet. Do not obstruct the air flow over a heat sink.
- If you have a high heat density, for example a board full of high-speed logic devices, consider using a thermally conductive ladder fixed on the board and in contact with the IC packages, brought out to the edge of the board and bonded to an external heat sink. PCB laminates themselves have a low thermal conductivity.
- If you have to use a case with no ventilation, for environmental or safety reasons, remember that cooling of the internal components will be by three stages of convection rather than one: from the component to the inside air, from the inside air to the case, and from the case to the outside. Each of these will be inefficient, compared to conductive heat transfer obtained by mounting hot components directly onto the case. But if you take this latter course, check that the outside case temperature will not rise to dangerously high levels.

Appendix: Standards

Standards are indispensable to manufacturing industry. Not only do they allow interchangeability or interoperability between different manufacturers' products but they also represent a distillation of knowledge about practical aspects of technology—how to make measurements, what tests to use, what dimensions to specify, and so on. Each standard is the result of considerable work on the part of a collection of experts in that particular field and is therefore authoritative; notwithstanding which, any standard in a fast-changing area will be subject to revision and amendment as the technology progresses.

This appendix lists a few of the more relevant British and international standards, both those which have been referenced in the text of this book and some which the author feels to be of particular interest. The catalogs of the various standards bodies, updated yearly, give a full list of the available and current standards and are essential for the library of any development department. Although only British Standards (BS) and International Electrotechnical Commission (IEC) publications are mentioned here, for the sake of brevity, there are of course many other standards sources that you may need to consult for a particular application.

BRITISH STANDARDS

These are available from:

British Standards Institution
Customer Services
389 Chiswick High Road
London W4 4AL
UK
Tel +44 (0)208 996 9001
Email cservices@bsi-global.com
Website www.bsi-global.com

Some BS are related to, equivalent to, or identical to other European or international standards. This is indicated where appropriate. Where a British Standard's numbering starts "BS EN" this means that it is identical to the equivalent European document produced by CENELEC, which itself may well be identical to the IEC source document, although this is not assured.

BS no.   Related, equivalent or identical standards
BS 613

Specification for components and filter units for electromagnetic interference suppression

BS 2316   **IEC 60096**

Specification for radio-frequency cables

BS 2488   **IEC 60063**

Schedule of preferred numbers for resistors and capacitors

BS 2754   **IEC 60536**

Memorandum: construction of electrical equipment for protection against electric shock

BS 4808   **IEC 60189**

Specification for LF cables and wires with PVC insulation and PVC sheath for telecommunications

BS 5783

Code of practice for handling of electrostatic sensitive devices

BS 6221   **IEC 60326**

Printed wiring boards

BS 6500

Electric cables. Flexible cords rated up to 300/500 V, for use with appliances and equipment intended for domestic, office and similar environments

BS EN 13602

Copper and copper alloys. Drawn, round copper wire for the manufacture of electrical conductors

BS EN 55014   **CISPR 14**

Electromagnetic compatibility. Requirements for household appliances, electric tools, and similar apparatus

BS EN 55022   **CISPR 22**

Information technology equipment. Radio disturbance characteristics. Limits and methods of measurement

BS EN 60062   **IEC 60062**

Specification for marking codes for resistors and capacitors

BS EN 60065   **IEC 60065**

Audio, video and similar electronic apparatus. Safety requirements

BS EN 60068   **IEC 60068**

Environmental testing

BS EN 60127   **IEC 60127**

Miniature fuses

BS EN 60182   **IEC 60182, IEC 60851**

Basic dimensions of winding wires. Specification for maximum overall diameters of enamelled round winding wires

BS EN 60269   **IEC 60269**

Low-voltage fuses (see also some parts of BS 88)

BS EN 60285   **IEC 60285**

Alkaline secondary cells and batteries. Sealed nickel-cadmium cylindrical rechargeable single cells

BS EN 60431   **IEC 60431**

Specification for dimensions of square cores (RM-cores) made of magnetic oxides and associated parts

BS EN 60529   **IEC 60529**

Specification for degrees of protection provided by enclosures (IP code)

BS EN 60617   **IEC 60617**

Graphical symbols for diagrams

BS EN 60950   **IEC 60950**

Safety of information technology equipment

BS EN 61951   **IEC 61951**

Secondary cells and batteries containing alkaline or other nonacid electrolytes (previously IEC 60509)

BS QC 300XXX   **IEC 60384**

Harmonized system of quality assessment for electronic components. Fixed capacitors for use in electronic equipment

BS 9940   **IEC 60115, QC 400XXX**

Harmonized system of quality assessment for electronic components. Fixed resistors for use in electronic equipment

INTERNATIONAL ELECTROTECHNICAL COMMISSION STANDARDS

The IEC is responsible for international standardization in the electrical and electronics fields. It is composed of member National Committees. IEC publications are available from the British Standards Institution, address as before, other national committees, or from the IEC webstore at www.iec.ch, postal address:

IEC Central Office
3 Rue de Varembé
1211 Geneva 20
Switzerland

Most IEC publications have related BS (see the list above). A few that do not are listed below.

IEC 60479

Effects of current passing through the human body

IEC 60647

Dimensions for magnetic oxide cores intended for use in power supplies (EC-cores)

INSTITUTE OF ELECTRICAL AND ELECTRONICS ENGINEERS STANDARDS

The IEEE (Institute of Electrical and Electronics Engineers) is originally an organization covering the United States of America, and was formed in 1963 from the union of the American Institute of Electrical Engineers and the Institute of Radio Engineers. The IEEE now has more than 390,000 members worldwide, in over 160 countries, and has a strong voice in defining standards relevant to electrical and electronic systems design. IEEE Standards are available from the IEEE Standards Association website (www.ieee.org), and the postal address is given below:

IEEE Operations Center
445 Hoes Lane
Piscataway, NJ 08854-4141 USA
Phone: +1 732 981 0060

The IEEE standard number is usually followed by the year in which the standard was ratified. The following is a brief list of some relevant standards, but for the most up-to-date listing, the reader should refer to the IEEE standards website http://standards.ieee.org.

295-1969—Electronics Power Transformers

This standard pertains to power transformers and inductors that are used in electronic equipment and supplied by power lines or generators.

388-1992—Transformers and Inductors in Electronic Power Conversion Equipment

In this standard transformers of both the saturating and nonsaturating type are covered.

393-1991—Test Procedures for Magnetic Cores

Test methods useful in the design, analysis, and operation of magnetic cores in many types of applications are presented.

484-2002—Recommended Practice for Installation Design and Installation of Vented Lead-Acid Batteries for Stationary Applications

Recommended design practices and procedures for storage, location, mounting, ventilation, instrumentation, preassembly, assembly, and charging.

937-2007—Recommended Practice for Installation and Maintenance of Lead-Acid Batteries for Photovoltaic (PV) Systems

Design considerations and procedures for storage, location, mounting, ventilation, assembly, and maintenance of lead-acid secondary batteries.

1159-2009—Recommended Practice for Monitoring Electric Power Quality

This recommended practice encompasses the monitoring of electrical characteristics of single-phase and polyphase ac power systems.

1515-2000—Recommended Practice for Electronic Power Subsystems: Parameter Definitions, Test Conditions, and Test Methods

This recommended practice defines many common parameters for AC-DC and DC-DC electronic power distribution components and subsystems.

1573-2003—Recommended Practice for Electronic Power Subsystems: Parameters, Interfaces, Elements, and Performance

A technical basis for implementation of electronic power subsystems is provided in this recommended practice. It is intended for electronics.

802.3.1-2011—Management Information Base (MIB) Definitions for Ethernet

The Management Information Base (MIB) module specifications for IEEE Std 802.3, also known as Ethernet, are contained in this standard.
(The IEEE Standards starting 802 define all the aspects of modern Ethernet systems).

139-1988—Recommended Practice for the Measurement of Radio Frequency Emission from Industrial, Scientific, and Medical (ISM) Equipment Installed on User's Premises

Describes equipment inspection and radio frequency (RF) electromagnetic field measurement procedures for evaluation of RF industrial, scientific, etc. equipment.

644-1994—Standard Procedures for Measurement of Power Frequency Electric and Magnetic Fields from AC Power Lines

In this IEEE Standard uniform procedures for the measurement of power frequency electric and magnetic fields from alternating current (AC).

1528-2003—Recommended Practice for Determining the Peak Spatial-Average Specific Absorption Rate (SAR) in the Human Head from Wireless Communications Devices: Measurement Techniques

IEEE Std 1528-2003 specifies protocols and test procedures for the measurement of the peak spatial-average SAR induced inside a simplified model.

1775-2010—Power Line Communication Equipment—Electromagnetic Compatibility (EMC) Requirements—Testing and Measurement Methods

Electromagnetic compatibility (EMC) criteria and consensus test and measurements procedures for broadband over power line (BPL) communication.

C63.4-1991—Methods of Measurement of Radio-Noise Emissions from Low-Voltage Electrical and Electronic Equipment in the Range of 9 kHz to 40 GHz

This standard sets forth uniform methods of measurement of radio noise emitted from low-voltage electrical and electronic equipment.

Bibliography

When writing a book of this nature, one is invariably indebted to a great many sources of information, which are often used in day-to-day work as a circuit designer. This reading list pulls together some of the references I have found most useful over the years, and/or which have been drawn on extensively for some of the chapters. Most are manufacturers' data books. I have also added some links on the Web, which has become an excellent source of information, whether technical or design oriented.

On Batteries

Guide for Designers, Duracell Europe, 1985
Short applications guide and data book covering most primary cell systems.
Yuasa little red book of batteries, Yuasa Battery (UK) Ltd., from the Yuasa website
Intended "to give the layman battery user an insight into lead–acid batteries."
Linden's Handbook of Batteries, ed. Thomas Reddy, 4th edition (Electronics). ISBN-13: 978-0071624213

On Digital Design

High-Speed CMOS Designer's Guide and Applications Handbook, Mullard, 1986 (now Philips Components)
Most manufacturers produce extensive applications literature on their standard logic families. This book is the most comprehensive guide I have found on HCMOS.
Transmission line effects in PCB applications, Motorola application note AN1051, 1990
Presents design analysis and examples for PCB transmission line effects in digital circuits. Includes Bergeron diagram analysis.
IEEE Std 1149.1 (JTAG) Testability Primer, Texas Instruments application note SSYA002C, 1997
Details boundary-scan techniques, including a clear and complete description of JTAG architecture and applications.
Logic and Computer Design Fundamentals, Mano and Kime, Prentice-Hall, 2008
M. Morris Mano has published numerous textbooks on the topic of logic and computer design, and this is a more recent edition. This book covers many of the important fundamentals that are useful for anyone interested in digital design.

On Discrete Devices

Philips Components Data Book 1 Part 1b, **Low-frequency power transistors and modules**, 1989
Contains information on transistor ratings, safe operating area, and letter symbols.
Thyristor & Triac Theory and Applications, in Motorola Thyristor Databook, 1985
Nine chapters cover the title's scope comprehensively and in detail.
Applications, in Low Power Discretes Data Book, Siliconix, 1989
Covers all aspects of JFET applications.

Power MOSFET Application Data, in HEXFET Databook, International Rectifier, 1982

Covers all aspects of power MOSFETs and is equally applicable to devices from other manufacturers.

IGBT Characteristics, International Rectifier AN-983, and **Application Characterization of IGBTs**, International Rectifier AN-990; **Application Considerations Using IGBTs**, Motorola AN1540

These together cover most aspects of IGBTs, and as above are equally applicable to devices from other manufacturers.

Mounting Considerations for Power Semiconductors, Bill Roehr, Motorola Applications Note AN1040, 1988

Useful summary of techniques and considerations for good thermal and mechanical design practice.

On Grounding and Electromagnetic Compatibility

Grounding and Shielding Techniques in Instrumentation, Ralph Morrison, Wiley-Interscience 1986

A rare guide to the fundamentals of grounding and shielding practice. Terse style but very useful.

Electromagnetic Compatibility, Jasper Goedbloed, Prentice Hall, 1992

Addresses fundamental interference problems, so that the reader, "starting out correctly, will be able to design in an EM-compatible way … insight is considered more valuable than the recipe for tackling a symptom." Cannot be too highly recommended.

EMC for Product Designers, 3rd Edition, Tim Williams, Newnes, 2001

A guide to EMC techniques and effects, as well as compliance regimes and tests. By the present author.

On Electromagnetics

Electromagnetics, 4th Edition, John Kraus, McGraw Hill, 1991

An excellent textbook that covers a lot of ground from fundamental electromagnetism to practical advice on impedance matching and transmission lines.

On Op-amps

Intuitive IC Op Amps, Thomas M Frederiksen, National Semiconductor, 1984

This is undoubtedly the most useful book on op-amps there is. Frederiksen designed the LM324 among other things.

Op Amp Orientation, in Analog Devices Databook Vol I, 1984

ADs are heavily committed to op-amp technology and this is a compact but comprehensive guide to specifying the devices. The data book also contains useful applications notes on interference shielding and guarding, grounding and decoupling.

On Passive Components

Ceramic Resonator (CERALOCK) Application Manual, Murata, 2002
Describes the application and characteristics of ceramic resonators.
Technical papers published between 1981 and 1988, Wima, 1989
A collection of informative articles on the technology of plastic film capacitors.
Introduction to Quartz Crystals, in ECM Electronics Ltd. components catalogue, 1989
All the information you need on specifying and designing circuits for crystals.
Soft Ferrites: General Information, in Neosid Magnetic Components data book, 1985
A clear introduction to the properties and manufacture of ferrite components.
The Trimmer Primer, Bourns, Inc., from the Bourns website
Everything you ever wanted to know about trimmer circuits.

On Power Supplies

Linear/Switchmode Voltage Regulator Manual, Motorola, 1989; **Voltage regulator Handbook**, National Semiconductor, 1982
Both these books, besides offering data sheets, contain chapters on aspects of power supply and regulator design.
Unitrode Applications Handbook, Unitrode, 1985
Unitrode specialize in power conversion semiconductors, and this handbook contains a comprehensive selection of application notes covering switching supply design.
The Essence of Power Electronics, Neil Ross, Prentice Hall, 1997
This is an introduction to the key concepts in power electronics and has some useful introductory material for DC/DC converters.
Switch Mode Power Conversion, K. Kit Sum, Marcell Dekker, 1984
This is an excellent introduction to switch-mode power converters, with very clear derivation of basic principles and useful design examples.

On Power Circuits

Power Electronics, Mohan, Undeland and Robbins, Wiley, 2003
This is an excellent text that covers many of the fundamentals of power design including conversion from AC−DC and vice versa.
Power Electronics, Lander, McGraw Hill, 1993
This is an excellent text that covers many of the fundamentals of power design in a very accessible way and also has clear explanations of many of the theoretical aspects of power systems design.

On Printed Circuits

An Introduction to Printed Circuit Board Technology, J A Scarlett, Electrochemical Publications Ltd., 1984

Introduces all aspects of standard, PTH, and multilayer board manufacture. Also covers design rules.

BS6221: Part 3: 1984, Guide for the design and use of printed wiring boards, British Standards Institution

Exactly as the title describes it. Other parts of BS6221 cover assembly and repair of PCBs, as well as specification methods for each type (equivalent to IEC326).

Printed Circuits Handbook, ed. Clyde F Coombs, 3rd edition, McGraw Hill, 1988

Well-established reference handbook.

Flex Design Guide, All Flex Inc., from the All Flex website

Technical information on the use and application of flexible circuits.

On Reliability

Electronic equipment reliability, J C Cluley, Macmillan, 1981 (2nd Edition)

Useful overview of statistical basis, prediction, component failure rate, and design.

On Thermal Design

Switchmode power supply handbook, Keith Billings, McGraw Hill, 1989

Although this is all about switching supplies, Chapter 16 is titled "Thermal Management" and gives a useful and concise review of the how and why of heat sinks.

On Transient Suppression

Transient voltage suppression manual, General Electric, 5th Edition, 1986

GE are major suppliers of varistor transient suppressors. This manual offers an overview of voltage transients and how to suppress them with varistors.

Application notes on Transient Suppression, Semitron

Similar to the above but concentrating on avalanche-type transient absorbers.

On Simulation

Inside SPICE — overcoming the obstacles of circuit simulation, Ron Kielkowski, McGraw Hill, 1994

This excellent book is really useful for understanding how the SPICE simulator actually works and is more accessible than many of the more numerically oriented analysis textbooks available.

From the Web

Now there are some excellent resources, including application notes, available on the web. In many cases, the application notes from manufacturers such as Texas Instruments, Analog Devices, and International Rectifier are excellent. In many areas, you can find application notes that apply regardless of the technology or device chosen, making this an extremely valuable resource.

Due to the fickle nature of the corporate world, companies are often acquired or merged, making web links particularly prone to becoming out of date, however I will provide some links that I have found extremely useful, which will hopefully persist for some time.

Texas Instruments: http://www.ti.com

Analog Devices: http://www.analog.com

International Rectifier: http://www.irf.com

ARM: http://www.arm.com

Ferroxcube: http://www.ferroxcube.com

In addition to the company information given above, there are many sources of technical data online, with the best ones being the published archives of the major international academic publishers including:

IEEE: http://ieeexplore.ieee.org

IET Digital Library: http://www.ietdl.org

Elsevier: http://www.elsevier.com

Universities worldwide also publish research, with often the use of open source repositories becoming widely used. An example of this is the "Eprints" project at the University of Southampton, where papers, theses, and reports can be accessed:

Southampton Eprints: http://eprints.ecs.soton.ac.uk

Index

'*Note*: Page numbers followed by "f" indicate figures, "t" indicate tables.'

Printed in the United States
By Bookmasters